Table of Contents

Chapter 15 : Turbochargers & Superchargers   329

Copyrights

Scope

- This book is a good complement to prepare the ABYC certification in diesel engines. Additionally gives to reader the basic learning to diagnosis and repair of marine diesel engines

- This course **is not equivalent** to the certification granted by the engine manufacturers (Caterpillar, Cummins, Volvo, MTU, etc)

- We will follow the recommendations of **ABYC** Standards P-1, H-2 , H-24 and H25 , the **USCG** regulations and the **CFR's** (code of federal regulations)

ABYC Certification

This book is a good complement for students interested to take the ABYC Diesel Certification

Recognition

This book is the result of many years of practice and study about diesel engines. Most of the items presented here are taken as reference, from research and articles of other authors. All I want to express is my gratitude because thanks to these works , this book will be usefull for boat owners and marine engineers interested in diesel engines service. In the same way some graphics are taken from public internet sites. To the people who have uploaded these images to the cloud , thank you very much for its contribution

Chapter 1
Mechanical Principles

Effects of Friction

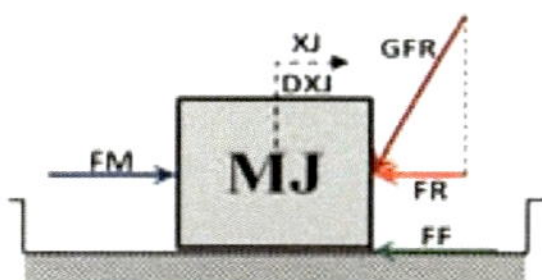

Friction is the resistance to motion of one object moving relative to another. It is not a fundamental force, like gravity or electromagnetism

- It is always present
- The type of surface has an effect on friction
- Rough surfaces will produce more friction than smooth surfaces

Fluid Friction

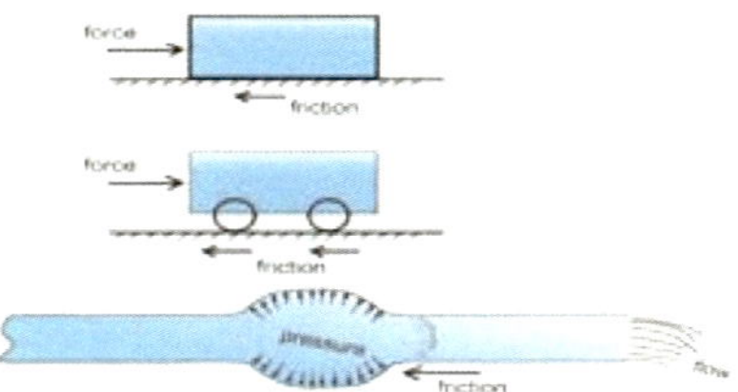

Fluid friction is the resistance to an object's motion through a liquid or gas. When the motion is occurring in a liquid, it is referred to as viscous resistance

The friction will not be caused by the surfaces being in contact, but from the oil in between them

Fluid Friction

- Fluid friction occurs in fluids that are moving relative to each other.
- Fluid friction contributes to aerodynamic **drag**, which is a resistance to the forward motion of a body through a fluid (the air)

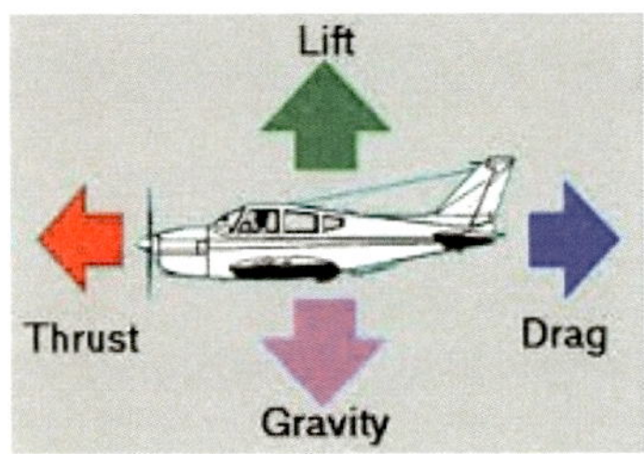

Viscosity & Boat resistance

- Viscosity can be considered as the resistance to flow or to shear within the fluid
- Wave resistance is a form of drag that affects the surface of a watercraft, such as boats and ships, and reflects the energy required to push the water out of the way of the hull
- The rougher the hull surface the more water is dragged with it
- In terms of diesel engines , the engine life time is directly related with the oil quality

Laminar Flow

- In fluid dynamics, **laminar flow** (or streamline **flow**) occurs when a fluid **flows** in parallel layers, with no disruption between the layers
- At low velocities, the fluid tends to **flow** without lateral mixing, and adjacent layers slide past one another like playing cards

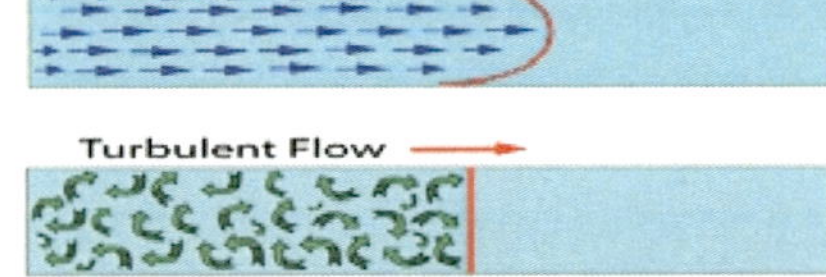

Metric Vs Standard

Due that the engine components are bolted with metric and standard fasteners. It is recommended that the readers try to memorize the equivalence values of wrenches and sockets of similar dimension

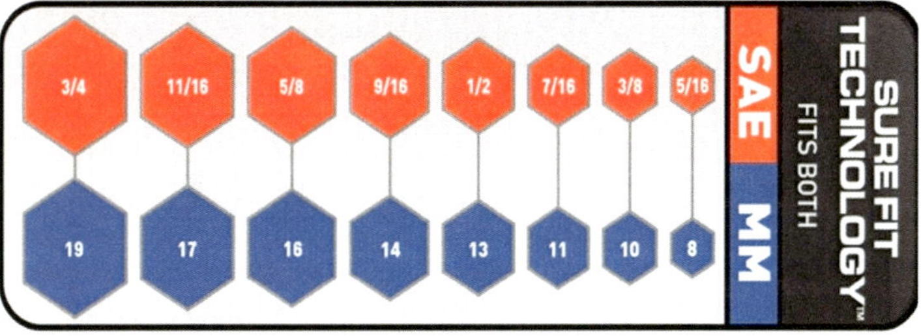

Mass

- Mass is the amount of matter a substance contains. It is measured in kilograms (kg) or grams (g) and it never changes
- Mass in Science is how much matter is inside of an object. being directly related to the number and type of atoms present in the object.
- And matter is just what kind of substance an object is made of. Mass is not the weight of an object

Weight

Weight is the gravitational force acting on a body (although for trading purposes it is taken to mean the same as mass)

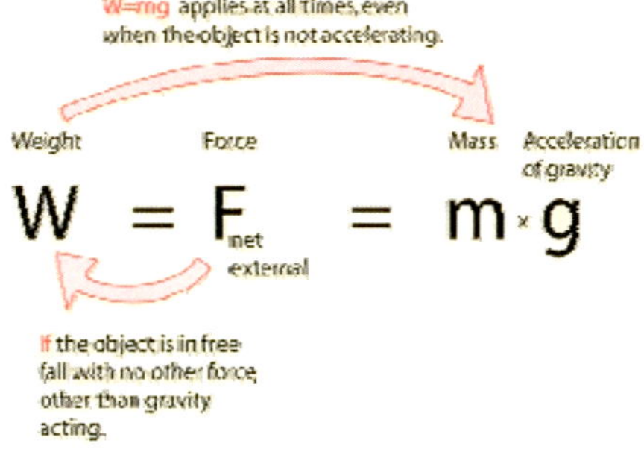

Force

A force acting on an object can cause it to accelerate. The relationship between force, mass and acceleration is given by the equation.

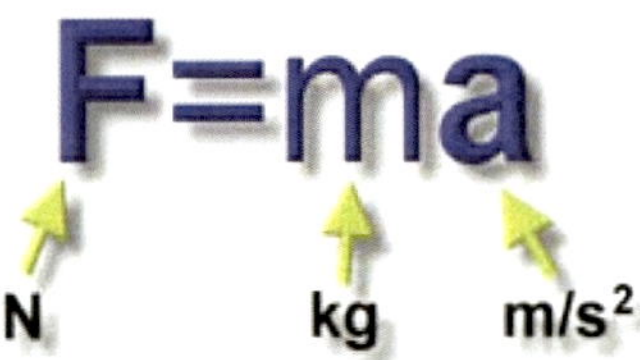

Second Law of Newton

- Newton's second law details the relationship between net force, the mass, and the acceleration
- If you push or pull an object in a particular direction, it accelerates in that direction.

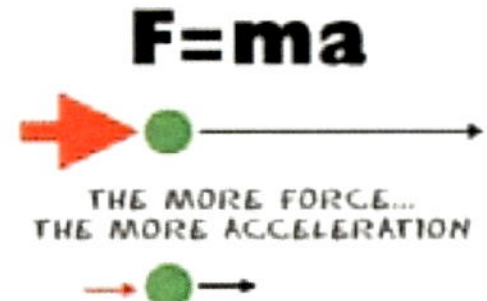

Force

Electricity = **Ohm** Law

Force = Voltage = I x R

Mechanical = **Newton** Law

Force = Mass x Acceleration

Hydraulics = **Pascal** Law

Force = Pressure x Area

Work

- It is dimensionally equivalent to a force of one newton acting over a distance of one meter
- The unit *newton meter* is properly denoted N·m or Joule

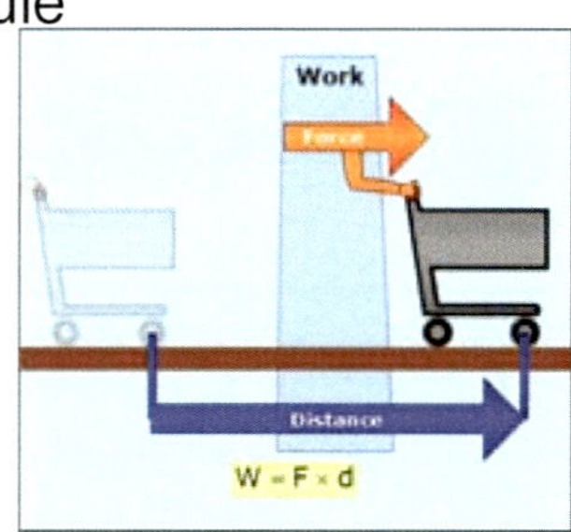

Torque

- Is the tendency of a force to rotate an object about an axis
- torque is defined as the cross product of the lever-arm and the force vector, which tends to produce rotation.

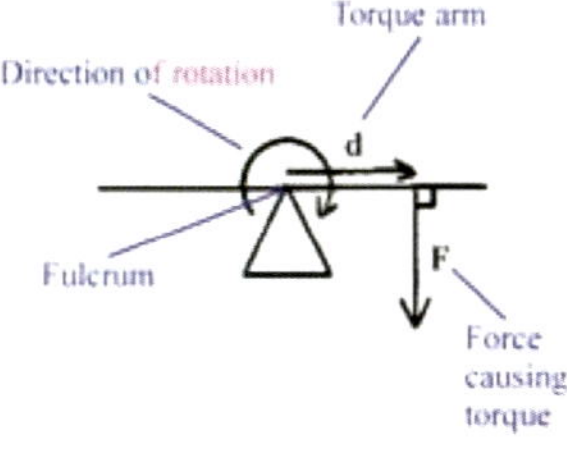

Torque

Torque is a measure of how much a force acting on an object causes that object to rotate. The object rotates about an axis, which we will call the pivot point

Torque

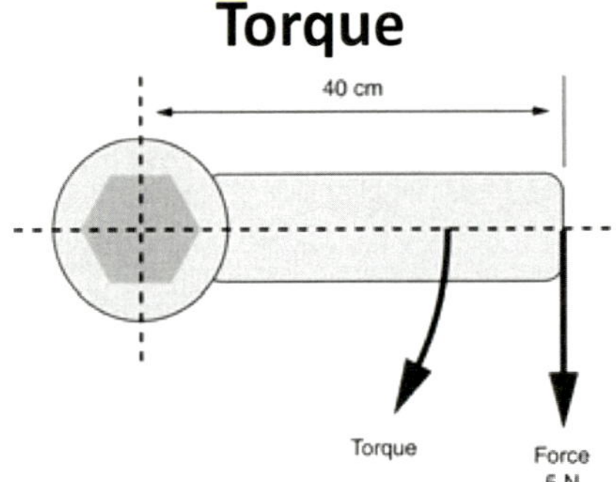

Torque **= Force (N) x Radius (m)**

In this example:

Torque = 5 N x .4 m = 2 Nm

Work and Torque

Torque is a vector which indicates the angular acceleration of an object around an axis, while work is a scalar which indicates the increase in energy of an object being accelerated along a path

Work and Torque

The component of distance involved in the two is different; torque involves the perpendicular distance to the axis involved, and work involves the component of the movement of the body parallel to the force

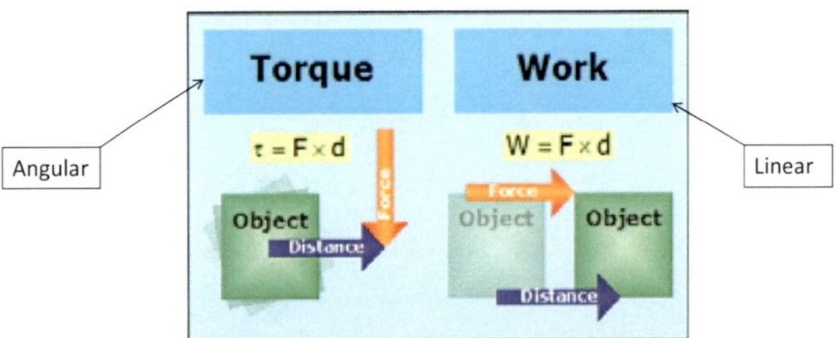

Power

- The unit of power is the ***Watt* (W)**
- Power is the rate at which work is done, that is, the work done per second
- One *Watt* (W) of power is one *joule* (J) of work done in one second
  - 1 Watt (W) = 1 joule (J) per second (J/s)
- The *Watt* (W) is used for both mechanical and electrical power
  - Electrical Power: 1 Watt (W) = 1 Volt (V) x 1 Amp (A)
- Engines are rated in kilowatts (kW)
- 1 kW = 1.34 HP   or   **1 HP = 746 Watts**

Marine Engine Torque Vs. Marine Engine Horsepower

- Most people make the common mistake of focusing on the marine engine horsepower rather than the marine engine torque.
- There is a common saying that "Horsepower sells a boat however Torque is what actually moves it".

Output Torque Measurement

The torque produced by the mechanical device attached to the crankshaft of the engine is 300 Foot per pound @ 2500 RPM

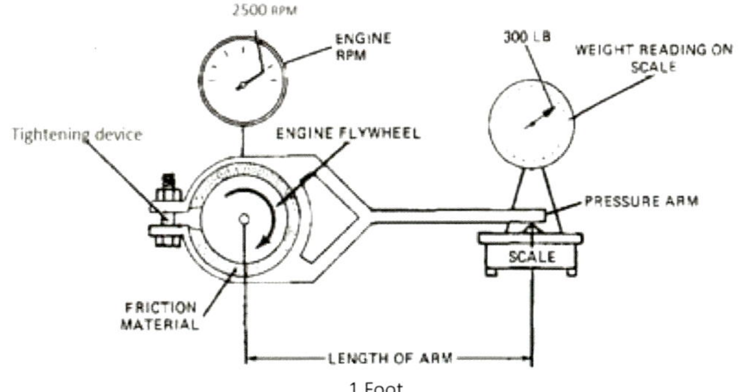

Marine Engine Torque Vs. Marine Engine Horsepower

- One should realize that horsepower is really a measure of the torque over a given period of time .
- This taken into account by the rpm variable in the specification. The following equation may help to shed some light as well.
  - Torque = Hp x 5252 / RPM   (5252 is a constant)

Engine Output Torque

$$Torque_{(Lb\text{-}Ft)} = \frac{5252 \ \times \ HP_{(Per\ Engine)}}{RPM(Engine)}$$

5252 = Constant

HP = Is the Brake Horsepower of the Engine

RPM= Revolutions per Minute

Note: The torque tends to grow when the revolutions decrease

Engine Efficiency

The efficiency of a machine is the ratio of output power to the input power or the actual output power (Brake Horsepower) compared to theoretical output power (Indicated Horsepower)

$$\frac{\underline{Actual\ Output\ Power}}{Theoretical\ Output\ Power} \ = \ Efficiency\ (\%)$$

Volumetric efficiency

- The theoretical maximum *volume* of air that each cylinder can ingest during the intake cycle is equal to the swept volume of that cylinder [(3.1416/4) x bore x bore x stroke]
- The actual amount of air the engine ingests compared to the theoretical maximum is called volumetric efficiency (VE)
- An engine operating at 100% VE is ingesting its total displacement every two crankshaft revolutions

Volumetric efficiency

- If life was perfect, we could fill the cylinders completely with air.
- If we had 17 psi boost in the intake manifold, we would open the intake valve and get 17 psi in the cylinder before the intake valve closed .
- Unfortunately, this doesn't usually happen. With some exhaust remaining in the cylinder and the restriction offered by the intake ports and valves the actual amount of air that flows into the cylinder is somewhat less than ideal .

Volumetric efficiency

- The amount that does flow divided by the ideal amount is called the volumetric efficiency.
- For your basic Cummins small block 6-cyl, this number is around 0.85 (or 85%).
- With tunnel rams some normally aspirated engines can get over 90% at certain rpm's due to the ram effect.
- To take this into account when we calculate flow into the engine (cfm), we multiply the ideal amount of air by the efficiency to get the actual amount of air

Chapter 2
Bearings , Gaskets & Seals

Introduction

- The wheel is often described as the most significant invention of all time
- The real innovation though was placing the wheel on an axle fitted with a bearing
- There is ancient evidence of the use of rounded surfaces to reduce effort of moving objects

Mechanical Bearings

- A **bearing** is any of various machine elements that constrain the relative motion between two or more parts to only the desired type of motion.
- Bearings may be classified broadly according to the motions they allow and according to their principle of operation, as well as by the directions of applied loads they can handle

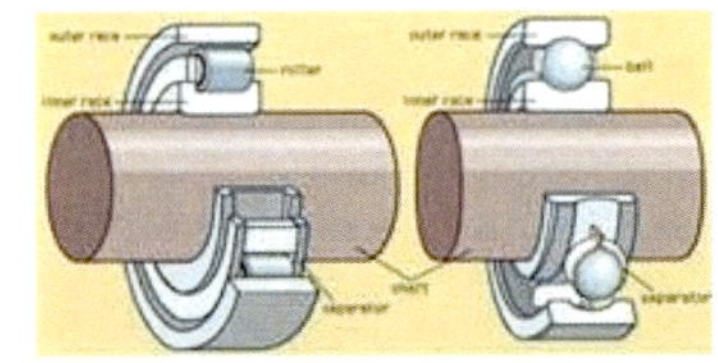

Application of Friction to Bearings

- When two surfaces rub together there is a concern about friction
- As Friction increases – so does wear
- Bearings are machined elements designed to **take loads** and **reduce friction**

Bearings Uses

- Many parts of a machine are fitted with bearings
- Where loads are light
  - Bearings are simple and lubricated by simple means
- Where loads are heavy and constant
  - Bearings are much more important and lubrication is critical

Bearing Functions

- Decrease friction, heat and wear

- Support static weight of shafts and machinery

- Support radial and thrust loads

- Allow tighter fitting tolerances

Types of Bearings

- Two main types of bearings:
  - Plain Bearings
    - Sliding bearings, journal bearings
    - Shafts run directly on the bearing surface
  - Anti-friction
    - Rolling element bearings, rolling contact bearings
    - Balls or rollers as part of the bearing

Plain Bearings

Journal or plain bearings consist of a shaft or journal which rotates freely in a supporting metal sleeve or shell (There are no rolling elements in these bearings)

Is a shell of steel covered with a layer of white metal or babbitt (Alloy of: Tin, Antimony, Copper , Lead)

Plain Bearings

Simple shell-type journal bearings accept only radial loading, perpendicular to the shaft, generally due to the downward weight or load of the shaft (i.e.: Crankshaft bearings)

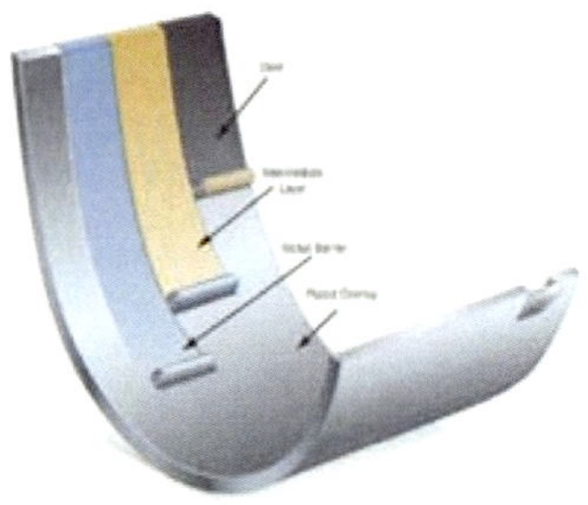

Antifriction Bearing Design

Types of loads on Bearings

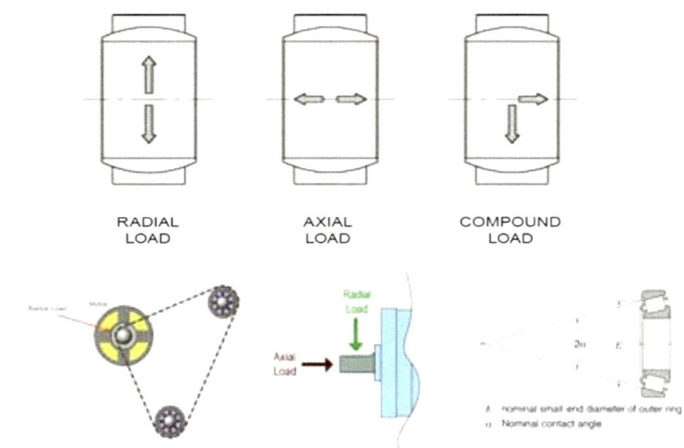

Bearing Loading

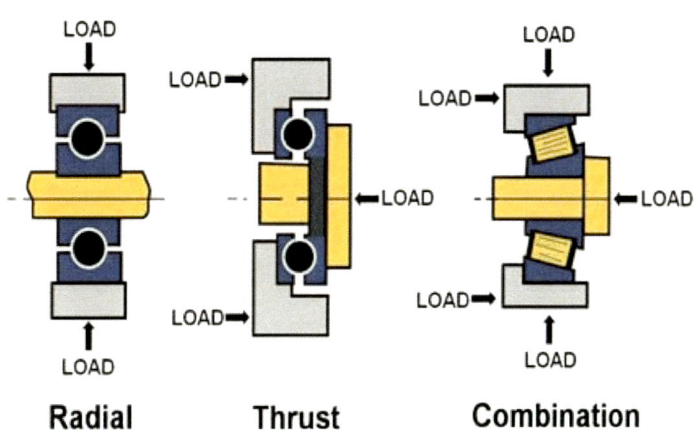

Ball Bearings

Ball bearings are used for high precision applications in a variety of markets, including aerospace , marine, agriculture, construction, health, machine tool and general industry

Radial ball Bearings

Radial ball bearings tolerate relatively high-speed operation under a range of load conditions. Bearings consist of an inner and outer ring with a cage containing a complement of precision balls

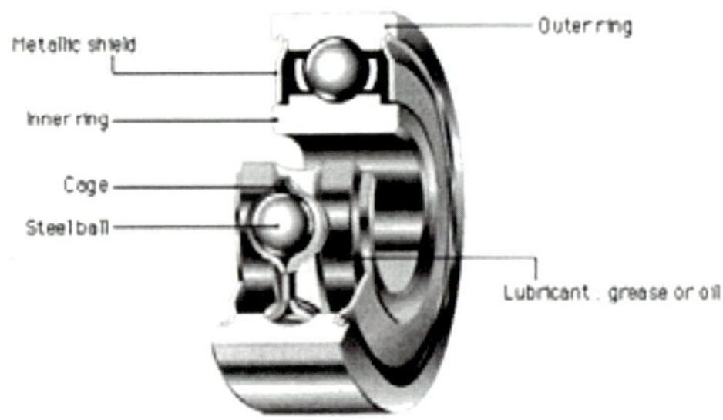

100 % Radial Bearing

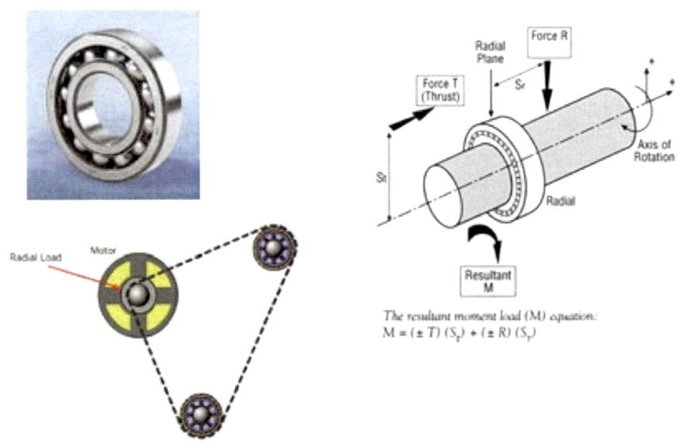

100 % Axial Load

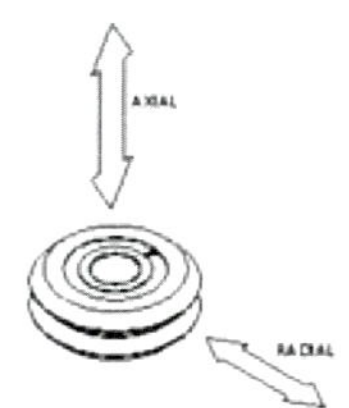

Compound Load

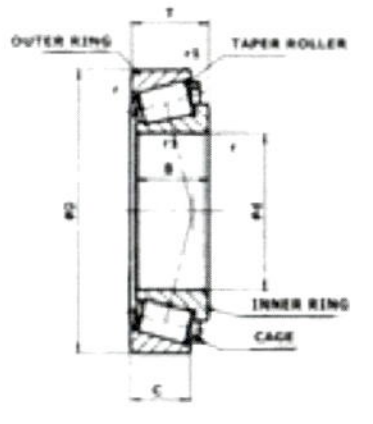  

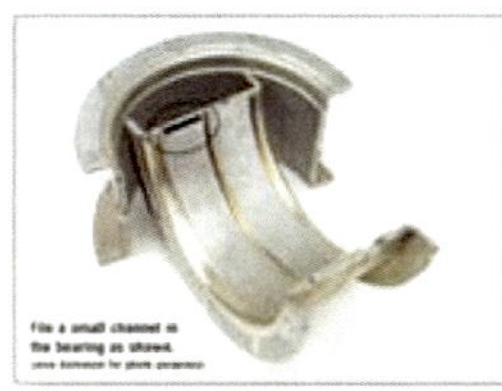 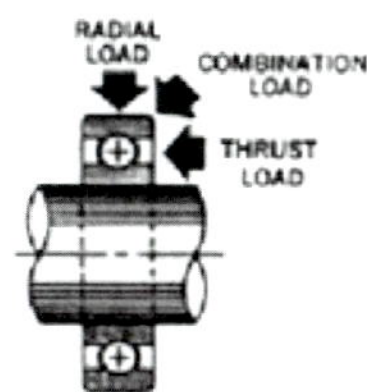

Angular ball bearing

- Angular contact bearings are designed such that a contact angle between the races and the balls is formed when the bearing is in use.

- The major design characteristic of this type of bearing is that one or both or the ring races have one shoulder higher than the other

 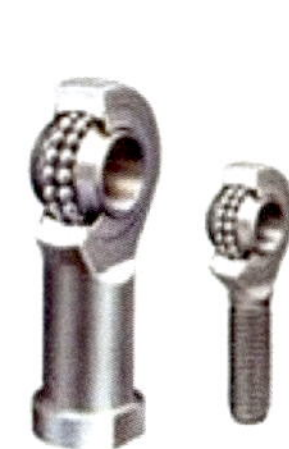 

Roller Bearings

- Three basic types:

  - Straight roller bearings 

  - Needle roller bearings 

  - Tapered roller bearings 

Roller Assembly, Caged Needle Rollers & Needle Thrust Bearing

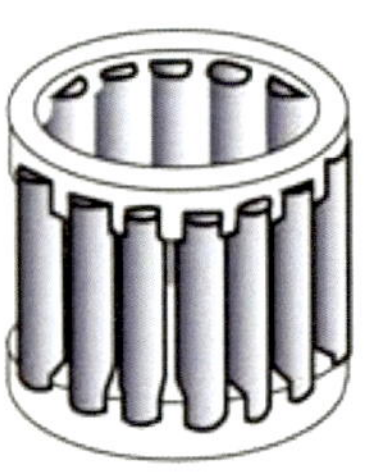

Roller Assembly   Caged Needle Rollers   Needle Thrust Bearing

Removing Bearings

Use appropriate tools to avoid damage the housing metal

Removing Bearing Tools

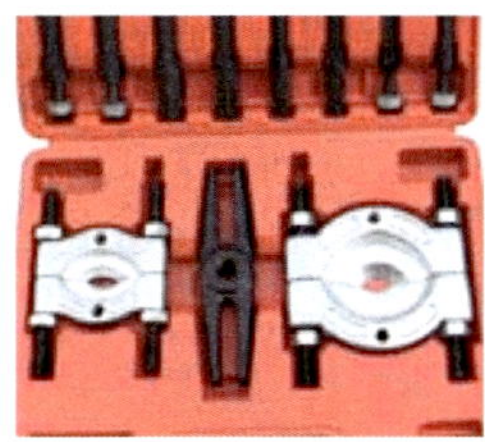
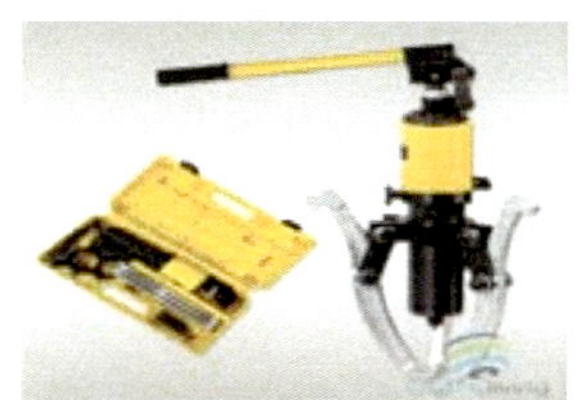

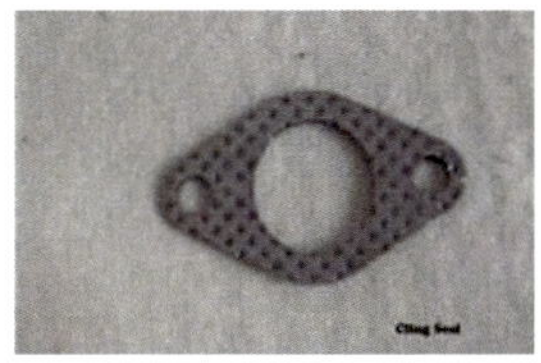
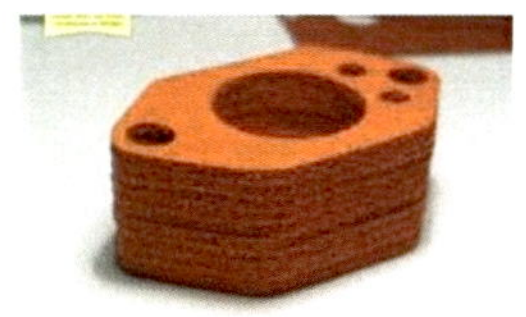

Gaskets

Materials and Types

Gaskets

- A **gasket** is a mechanical Seal which fills the space between two or more mating surfaces, generally to prevent leakage from or into the joined objects while under high compression
- Gaskets are commonly produced by cutting from sheet materials, such as gasket paper, rubber, silicone, metal, cork, felt and neoprene

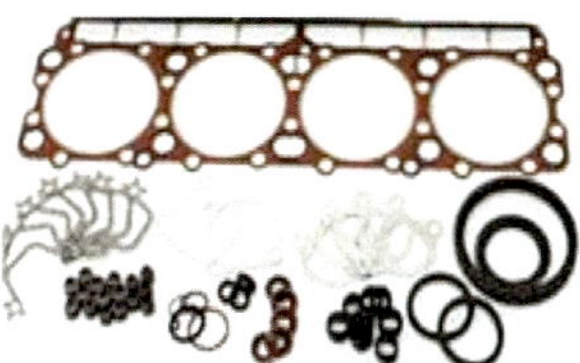

Gasket Material

- Cork and Cork composition
- Special joint materials
- Fiber and Nylon
- Synthetic Rubber
- Asbestos and Asbestos composition
- Copper or Steel and Asbestos
- Steel core and composition
- Corrugated Steel

Gasket Installation

The required compression for your gasket material will depend many factors including

1) Surface area
2) Pressure being sealed
3) Size of bolts (assuming bolts are being used)
4) Number of bolts
5) Condition of the bolts
6) Lubrication on the bolts

Gasket Installation

- Treat gasket with care

- Clean surfaces

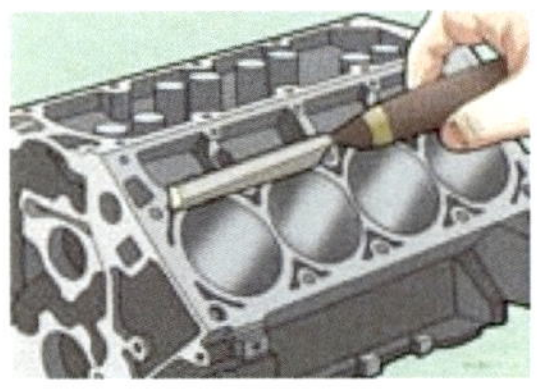

- Light coating of grease

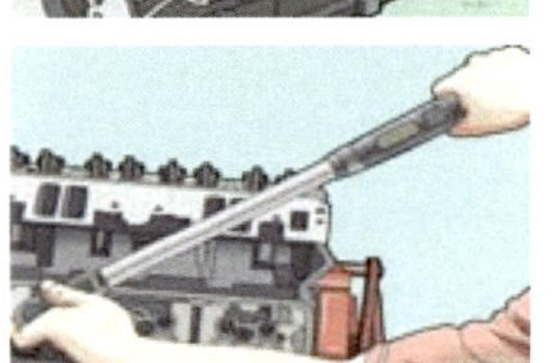

- Bolts tightened gradually and evenly

Never Reuse a Gasket

Go back to disassemble all parts to replace a seal in poor condition is more expensive than always to use original seals

Handling and Storing Gaskets

- Store gaskets flat

- Gasket kits
  - Some gaskets kits are made for more than one model

- Try to use original gaskets

- Some gaskets may shrink

Manifold Gaskets

- Intake manifolds – seal against air leaks

- Exhaust manifolds – seal against hot gas leaks (Stainless)

- Tightened:
  - Gradual
  - In the correct sequence
  - Correct torque specifications

Metal Gaskets

Some (piping) gaskets and Exhaust Manifold gaskets are made entirely of metal and rely on a seating surface to accomplish the seal; the metal's own spring characteristics are utilized (up to but not passing σ_y, the material's yield strength)

Gaskets / Sealants

- It is usually desirable that the gasket be made from a material that is to some degree yielding such that it is able to deform and tightly fills the space it is designed for, including any slight irregularities
- A few gaskets require an application of sealant directly to the gasket surface to function properly

O-Ring Seals

O-Rings Materials

- **Propylene**. It has fair resistance to brake fluids and phosphate esters while exhibiting good resistance to petroleum oils (Diesel Fuel)
- **Ethylene Propylene .** This material was introduced in 1961 and found broad acceptance in application requiring excellent resistance to Skydrol (ATF) and other phosphate ester fluids at higher temperatures

O-Rings Materials

- **Fluorocarbon**. They are especially good for hard vacuum service (Surgery equipment) and low gas permeability (Fuel Pumps).
- **Fluorosilicone.** They are used in aerospace applications for fuel systems (Fuel Injectors and Fuel Injection Pumps for Diesel Engines).
- **Neoprene**. Due to the excellent resistance to refrigerants such as Freon, neoprene compounds are used in refrigeration systems

How to Specify O-Rings

When you order an O-ring, the manufacturer needs to know the inside diameter (I.D.), the cross sectional diameter (W), and the compound (elastomer formula) from which it is to be made

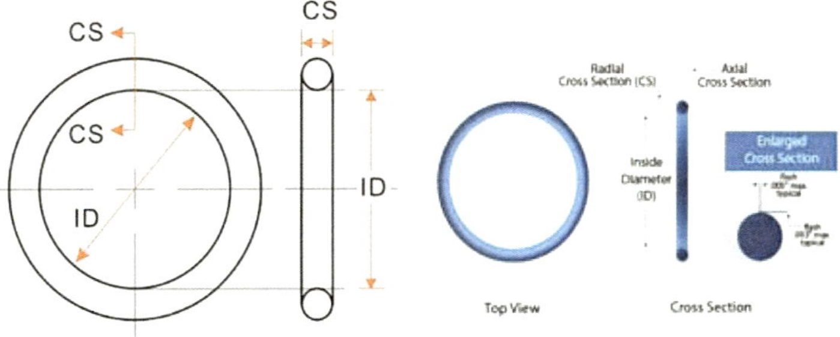

How to Select O-Rings

Selection of the appropriate O-ring is based on such factors as chemical compatibility, the temperature range of the given application, sealing pressures, lubrication needs and durability

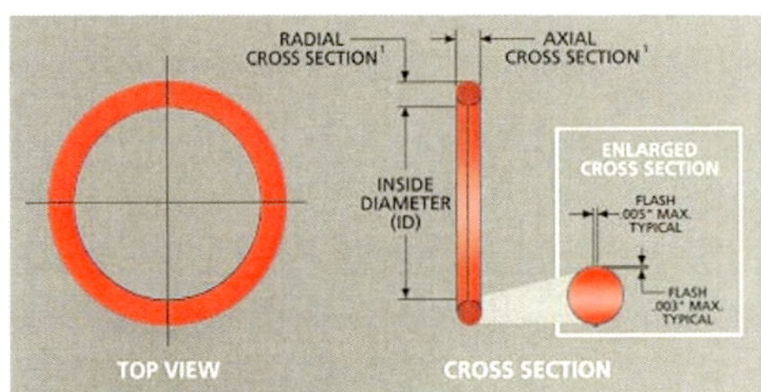

O-Ring Color Codes

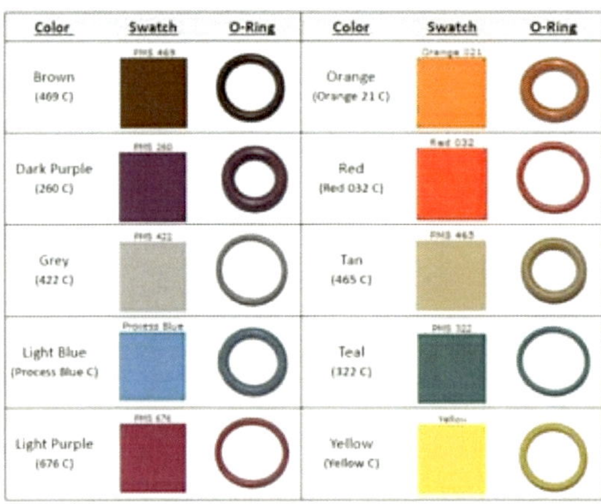

Color	Swatch	O-Ring	Color	Swatch	O-Ring
Brown (469 C)	PMS 469		Orange (Orange 21 C)	Orange 021	
Dark Purple (260 C)	PMS 260		Red (Red 032 C)	Red 032	
Grey (422 C)	PMS 422		Tan (465 C)	PMS 465	
Light Blue (Process Blue C)	Process Blue		Teal (322 C)	PMS 322	
Light Purple (676 C)	PMS 676		Yellow (Yellow C)	Yellow	

Oscillating Seal

OSCILLATING SEALS, in which the inner or outer member of the assembly moves in an arc relative to the other, rotating one of the members in relation to the O-ring

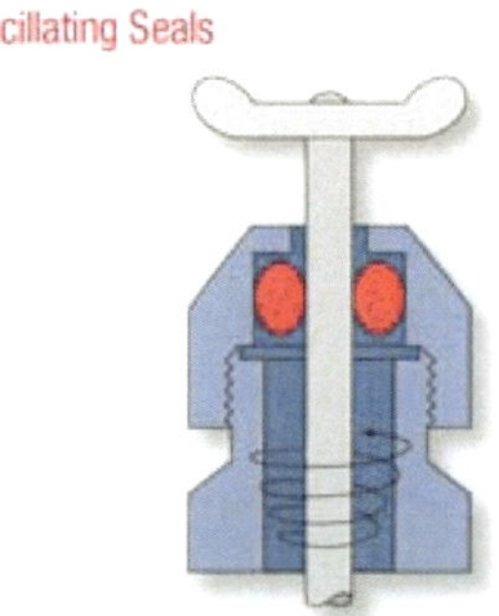

O-Rings or Rotary Seals

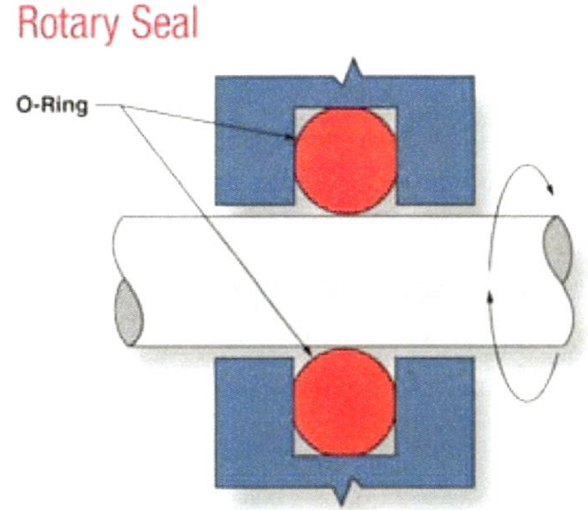

Rotary Seal

ROTARY SEALS, where an inner or outer member of the sealing assembly revolves around the shaft axis in only one direction

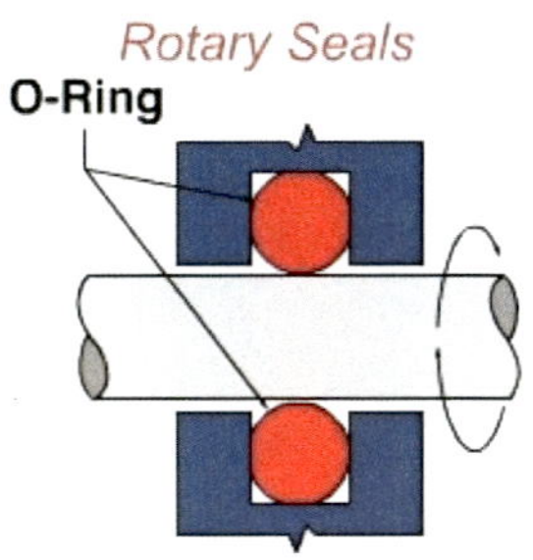

Seat Seals

Utilize an O-ring to close a flow passage by distorting the face of the O-ring against the opposite contact face Closing the passage distorts the seal element to create the closure

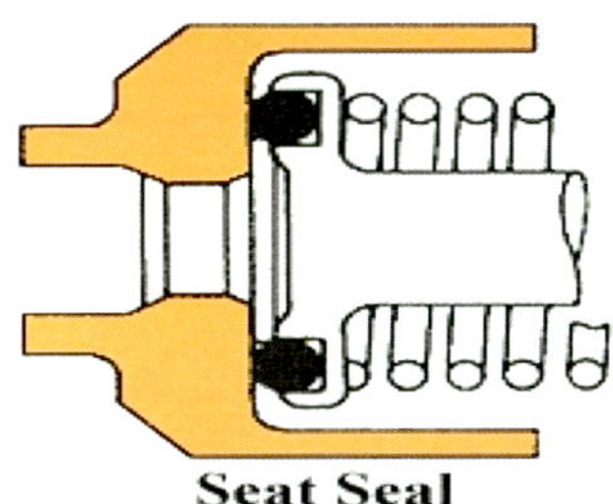

CRUSH SEALS

A variation of the static seal. literally "crushes" an O-ring into a space with a cross section different from the standard gland. Although often an effective seal. the O-ring is permanently deformed and must be replaced if the unit is opened.

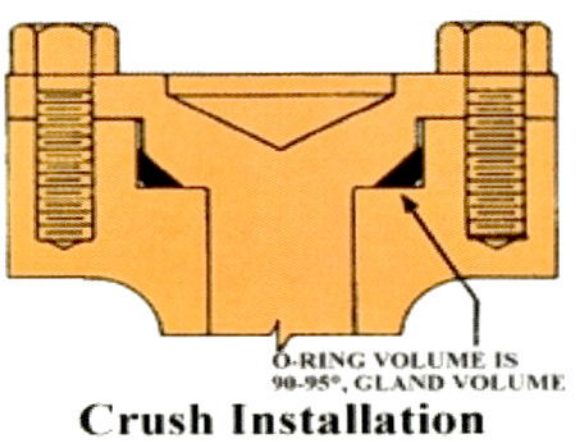

Crush Installation

How to Specify O-Rings

- When you order an O-ring, the manufacturer needs to know the inside diameter (I.D.), the cross sectional diameter (W), and the compound (elastomer formula) from which it is to be made
- Selection of the appropriate O-ring is based on such factors as chemical compatibility, the temperature range of the given application, sealing pressures, lubrication needs and durability

O-Ring Failure Diagnosis

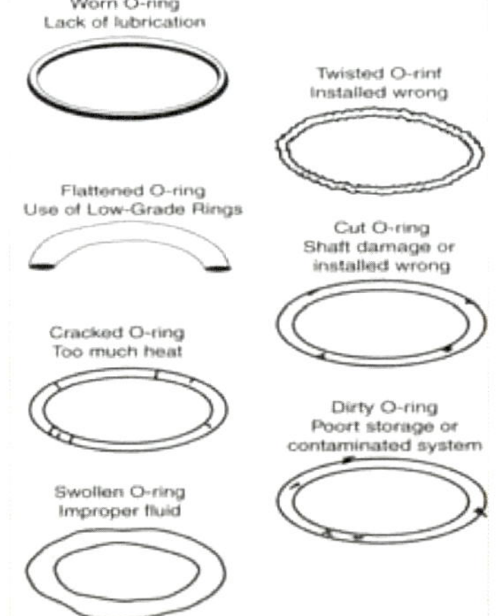

Correct Size O-Ring

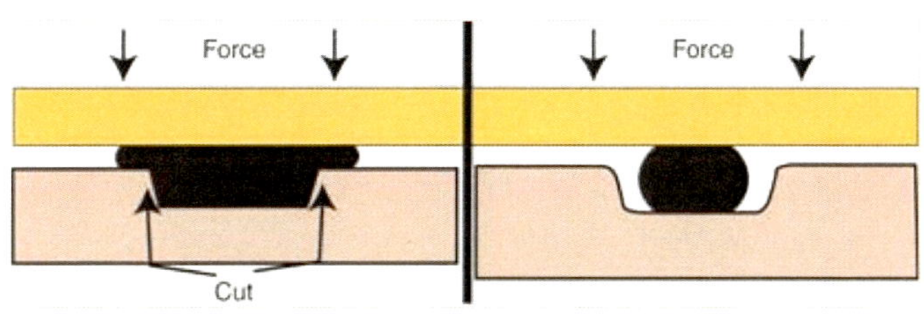

O-Ring Installation

- Ensure O-ring is compatible with fluid
- Clean the entire area
- Inspect O-ring grooves
- Inspect shaft or spool
- Lubricate the O-ring
- Install
- Align parts accurately

Check O-Ring After Installation

- Static O-ring:
  - Torqued again after unit is warmed and cycled several times

- Dynamic O-ring:
  - Cycled through their normal pattern
  - Allow to assume neutral position

- All dynamic rings should pass a very small amount of fluid when rotating

Hydraulic Seals

Uses, Applications and Installation Tips

Seals

- A piece of material or a method that prevents or decreases the flow of fluid or air between two surfaces

- The sealed surface may be stationary or have movement between them

Duties of a Seal

- Prevent lubrication leaks

- Keep out dirt and foreign bodies

- Keep different fluids apart

- Remain flexible

- Seal rough surfaces

- Wear faster than more expensive parts

Types of Hydraulic Seals

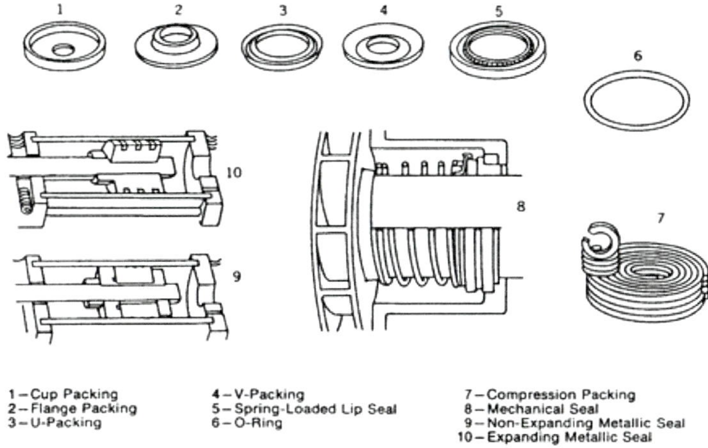

Hydraulic Seals

Hydraulic seals are devices that are designed to keep fluids from escaping from engines, transmissions, cylinders or pumps, and to prevent foreign contaminants from entering them

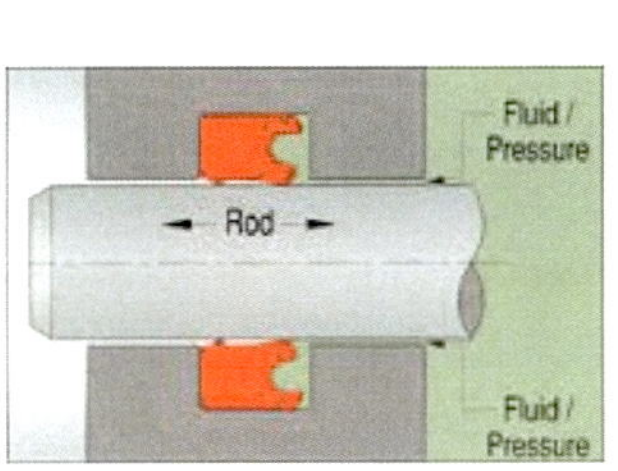

Spring-Loaded Lip Seal

Spring Loaded Seals are used for linear or reciprocating rod and piston sealing applications, for rotating shafts as rotary seals, and for static and face sealing applications

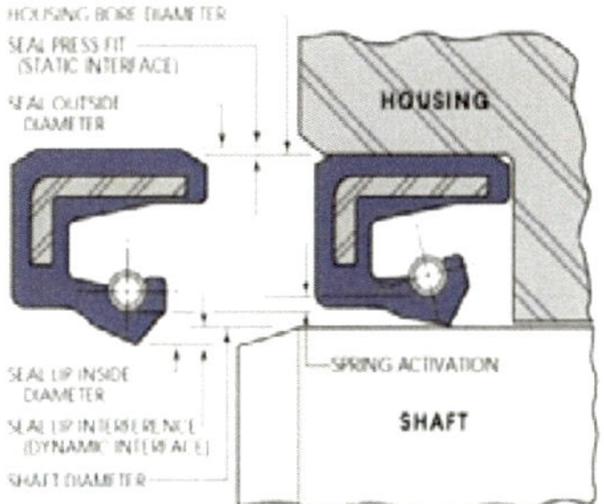

Cased Oil Seals

They are typically used in areas where elastomeric seals cannot meet the frictional, temperature, or chemical resistance requirements of the application

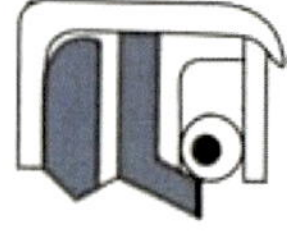

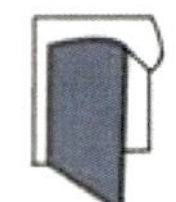
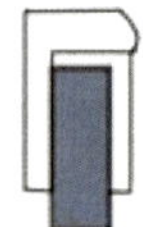

Standard Loaded Lip Seal

Loaded Lip Seals are very versatile and can be used as rod seals or piston seals. They are considered to be multi-purpose hydraulic seals and are designed for significantly improved performance

Rotary Shaft Seal

Hydraulic seals include shaft outer diameter **(OD)** or seal inner diameter **(ID)**, housing bore diameter or seal outer diameter, axial cross section or thickness **(t),** and radial cross section. Common features for hydraulic seals include spring loaded, integral wiper, and split seal

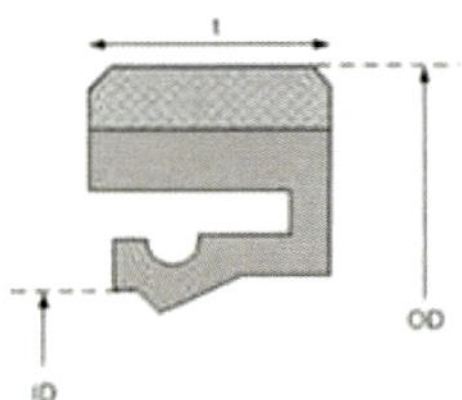

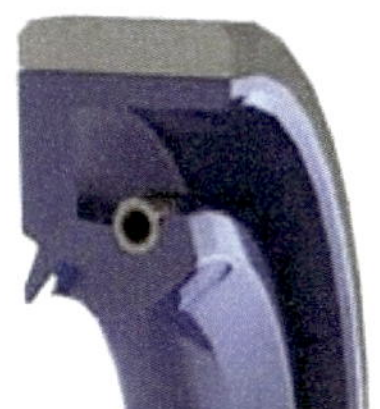

Servicing Oil Seals

- If a seal is removed for whatever reason – it should be replaced
- When installing a seal:
  - Ensure it is not damaged or distorted
  - Install the correct way
  - Lip-type seals must have the lips pointing in the right direction

Servicing Oil Seals

- Protecting seals while installing

- Lubricating seals before installing

- Checking sealing surfaces

- Checking boots

- Hydraulic brake seals

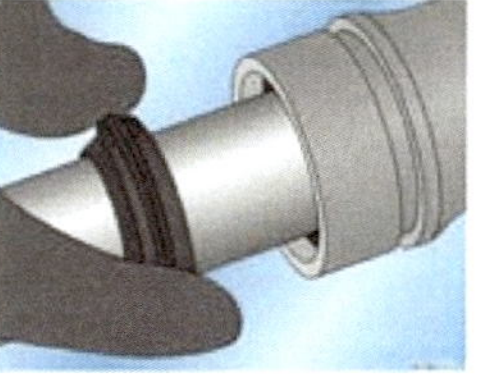

Checking Seals for Leakage

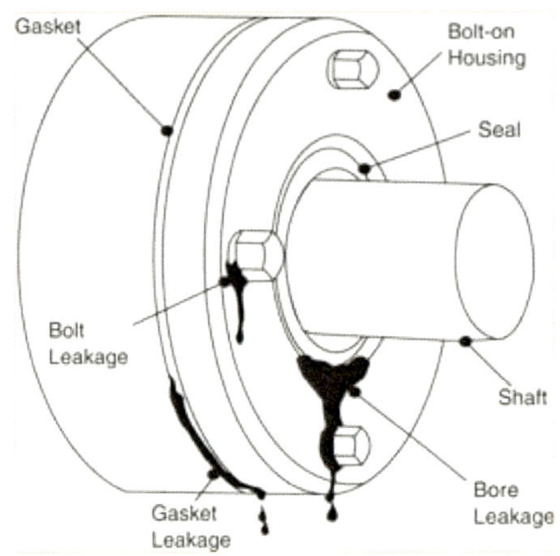

Removing Seals

To avoid damage the metal try to use appropriate tools to remove gaskets and seals

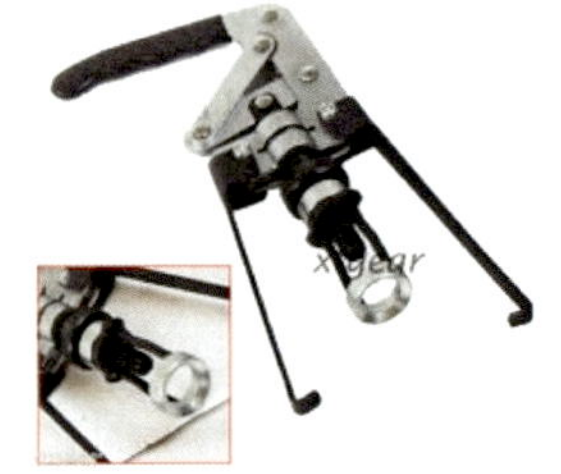

Removing Seals

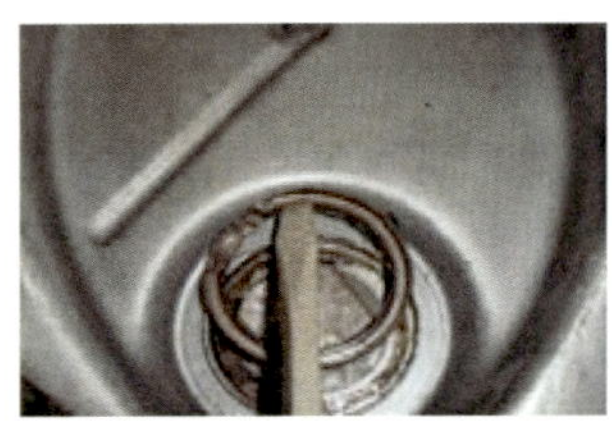

Removing seals

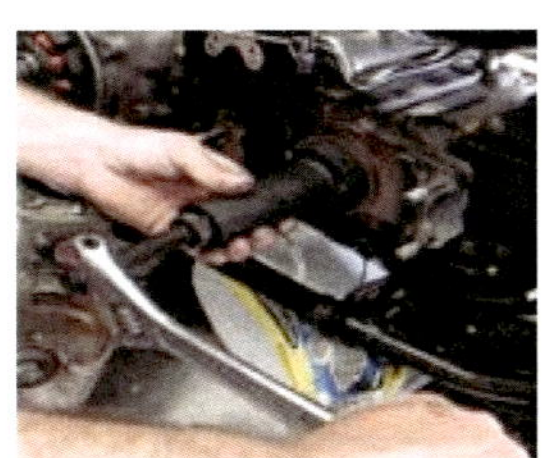

Seal Puller

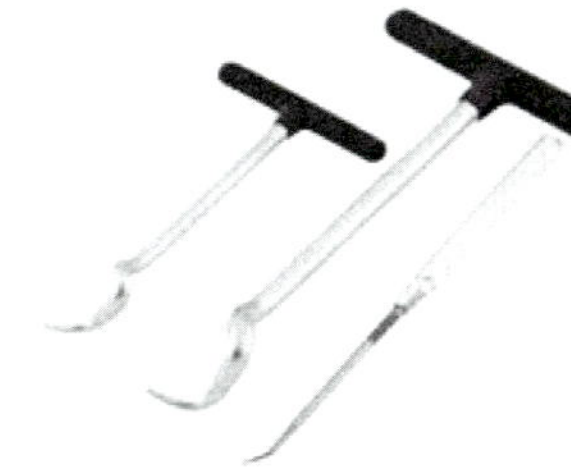

Installing Seals

Chapter 3
Marine Engines Classification

Marine Engines Classification

- **Inboard Engines**
  - Electric Motors
    - Alternating Current.
    - Direct Current
  - Diesel (2 &4 Strokes).
  - Gasoline
    - 4 Strokes

- **Outboard Engines**
  - 2 Strokes
  - 4 Strokes
  - Electrical Motors.
    - DC 12/24 Volts

Drive Systems

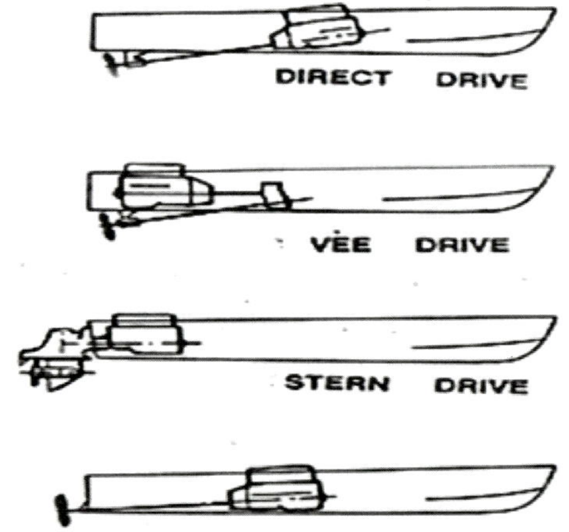

Inboard Engines

An inboard is a four-stroke automotive engine adapted for marine use. Inboard engines are mounted inside the hull's midsection or in front of the transom

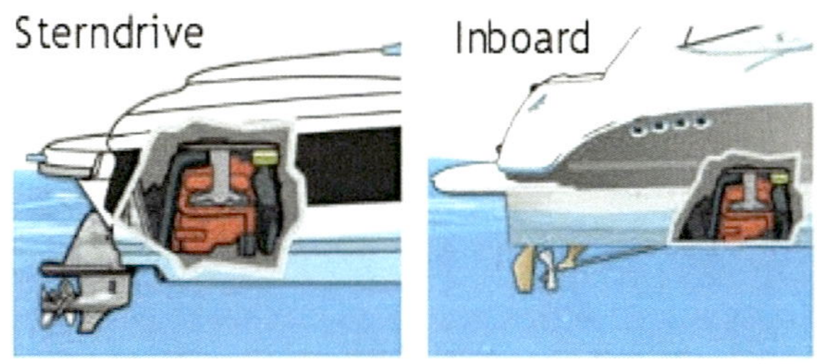

Inboard Engines

The engine turns a drive shaft that runs through the bottom of the hull and is attached to a propeller at the other end

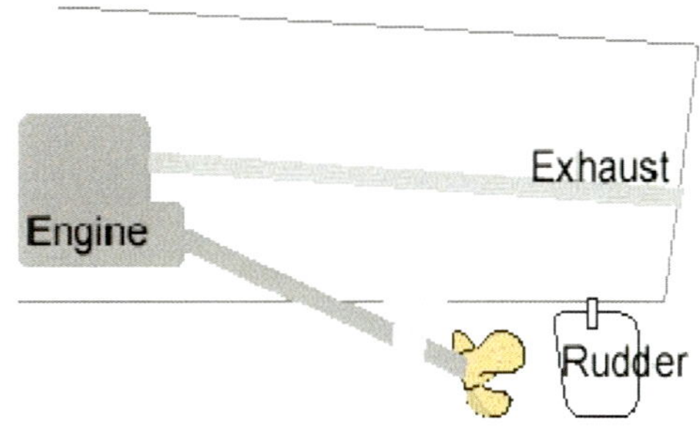

Inboard Engines

Steering of most inboard vessels, except PWCs and jet-drive boats, is controlled by a rudder behind the propeller

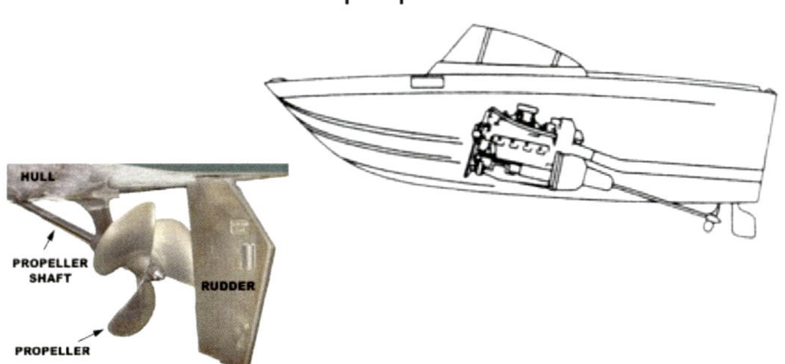

Diesel Engines

A **diesel engine** is an internal combustion engine that uses the heat of compression to initiate ignition to burn the fuel, which is injected into the combustion chamber during the final stage of compression

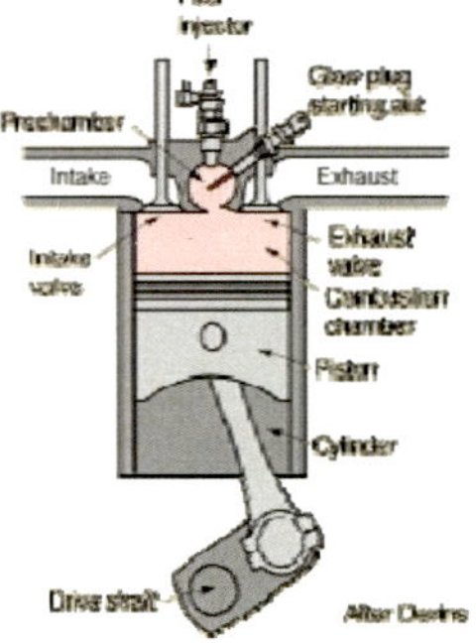

Diesel Engines

This is in contrast to a gas engine , which uses the Otto cycle , in which a fuel/air mixture is ignited by a spark plug

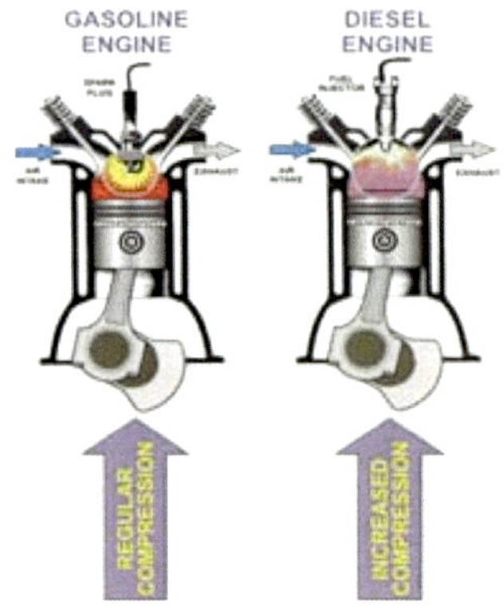

Diesel Engine Systems

• Fuel System

• Intake & Exhaust System

• Lubrication System

• Cooling System

• Ignition System

Fuel System

- Carburetors

- (EFI) Electronic Fuel Injection

- (DFI) Direct Fuel Injection

- Mechanical Fuel Injection

Cooling System

- Coolant Pump

- Salt Water Pump

- After Cooler

- Inter Cooler

Lubrication System

- Oil Pump.
  - Mechanical Pumps.
  - Hydraulic Pumps.
  - Electro-Mechanical Pumps

- Oil Pan

Diesel Engines Efficiency

Diesel engines have the highest thermal efficiency of any internal or external combustion engine, because of their compression ratio. Low-speed engines diesel engines thermal efficiency exceeds 50%

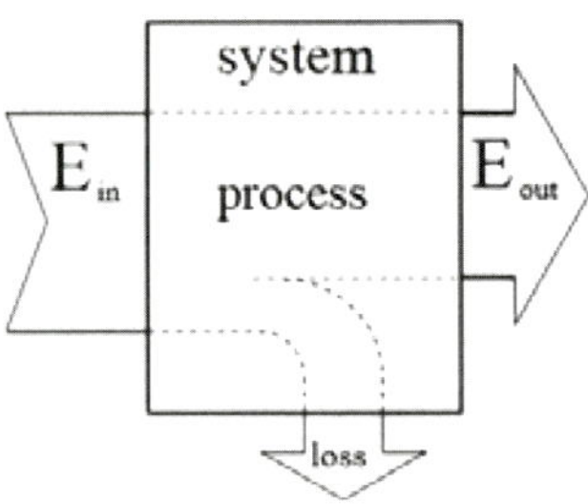

Diesel Engines Efficiency

Diesel engines have the highest thermal efficiency of any internal or external combustion engine, because of their compression ratio. Low-speed engines diesel engines thermal efficiency exceeds 50%

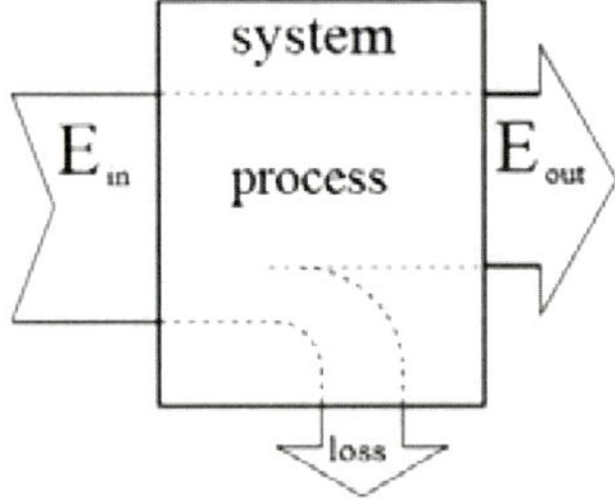

Diesel Engines

Diesel engines are manufactured in **two stroke** and **four stroke** versions

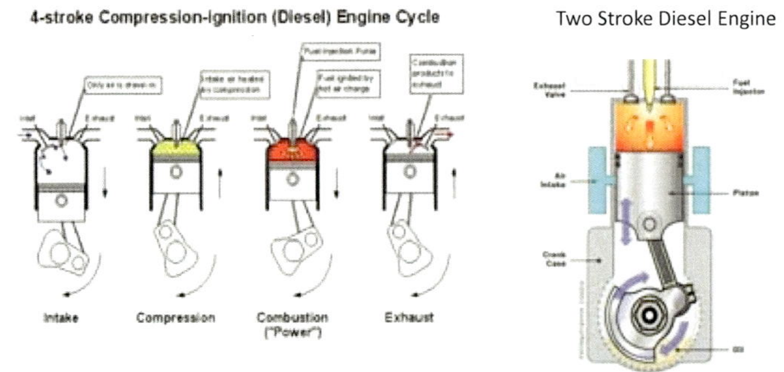

How Diesel engines work

In a diesel engine, only air is injected into the combustion chamber , during the intake stroke

The fuel is sprayed through the injector (2) at the end of the compression stroke just before the TDC

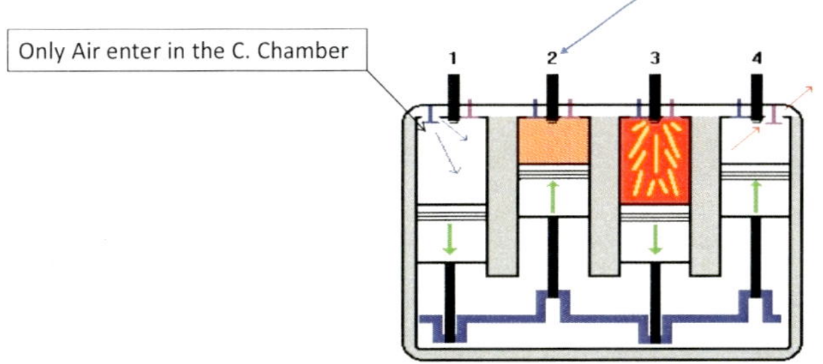

Combustion Temperatures

This high compression heats the air to 550 °C (1,022 °F). At about this moment fuel is injected directly into the compressed air in the combustion chamber

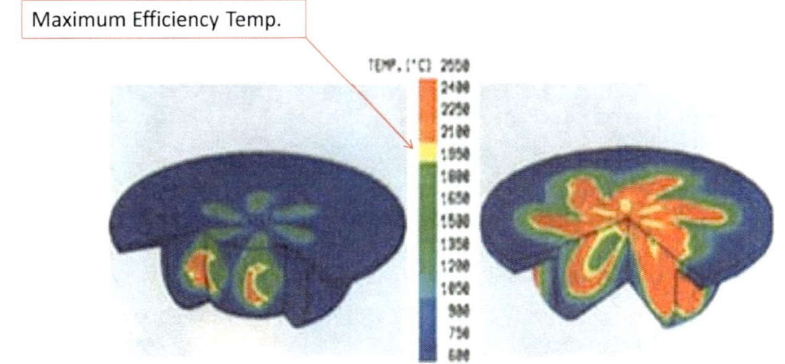

Engine Components

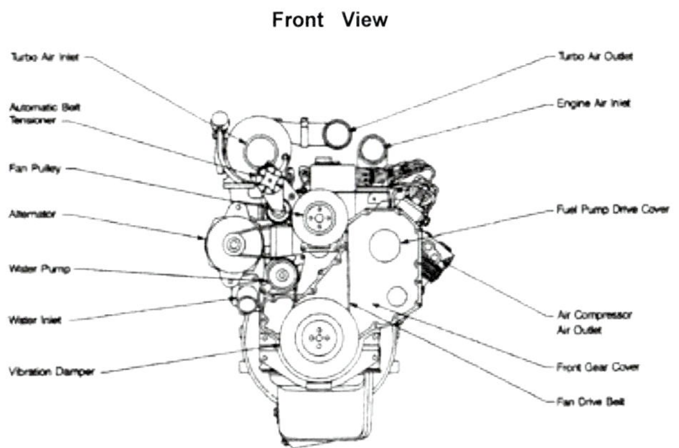

Engine Components

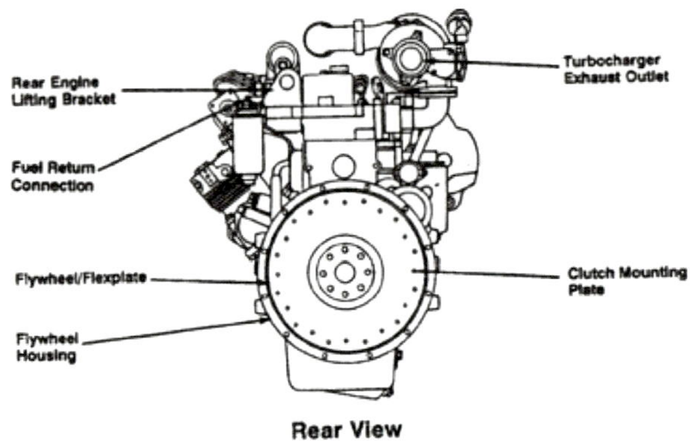

Rear View

Engine Components

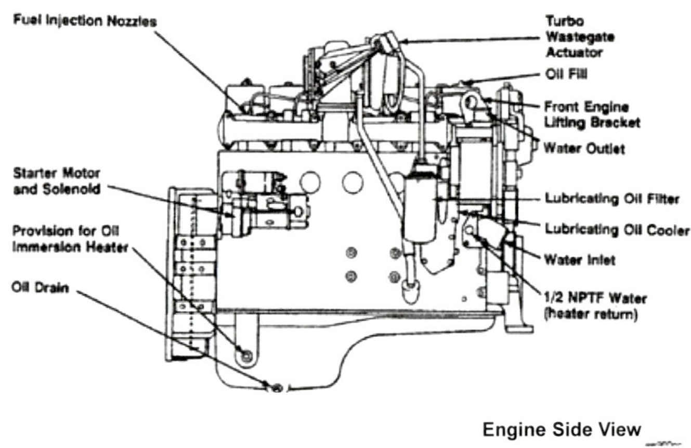

Engine Side View

Cylinder Arrangement

Common cylinder arrangements are from 1 to 6 cylinders in line or from 2 to 16 cylinders in V formation

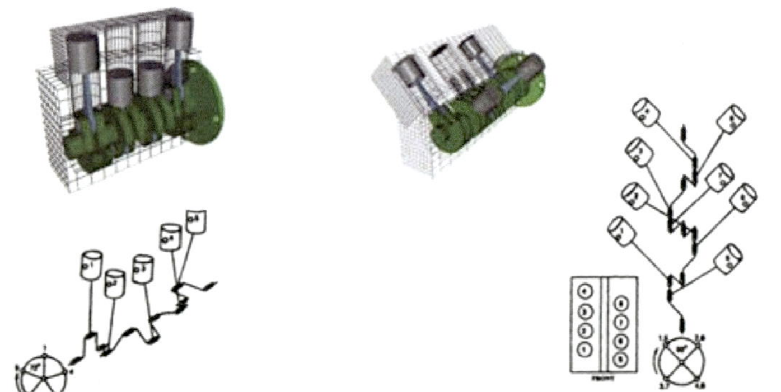

Cylinder Arrangement

Flat engines -- like a V design flattened out-- are common in small airplanes and motorcycles and were a hallmark of Volkswagen automobiles into the 1990s. Flat 6s are still used in many modern Porsches, as well as Subarus.

Cylinder arrangement

In flat engines a weak lubrication can produce catastrophic consequence, especially in the lower part of the cylinder

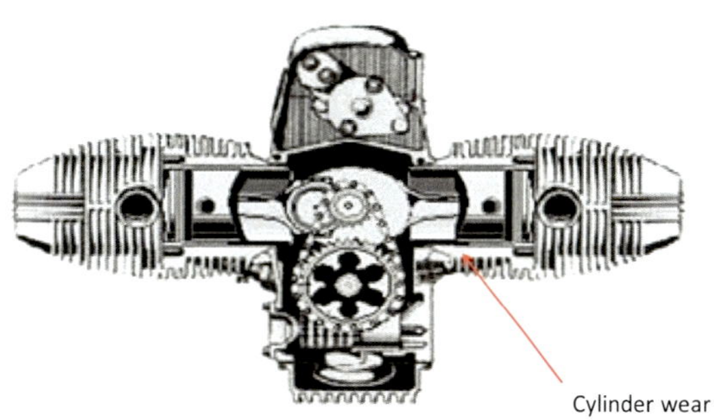

Cylinder Arrangement

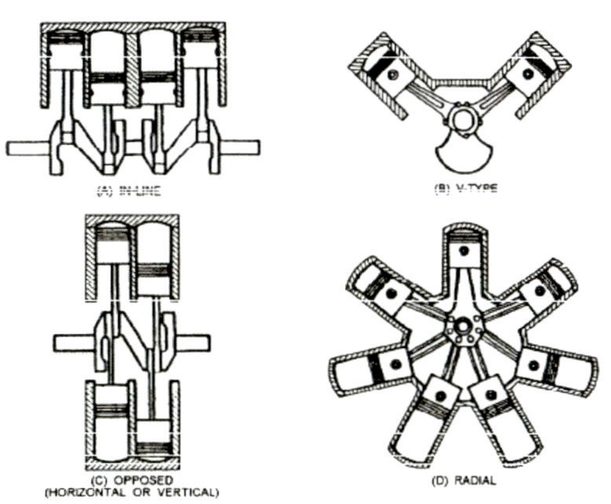

Figure 2-10.—Typical cylinder arrangements

Straight engine

- **The cylinders are arranged in a line in a single bank**
- the **straight engine** or **in-line engine** is an engine with all cylinders aligned in one row, with no or only minimal offset
- The straight-4 is by far the most common 4 cylinder configuration, whereas the straight-6 is slowly giving way to the V6, which requires less space

Straight Block Engine

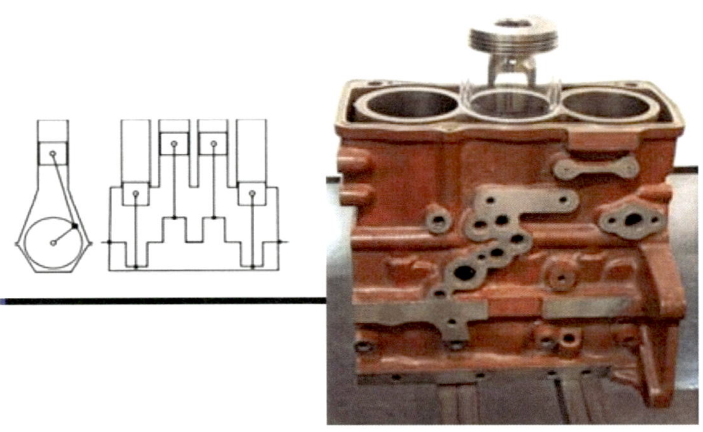

Flat configuration

- The cylinders are arranged in two banks on opposite sides of the engine
- flat engines or boxer engines or horizontally opposed engines the pistons lie horizontally oppo:

V Block Configuration

The most common V-block configurations are at 90 or 120 degrees

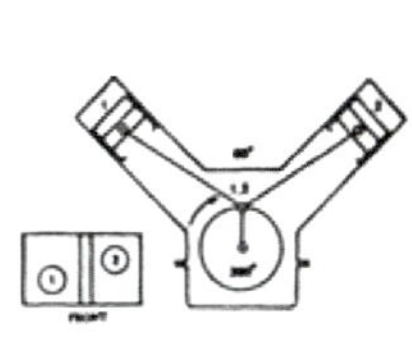

Two Stroke engines

- A **two-stroke** engine is a combustion engine that completes the cycle in two movements of the piston compared to twice that number for a four-stroke engine
- This increased efficiency is accomplished by using the beginning of the compression stroke and the end of the combustion stroke to perform simultaneously the intake and exhaust (or scavenging) functions

Two Stroke Diesel Engine

The four "cycles" (intake, compression, power, exhaust) occur in one revolution, while in a Two-stroke engine it occurs in two complete revolutions

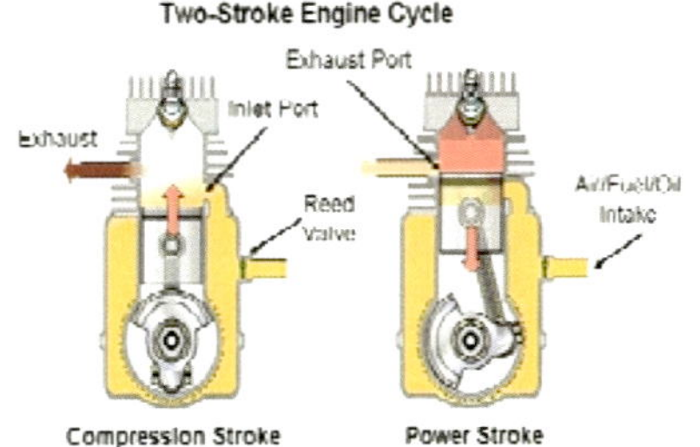

2 Stroke Cycle (Diesel)

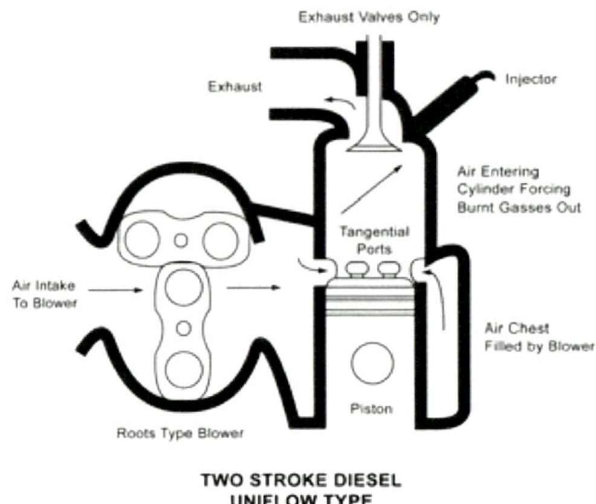

INTAKE & EXHAUST

- First Stroke : Intake + Exhaust for Gas engine
- In a two stroke gas engine , the mixture (Air, Gasoline + oil), is injected in the combustion chamber by differential pressure at the first stroke

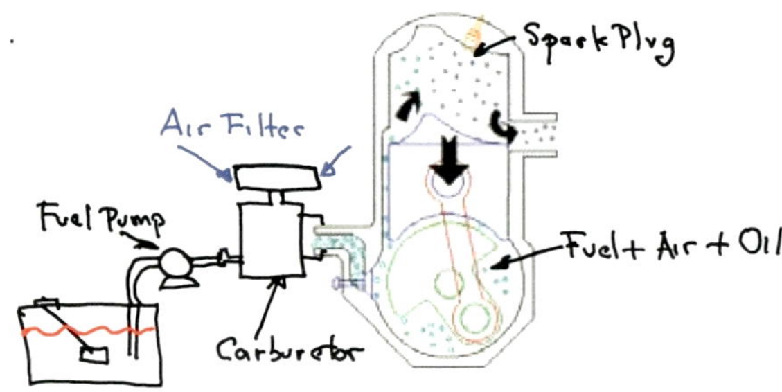

Two Stroke Engine (Diesel)
[Exhaust Valves on the Head]

- Complete the same cycle in only two strokes
- **Intake + Exhaust** = <u>First Stroke</u>

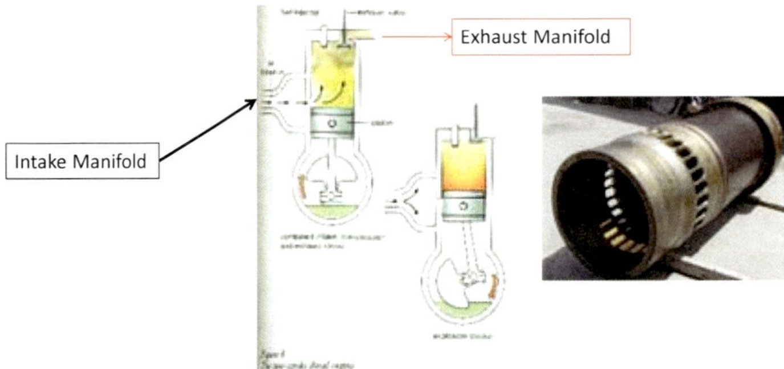

INTAKE   &   EXHAUST (Diesel)

- Toward the end of the stroke, the piston exposes the intake port, allowing the compressed fuel/air mixture in the crankcase to escape around the piston into the main cylinder. This expels the exhaust gasses out the exhaust port, usually controlled for the exhaust valves
- Unfortunately, some of the fresh fuel mixture is usually expelled as well.

Second Stroke (Gas Engine)

- Second Stroke : Compression + Power , with piston going down (Gas Engines)

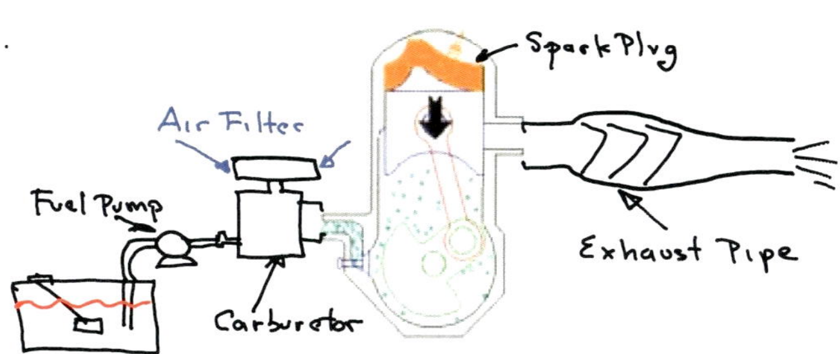

Two Stroke Engine

- **Power + Compression** = <u>Second Stroke</u>.
- This allows a power stroke for every revolution, instead of every revolution as in a four stroke

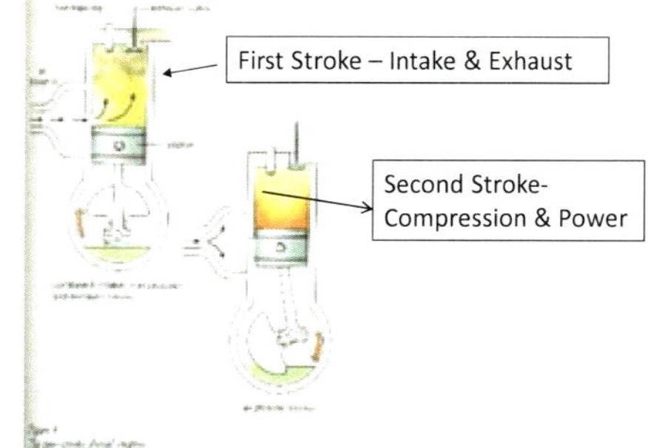

The Four Stroke Cycle
(Diesel Engines)

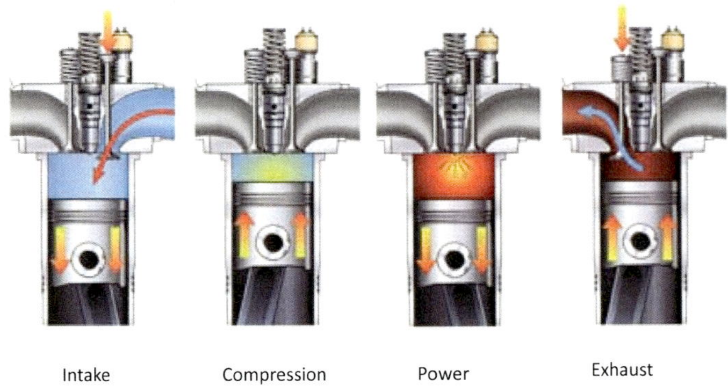

Intake Stroke

<u>Intake Valves</u> open and <u>Exhaust valves</u> closed
Air enter during the intake stroke when the piston is going down

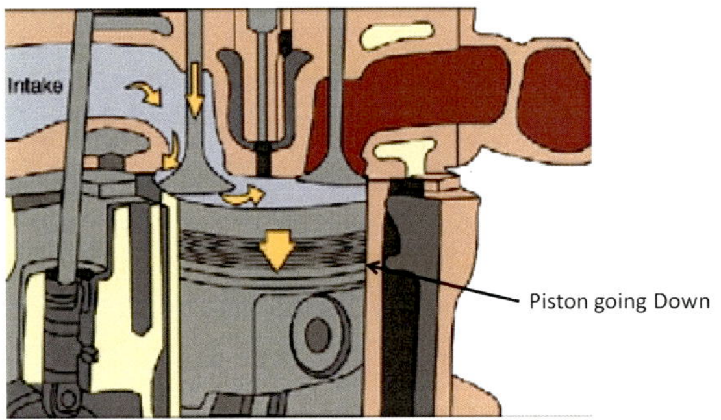

Compression Stroke

Intake & Exhaust Valves remain closed
Air is compressed and just before the piston
reaches the top dead center , the fuel is injected

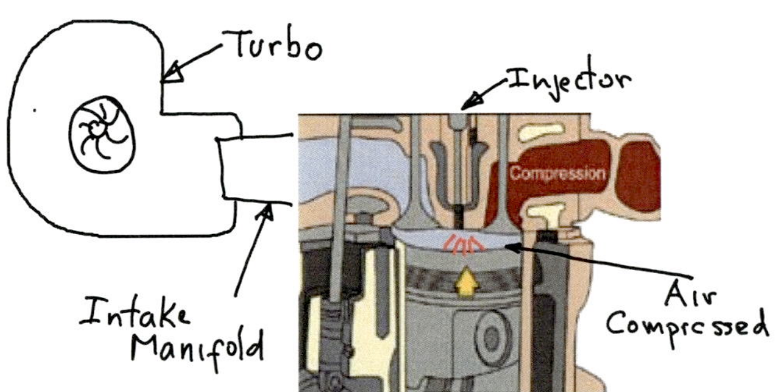

Power Stroke

Intake & Exhaust Valves remain closed
The intake and exhaust valves remain closed to seal
the combustion chamber

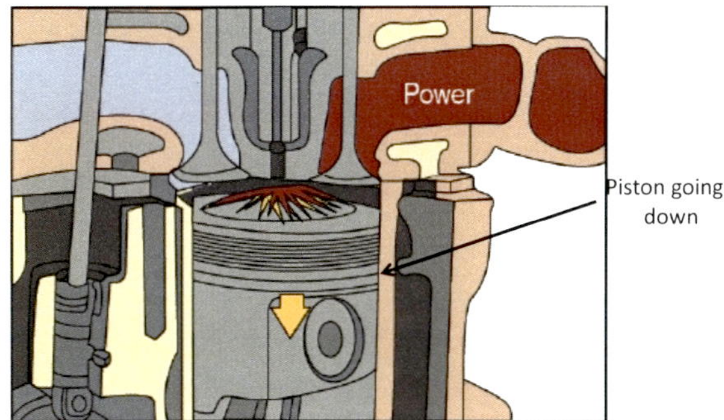

Exhaust Stroke

Intake Valves closed and Exhaust Valves open
The exhaust valve opens as the piston moves up and forces the
combustion gases out of the cylinder. Near TDC the exhaust
valve closes, the intake valve opens, and the cycle starts again

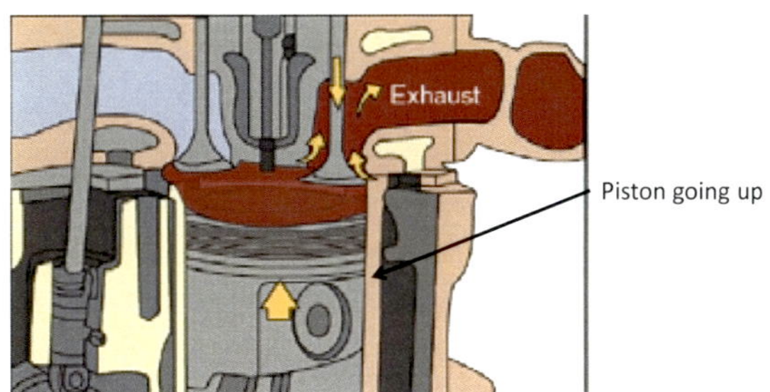

Chapter 4
The Combustion Chamber

Combustion Chamber

- The space at the head end of an internal combustion engine cylinder where most of the combustion takes place
- In Diesel engines, combustion is initiated when the fuel enter in the combustion chamber at the end of the compression stroke

Combustion Chambers

Direct Injection (Open Chamber)

Indirect Injection (Pre Combustion Chamber)
- Swirl Chamber
- Pre-combustion Chamber

Combustion Chamber Design

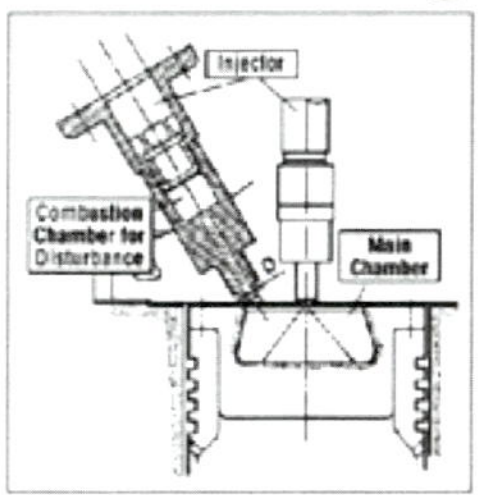

- Effects
  - Fuel efficiency
  - Engine performance
- Design of piston and method used to inject fuel determines how quickly and completely fuel burns

Combustion Chamber Design

- Pre-combustion design:
  - Use in earlier engines
  - No longer used by most manufacturers
- All current engines produced by Caterpillar & Cummins use Direct Injection

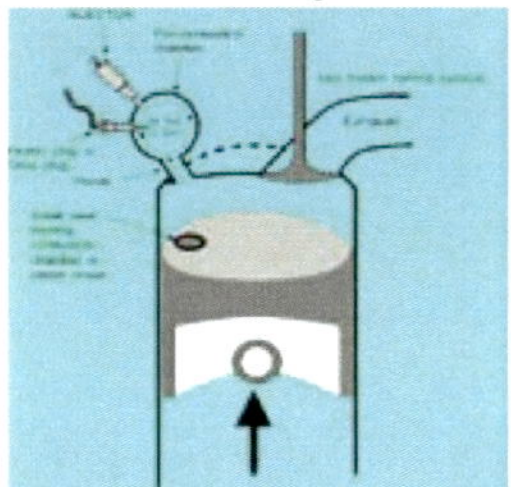

Wedge-Shaped Chambers

This type of chamber resembles an inclined basin recessed into the deck of the head. Inline valves are normally tilted to accommodate the sloping roof of this design

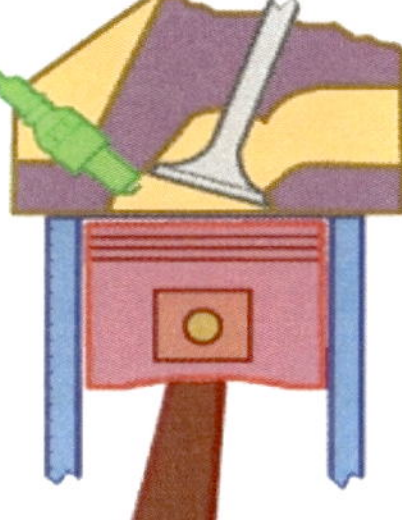

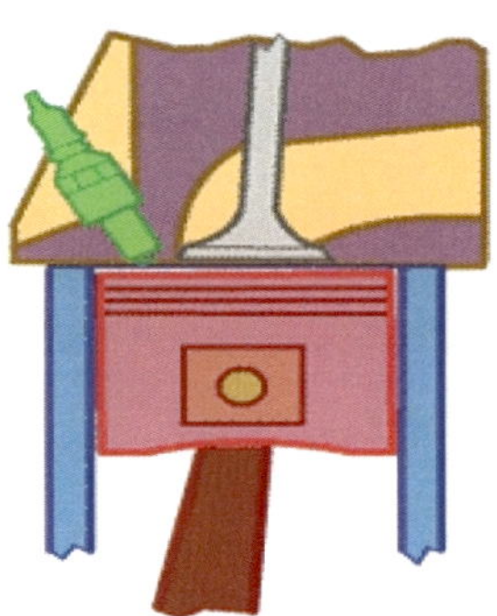

Hemi Spherical Chamber

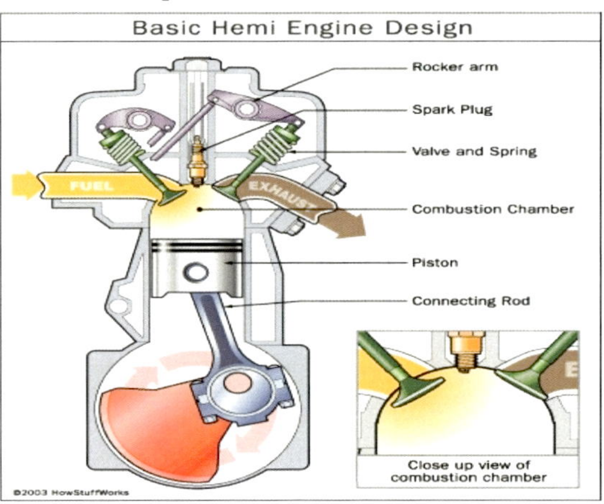

Fuel & Air Mixture

- The combustion process in the diesel engine is initiated by spontaneous ignition of the fuel when it is injected into a highly compressed charge of air, which has reached approximately **1,400 degrees Fahrenheit**

- In the diesel engine, the fuel is injected into the combustion chamber near the end of the compression stroke and ignites spontaneously

50

Fuel & Air Mixture

- The liquid fuel is usually injected at high velocity as one or more jets through small orifices or nozzles in the injector tip. It atomizes into small droplets and penetrates into the combustion chamber
- The ideal or stoichiometric Air/Fuel ratio for many gasoline type hydrocarbon fuels is very close to 15:1, for diesel engines its vary to 18 to 70

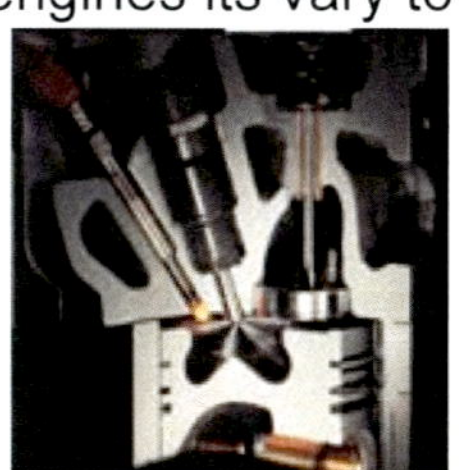

Fuel / Air Ratio (Diesel)

Diesel combustion is characterized by lean overall A/F ratio. The lowest average A/F ratio is often found at peak torque conditions. To avoid excessive smoke formation, A/F ratio at peak torque is usually maintained above 25:1, well above the stoichiometric (chemically correct) equivalence ratio. In turbocharged diesel engines the A/F ratio at idle may exceed 160:1

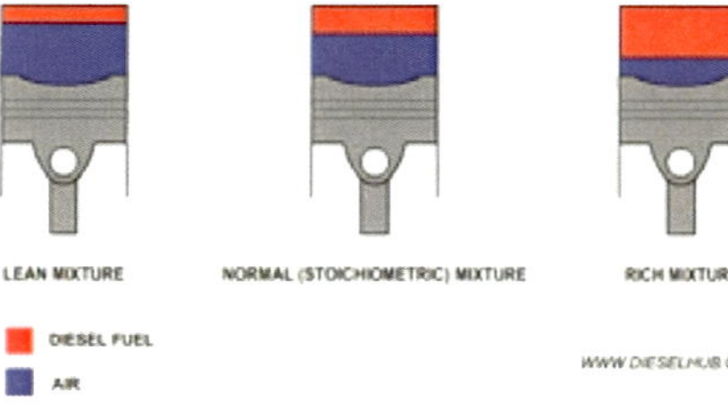

Air / Fuel Ratio (Diesel)

- The lowest average A/F ratio is often found at peak torque conditions. To avoid excessive smoke formation, A/F ratio at peak torque is usually maintained around 60:1
- In turbocharged diesel engines the A/F ratio at idle may exceed 160:1

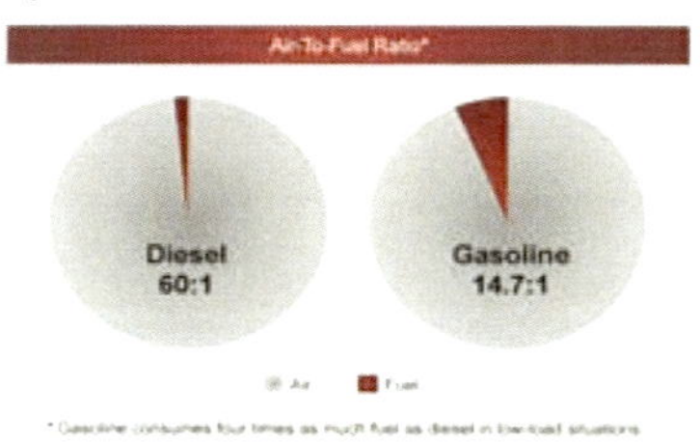

Admission of Fuel
Diesel Engine

- Only the air is compressed

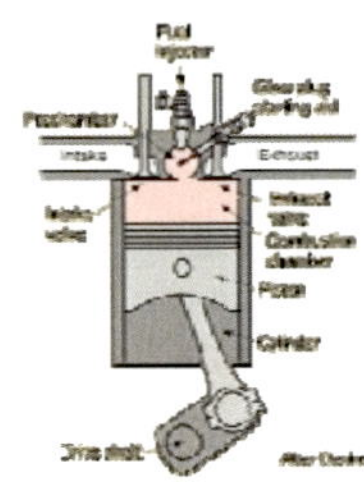

- Fuel
  - Metered by Fuel Injection Pump
  - Sprayed into combustion chamber
  - Mixes with hot compressed air and ignites

Admission of Fuel
(Gasoline Engine)

- Fuel and air are mixed prior to entering cylinder
- Ignited by a spark after it has been compressed

Engine Measurement Terms

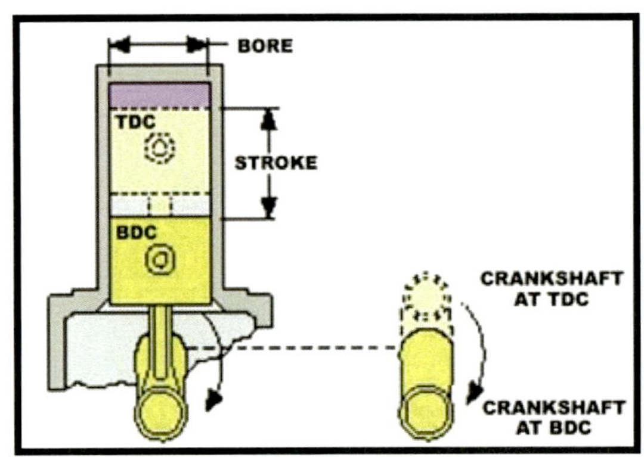

Top Dead Center

Top Dead Center (TDC) is a term used to describe the position of the piston when it is at its highest point in the cylinder

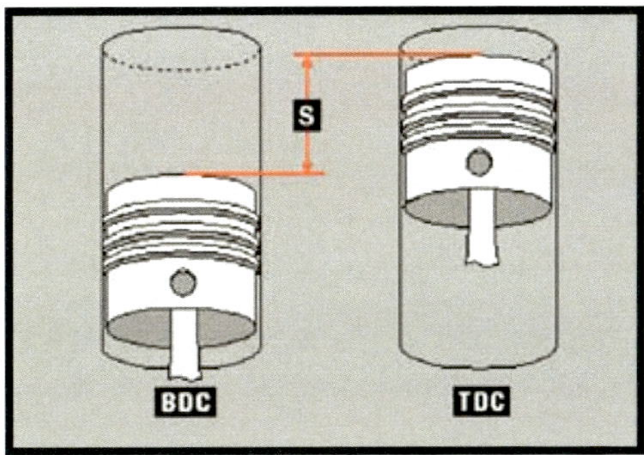

Bottom Dead Center

Bottom Dead Center (BDC) is a term used to describe the position of the piston when it is at its lowest point in the cylinder

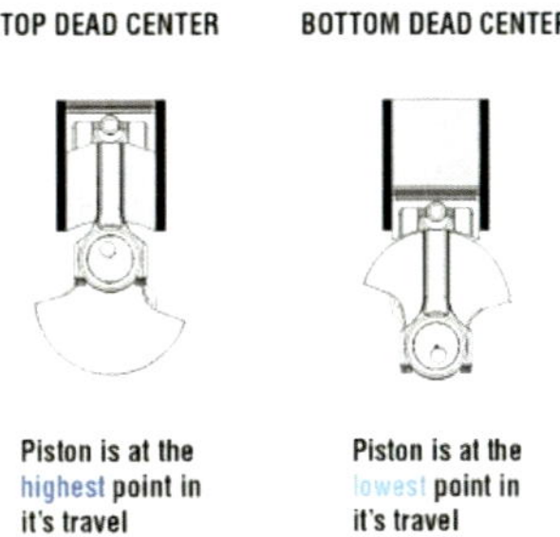

Piston displacement

- Is the Volume of Air displaced by a Piston as it moves from Bottom center to top center.
- To determine a Piston's displacement, you must multiply the Area of a Piston head by the length of the piston Stroke
- Piston Dis = [Pi X (BoreXBore)/4]X Stroke

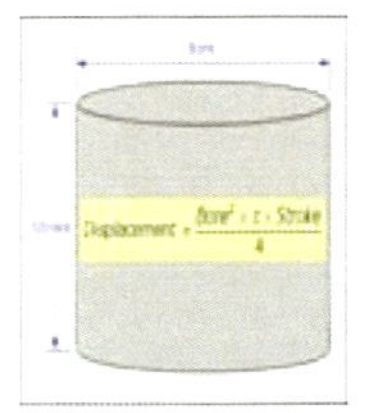

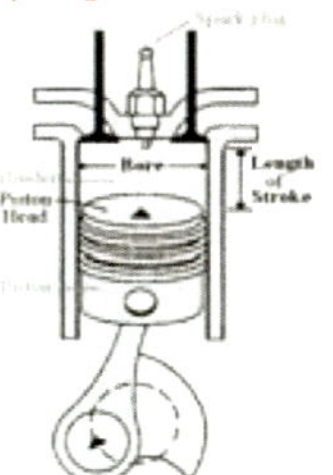

Engine displacement

Engine displacement is the volume swept by all the pistons inside the cylinders of an internal combustion engine in a single movement from *top dead center* (TDC) to *bottom dead center* (BDC)

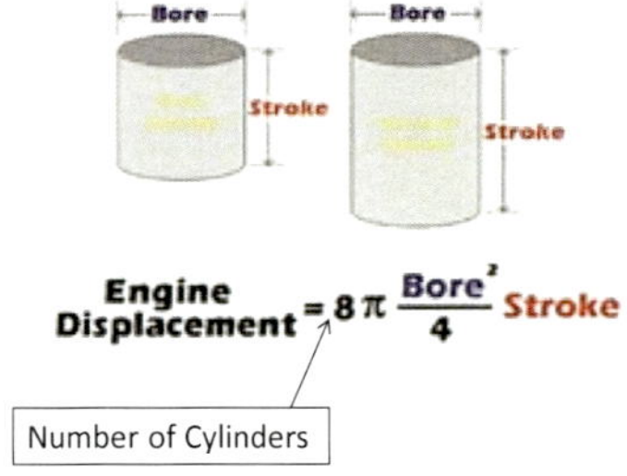

Engine displacement

- Engine displacement is determined from the bore and stroke of an engine's cylinders. The bore is the diameter of the circular chambers cut into the cylinder block.
- Eng Displacement = Pi * Bore2/4* Stroke * No de Cyl
- Example: The 427 Chevy bore is 4.312 in, and the stroke is 3.65 in, therefore the displacement for this 8-cylinder engine is
- (3.141 / 4) * 4.312 in * 4.312 in * 3.65 in * 8 = 426.3 in^3.

Compression ratio

In a piston engine it is the ratio between the volume of the cylinder and combustion chamber when the piston is at the bottom of its stroke, and the volume of the combustion chamber when the piston is at the top of its stroke

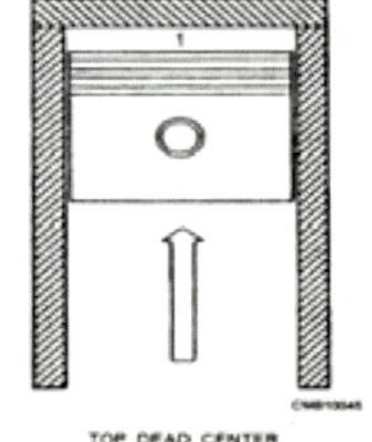

Compression Ratio

The cylinder displacement and the combustion chamber volume determine the compression ratio of an engine . In order to calculate the compression ratio, use the following formula:

CR = Total Cylinder Volume / Combustion Chamber

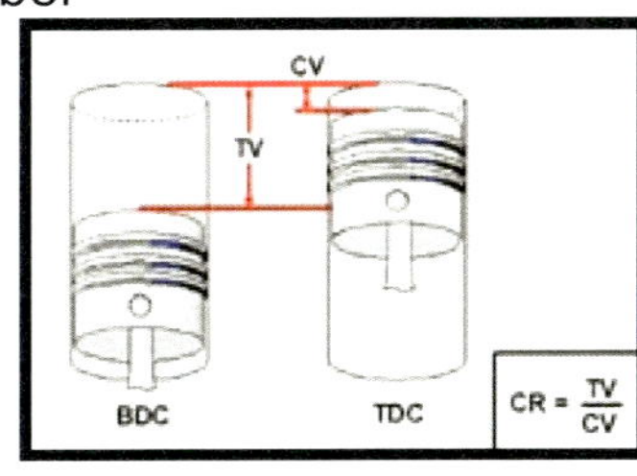

Compression Ratio Formula

- The ratio is calculated by the following formula

$$CR = \frac{\frac{\pi}{4}b^2 s + V_c}{V_c}$$

- b= cylinder bore (diameter)
- s = piston stroke length
- Vc = clearance volume. It is the volume of the combustion chamber

Compression Ratio

- Typical compression ratios of diesel engines range from 11:1 to 22:1; compared with the compression ratio of gasoline engines from 6:1 to 11:1

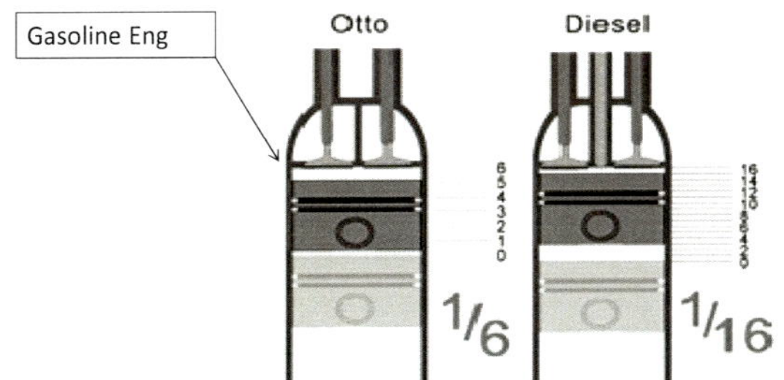

Chapter 5
Cylinder Head & Valves

Cylinder Head

The **cylinder head** (often informally abbreviated to just **head**) sits above the cylinders on top of the cylinder block. It closes in the top of the cylinder, forming the combustion chamber

Types of Cylinder Head

- Flat-Head.

- Overhead Valve (OHV)

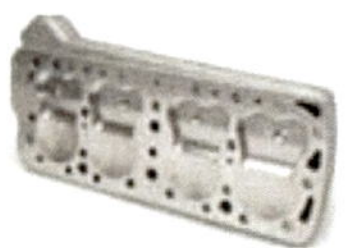

- Overhead Camshaft (OHC)

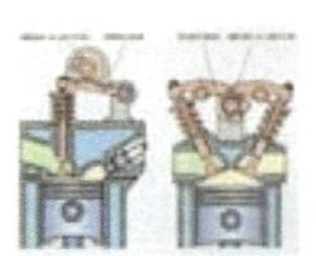

Flat Head

This engine head featured the valves themselves built on sides of the cylinders rather than in the top section, with the lower part of the head containing chambers for the valves to rise into to enable intake and exhaust

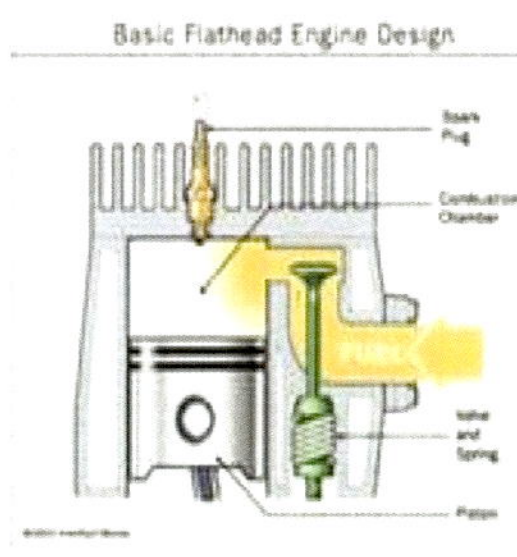

Flat Head

This design also offered simpler and better cooling mechanisms, but eventually lost popularity due to severe performance limitations as the airflow required a 90-degree turn to enter the combustion chamber resulting in inefficient compression and poor combustion

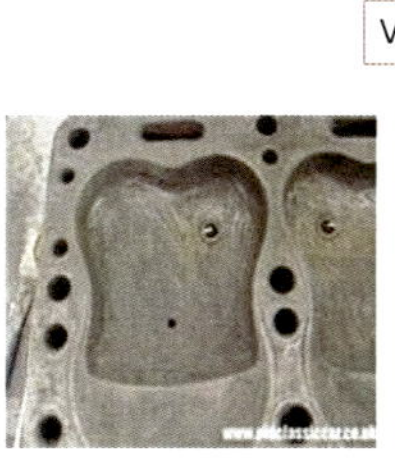
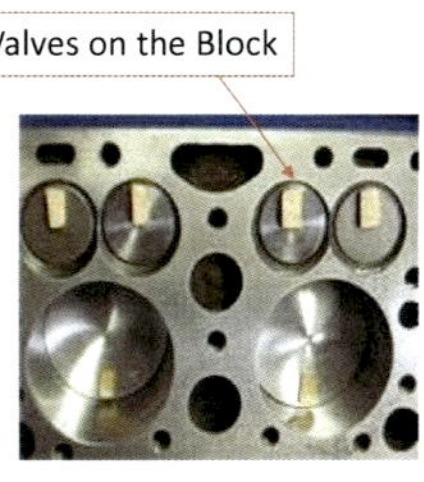

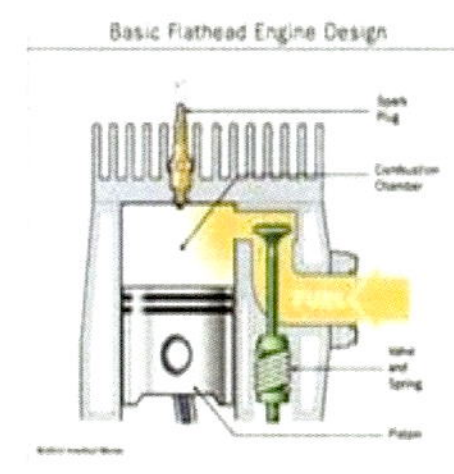

Cylinder Head Components

- (1).- Head
- (2).- Cover Valve
- (3).- Bridge Valve
- (4).- Spring Comp.
- (5).- Valve Guide
- (6).- Seat Valve
- (7).- Body Valve
- (8).- Rocker Arm

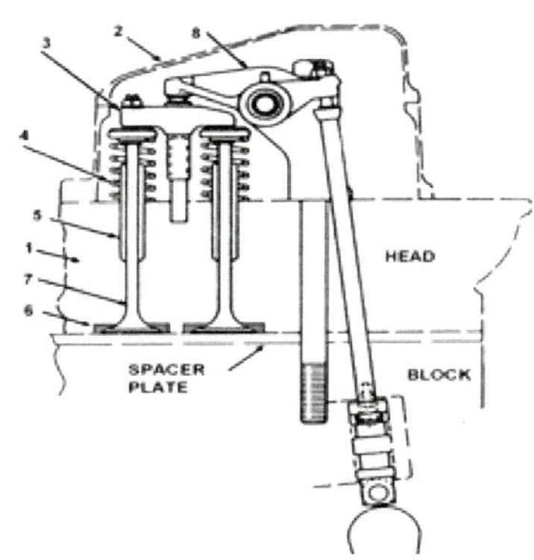

Overhead Cam Engine

Overhead Cam Engines have a camshaft (1) in the cylinder head. Valve lifters (2) are linked to the top of the valve stem. As the cam lobe rotates, the lifter follows the motion and opens the valve. As the cam continues to rotate, the valve spring (3) closes the valve.

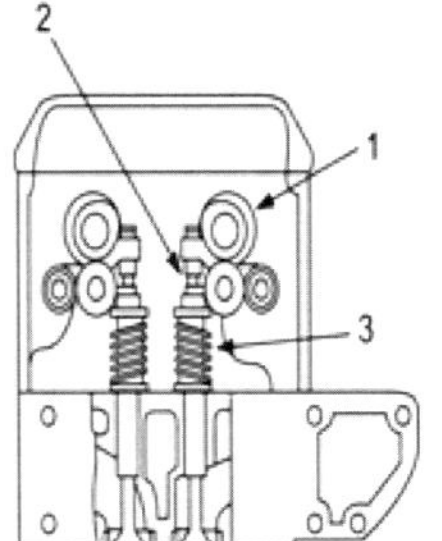

1. Overhead Camshaft
2. Valve Lifters
3. Valve Spring

Cam in Head Engine

As the camshaft turns, the rockers push the valves open

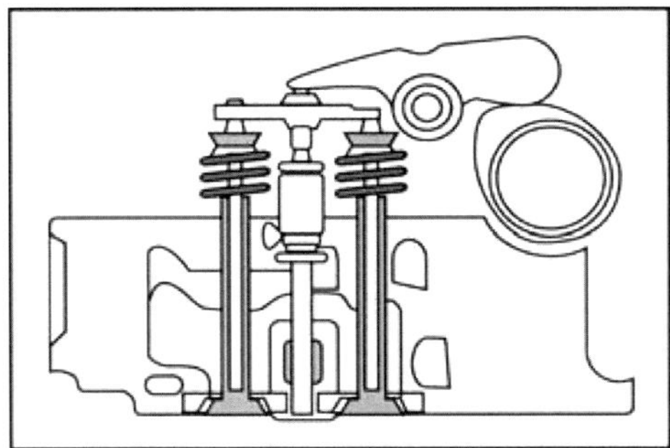

Overhead Valve (OHV)

This design helps meet some of the limitations of the flathead design, resulting in better performance while keeping the engine compact enough.

Overhead Valve (OHV)

The complexity of the drive timing system is also quite low due to the camshaft being located close to the crankshaft, a small chain—or more efficiently—a direct gear mechanism connecting them

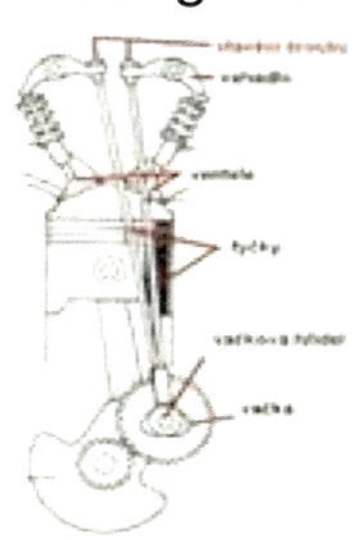

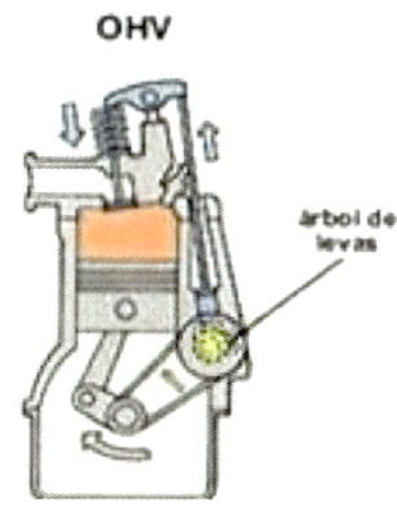

Overhead Camshaft (OHC)

This design eliminates the use of pushrods to actuate valves as the camshaft is located directly above the valves and can actuate them directly

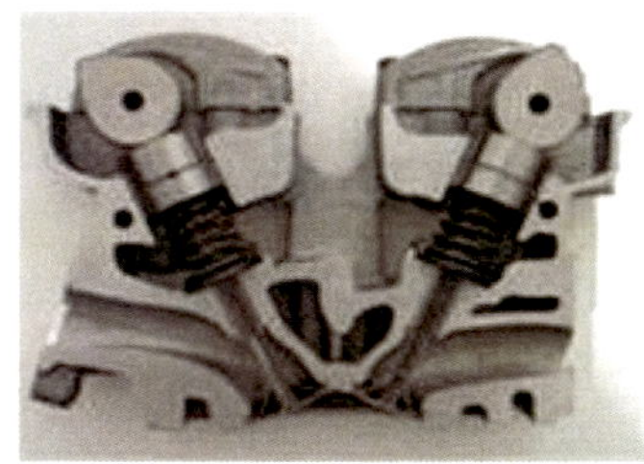

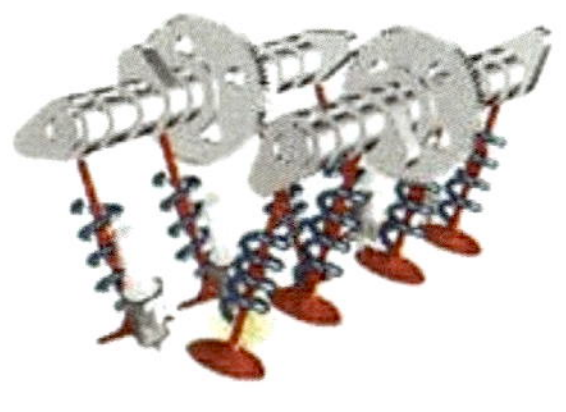

Single and Double Cam

These heads come in two variants: one for single overhead camshaft (SOHC) engines, featuring one camshaft built into the head, and another for double overhead camshaft (DOHC) engines with two camshafts in the head

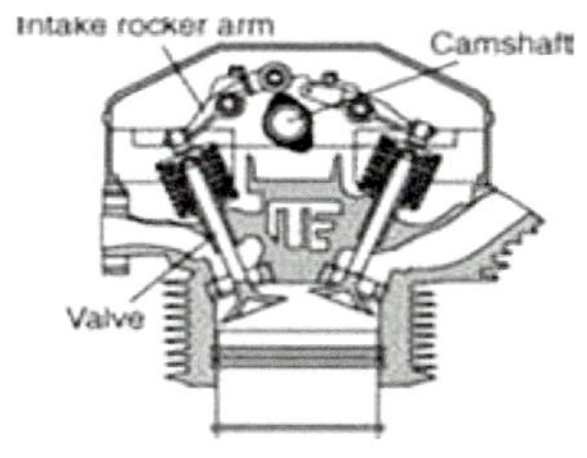

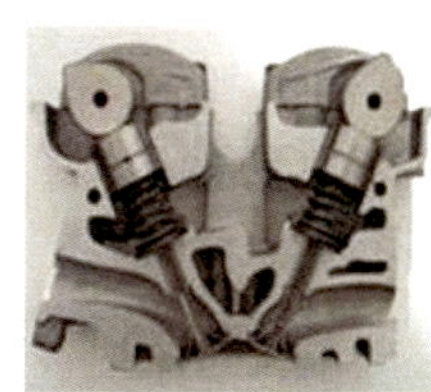

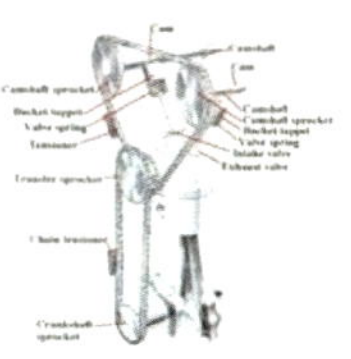

Single and Double Camshaft

In the latter version, one of the camshafts is responsible for controlling the inlet valves while the other one operates the exhaust valves. OHC heads also feature multiple valves per cylinder and modern OHC engine heads feature integrated variable valve timing systems for improved performance.

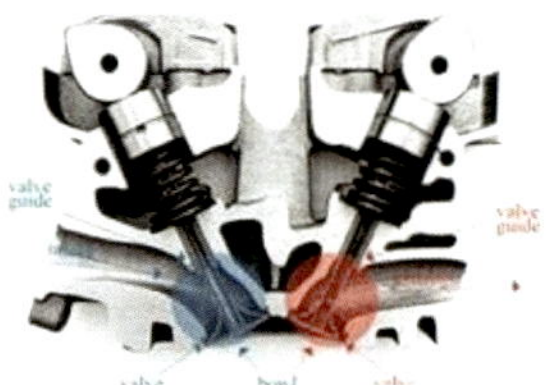

Double Cam

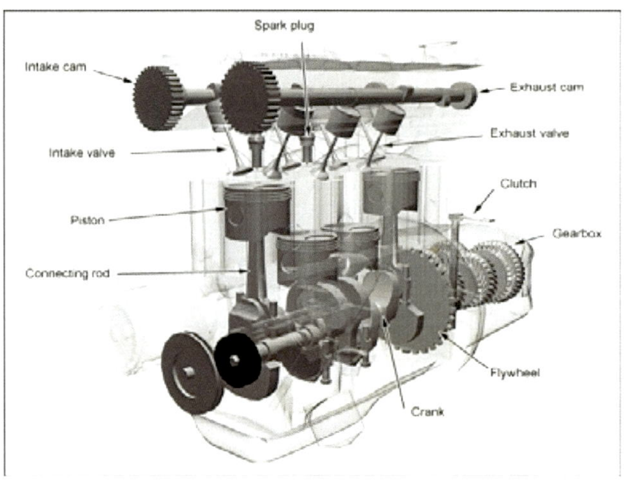

Valve Train Design
Pushrod Engine

Push Rod Engines uses camshaft, valve lifter, push rod and rocker arm

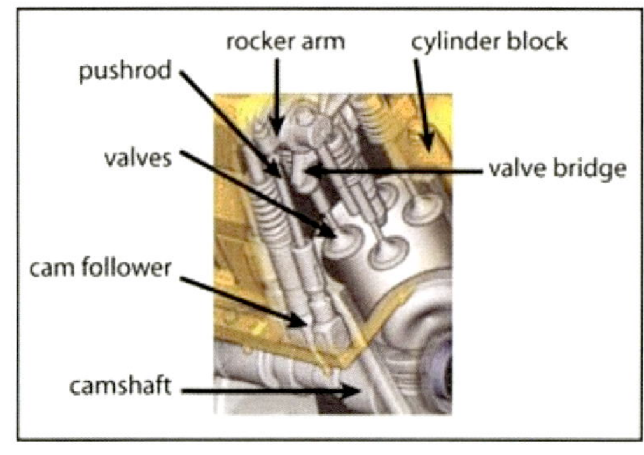

The Valve Cover

The **Valve Covers** fit on valve cover bases
(rocker boxes) that are fastened to the top of
the cylinder head

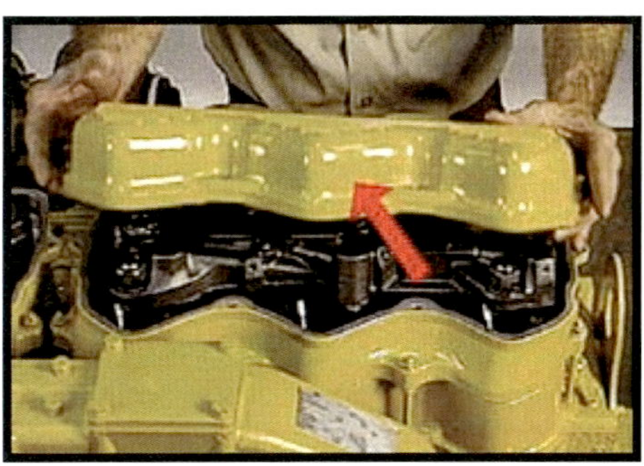

The Rocker Arm

Is an oscillating lever that conveys radial
movement from the cam lobe into linear
movement at the poppet valve to open it

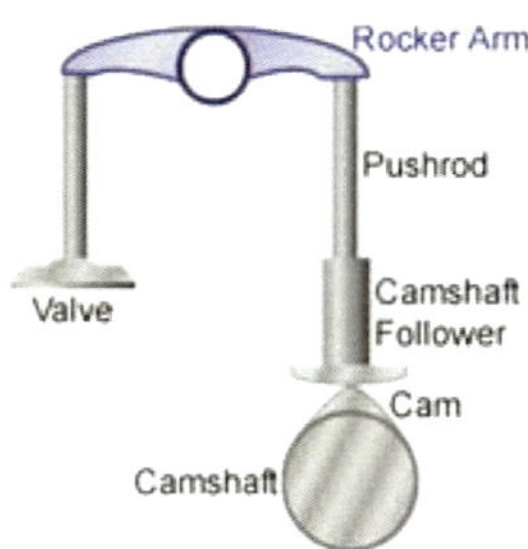

The Valve Bridge

The valve bridge enables the rocker arm to
operate two valves simultaneously. The valve
bridge in this engine is made of forged steel and
has a hardened ball socket into which the ball end
of the rocker arm adjusting screw fits

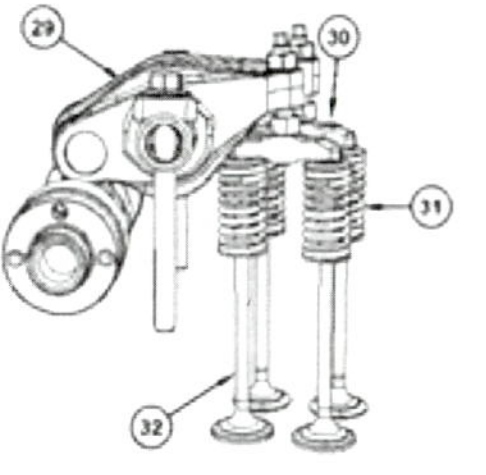

The Poppet Valve

A **poppet valve** (also called mushroom **valve**) is a **valve** typically used to control the timing and quantity of gas or vapour flow into an engine

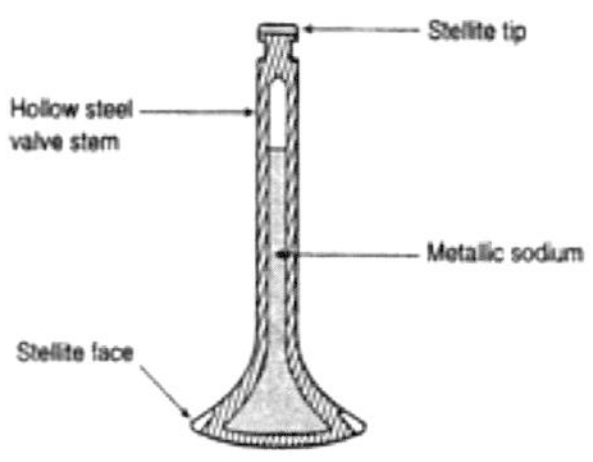

Head Valves

The shape and position of the cam determines the **valve lift** and when and how quickly (or slowly) the valve is opened

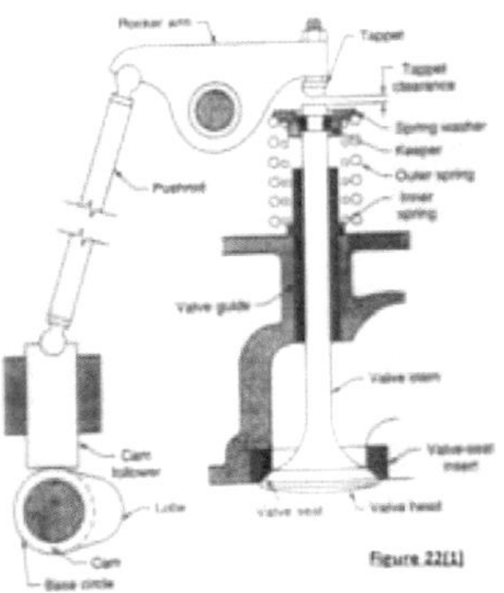

Head Valves /Poppet Valves

- The valve is usually a flat disk of metal with a long rod known as the *valve stem* attached to one side
- The stem is used to push down on the valve and open it, with a spring generally used to return it to the closed position when the stem is not being depressed

Valve Spring Assembly

Valve Springs keep the valves closed. They fit over the valve and are held in place by a keeper (1) and either a retainer (2) or rotator.

1. Keeper
2. Retainer (Rotator)

Valve Rotator

Each valve has a **Rotator**, which moves the valve face 3° relative to the valve seat insert during each valve actuation

Valve Seat

The valve seat is a critical component of an engine in that if it is improperly positioned, oriented, or formed during manufacture, valve leakage will occur which will adversely affect the engine compression ratio and therefore the engine efficiency

Valve Seat

Valve seats are often formed by first press-fitting an approximately cylindrical piece of a hardened metal alloy, such as Stellite, into a cast depression in a cylinder head above each eventual valve stem position and then machining a conical-section surface into the valve seat that will mate with a corresponding conical-section of the corresponding valve

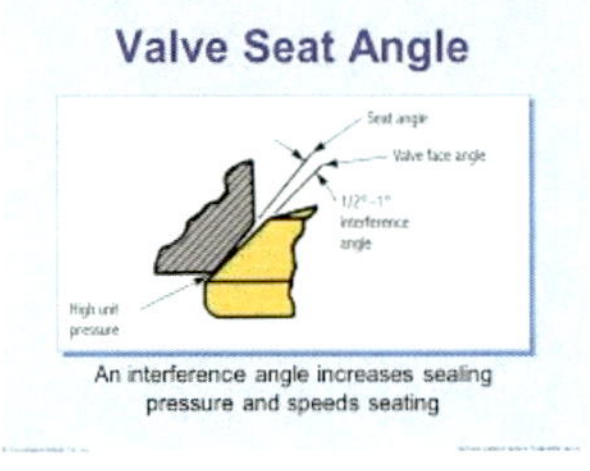

Seats and Springs

- If a seat is cracked or too badly worn to be refaced, the seat must be replaced .
- All aluminum heads have hardened steel seats that can be replaced
- The valve springs are all inspected and tested to make sure they are still capable of maintaining proper pressure

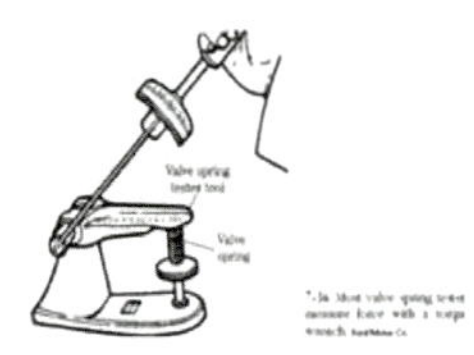

Valve Guide

Valve guides keep valves moving in a straight line and help carry heat away from the valve. The valve stem extends out of the guide on top of the cylinder head .

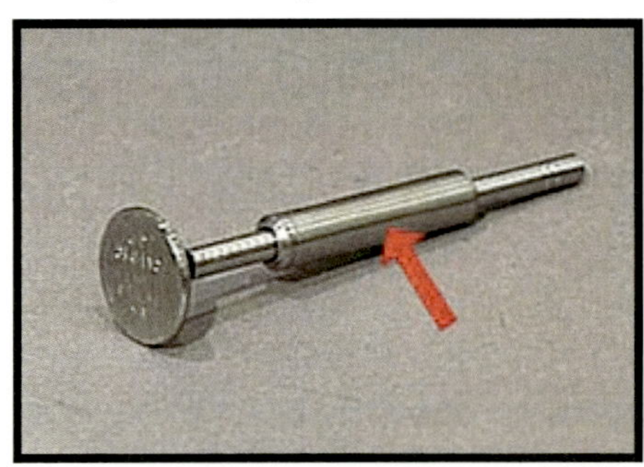

Valve Guides

A valve guide is a tube-shaped piece of metal, pressed into the cylinder head, with the valve reciprocating inside it.

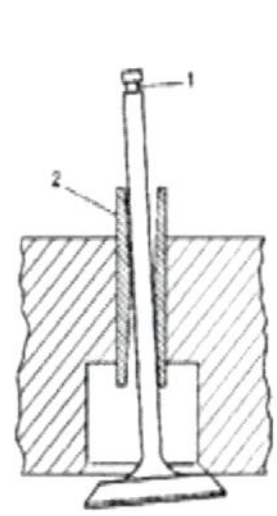 

Valves Guides and Seats

- Some shops install new guides, guide liners or bore out the old guides to accept new valves with oversized stems
- Aluminum heads have cast iron or bronze guides that can be replaced but most cast iron heads do not

 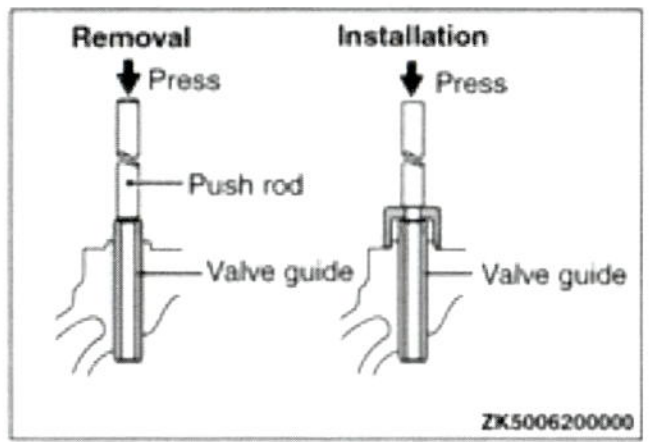

Valves Guides and Seats

- If the valves are to be reused, they will be inspected, checked for straightness then refaced
- Many shops automatically replace all the exhaust valves to reduce the risk of failure (exhaust valves run much hotter than intakes and are much more likely to fail)

Valve Guides

As valve guides wear, their ability to positively locate the valve to the valve seat decreases

 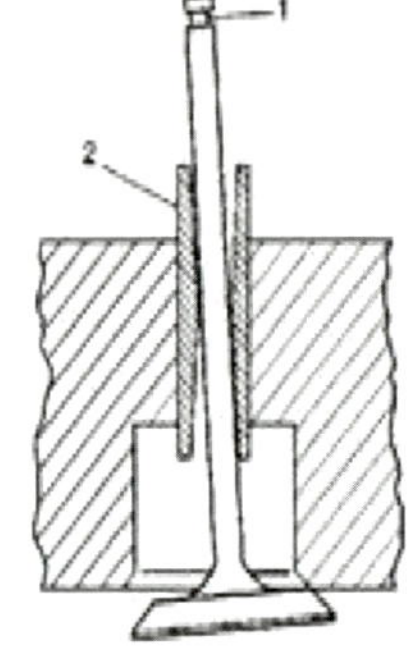

Valve Guides

As the valves lose their ability to seal the combustion chamber properly, the engine can lose performance and start to burn oil, leaking from the top of the cylinder head into the intake and exhaust manifolds

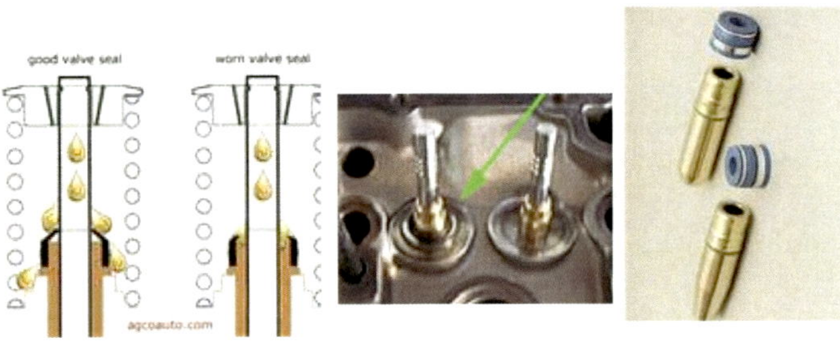

Hydraulic Valve Lifters

- Is a device for maintaining zero valve clearance in an internal combustion engine.
- The hydraulic lifter consists of a hollow expanding piston situated between the camshaft and valve

 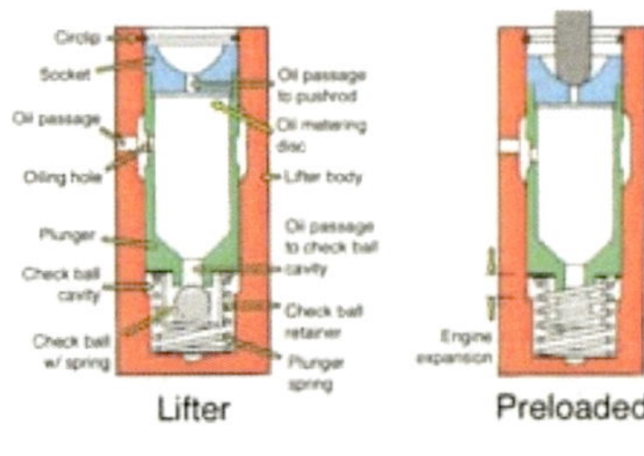

Hydraulic Lifter

It is operated either by a rocker mechanism, or in the case of one or more overhead camshafts, directly by the camshaft

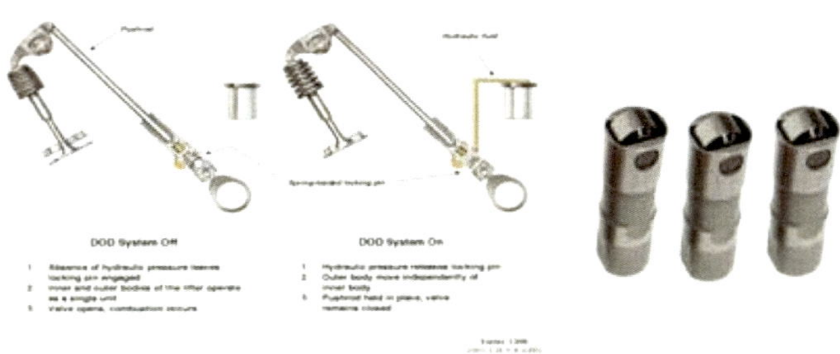

Head Porting

Cylinder head porting refers to the process of modifying the intake and exhaust ports of an internal combustion engine to improve the quality and quantity of the air flow

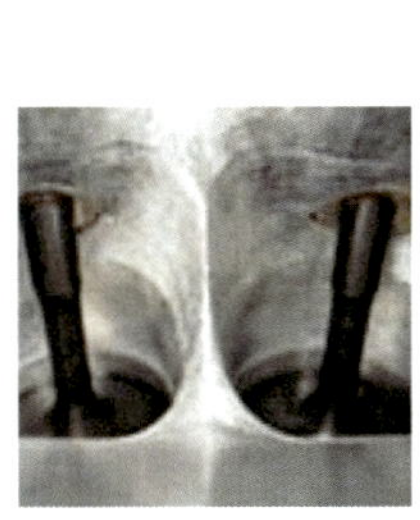

Lapping head valves

Valve Lapping is basically sanding down the surface of the Valves and the Head where they make contact with each other

Lapping Head Valves

When the valve is closed and there's not a perfect seal you loose compression. Lapping the valves restores this

Valve Spring

Valve springs play a vital role in how well an engine performs and have specific characteristics which must properly match the rest of the components in the valve train

Free Length

Free length is measured by the total length of the unloaded spring from one end to the other. The total length of a spring has some affect on the amount of pressure it exerts against the valve seat when installed

Open Load

Open load is the amount of pressure the valve spring exerts when the valve is at its maximum open position. The amount of pressure a valve spring exerts at open load is dependent upon the valve springs overall design

Engine Valve Failure

- Every place the valves touch another part during operation has the potential for showing us a cause of failure.
- One of the most important things that should be done in any valve failure analysis is to look in detail at every area of the valve before drawing conclusions

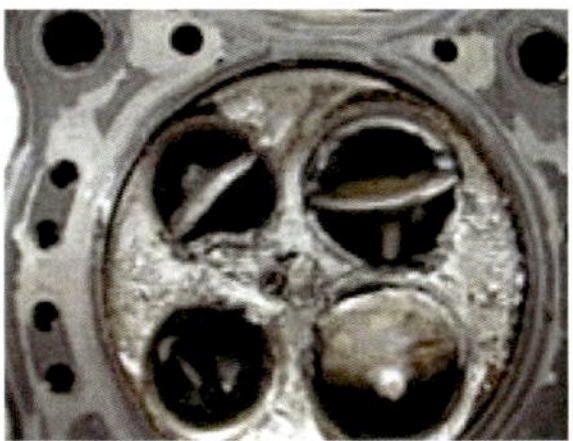

Head Fracture

A piece broken away from the valve head roughly in the shape of the chord of a circle due to very high cylinder pressure and valve temperatures; incorrect valve material; etc

Wrong Installation

The valve lock pins were installed improperly

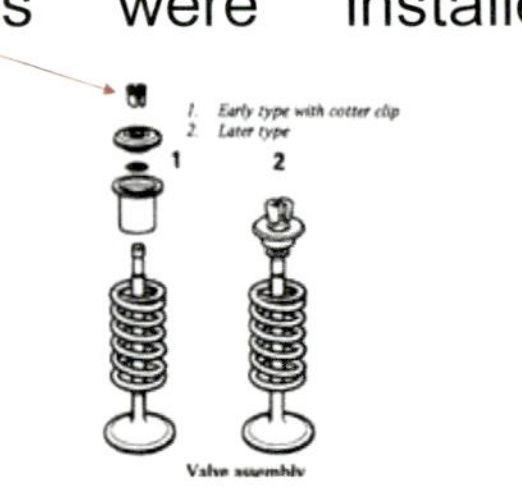

Engine Valve Failure

Valve stem sticking or seizing due to lack of lubrication; not enough stem to guide clearance; bent valve stem; carbon build up at the bottom of the valve stem, incompatible valve to guide material

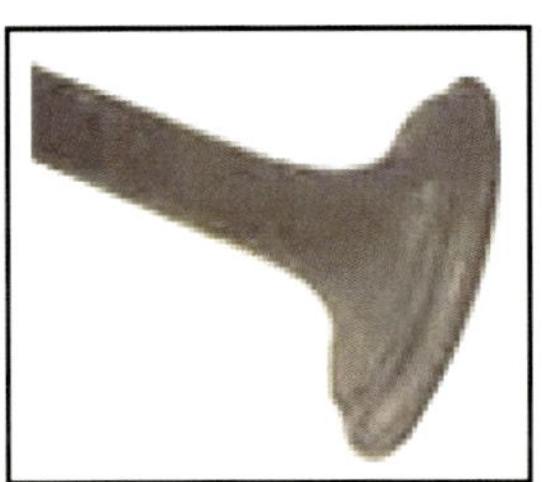

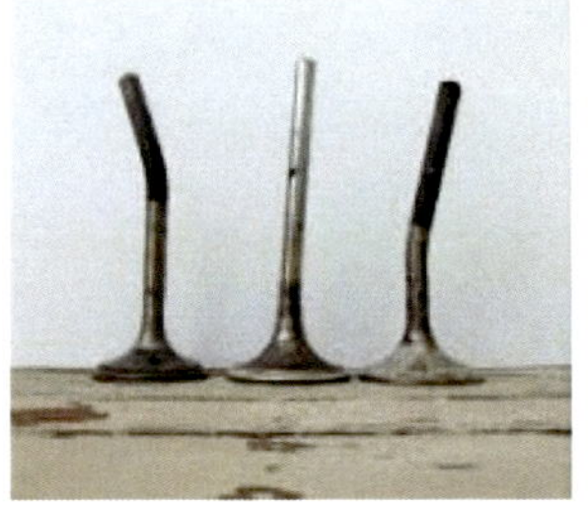

Engine Valve Failure

Badly pitted valve and valve seat faces, discovered on early overhaul due to power loss or a burn out due to excessive oil consumption, incorrect mixture setting producing a high level of solid particles or long periods of low power cold running

Valve Spring Broken

Head Valve Job

A valve job is removing the cylinder head(s) from the engine so the valves, guides and seats can be refurbished to restore compression and oil control

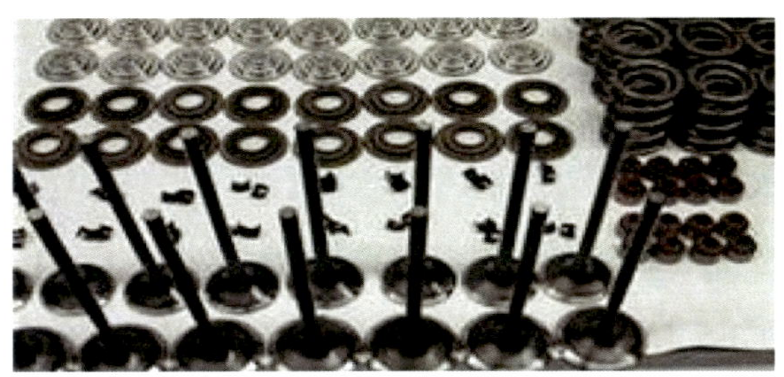

Head Valve Job

- A valve job typically begins by disassembling, cleaning and inspecting the cylinder head
- Cast iron heads are "Magnafluxed" to check for hairline cracks

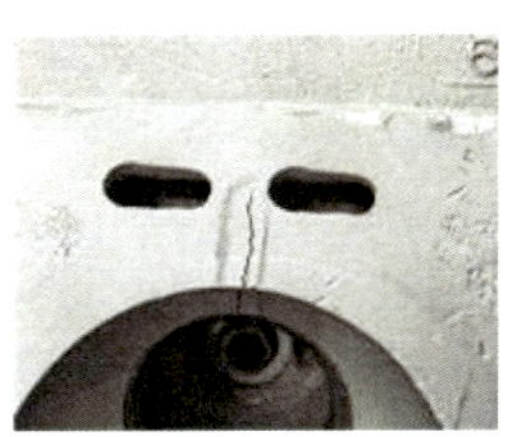

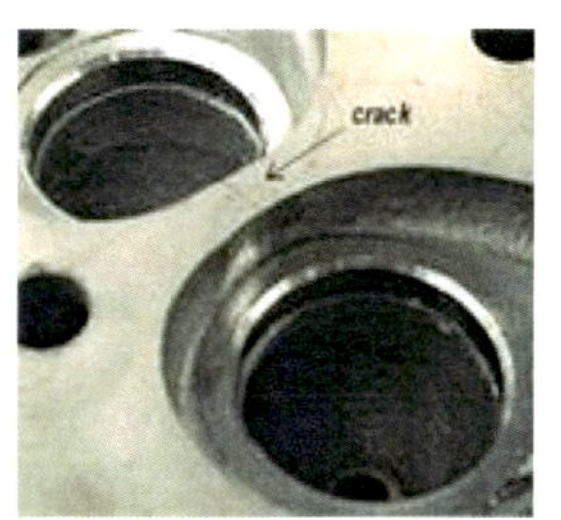

Head Valves Cracks

- Cracks are bad news because they can leak coolant into the combustion chamber damaging the cylinders and/or causing the engine to lose coolant and overheat
- If cracks are found in any critical areas of the head, the head must either be repaired or replaced

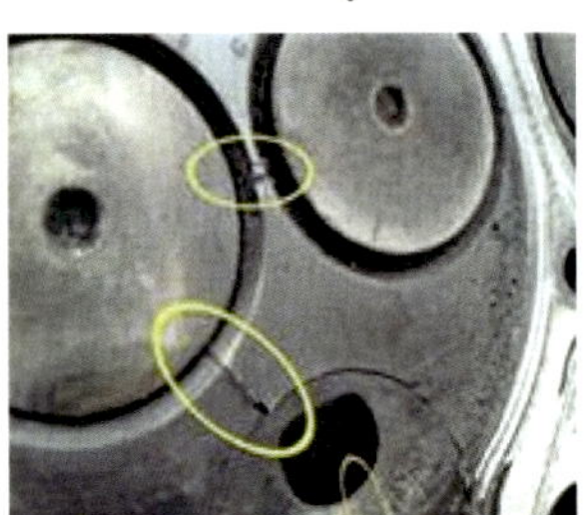

Cracks Inspection Methods

- The two most common methods of surface crack detection, **Liquid Penetrant Testing (PT)**, sometimes referred to as Dye Penetrant Inspection, and **Magnetic Particle Testing (MT)**
- These methods of inspection are specialized and should be carried out by suitably trained and qualified inspection personnel

Magnaflux Crack detection

- This process involves a Electro-magnet and Magnaflux powder
- Electro-magnet is placed on the cast cylinder head and magnaflux powder is applied to area between magnet. If there is a crack in the casting, the powder will adhere to the crack

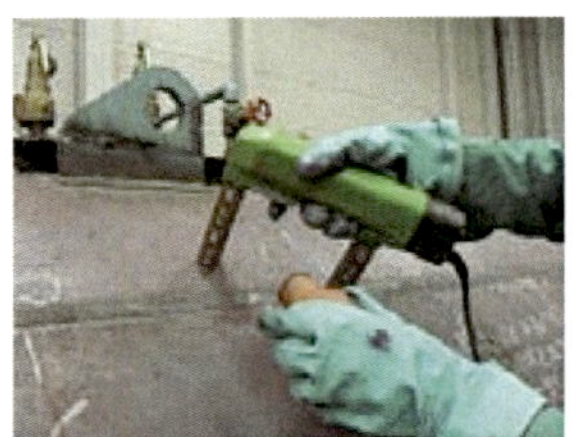

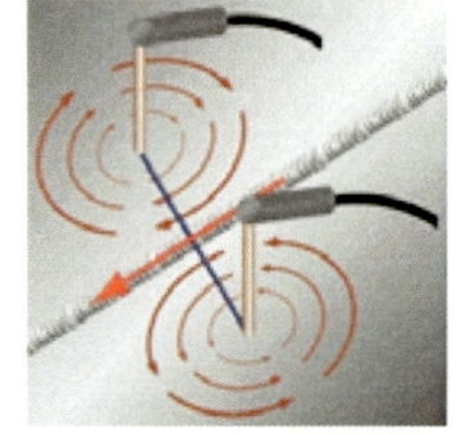

Magnetic crack detection

As long as the part has no cracks or other discontinuities, magnetic particles will not be attracted. When a crack or other discontinuity is present in the part being tested, north and south magnetic poles are set up at the edge of the discontinuity

Magnetic Crack Detection

Only ferromagnetic materials can be tested by this method. Ferromagnetic parts that have been magnetized during testing may retain a certain amount of residual magnetism. Certain parts may require demagnetization if they are to function properly in service

Liquid Penetrant Method

- **Dye penetrant inspection (DPI)**, also called **liquid penetrant inspection (LPI)** or **penetrant testing (PT)**, is a widely applied and low-cost inspection method used to locate surface-breaking defects in all non-ferrous materials and ferrous materials

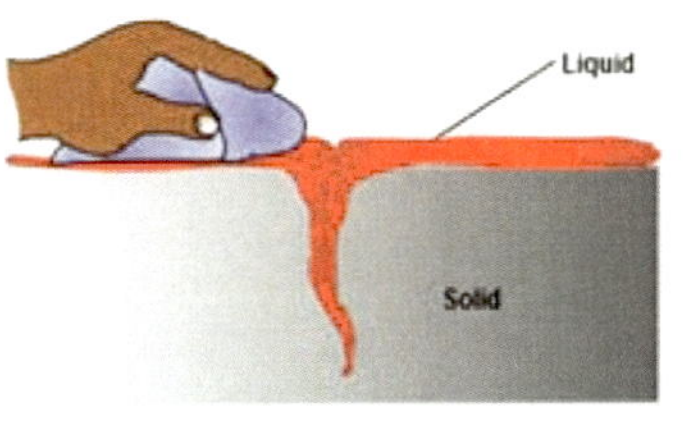

73

Liquid Penetrant Method

- The test surface is cleaned to remove any dirt, paint, oil or grease
- The penetrant is then applied to the surface of the item being tested

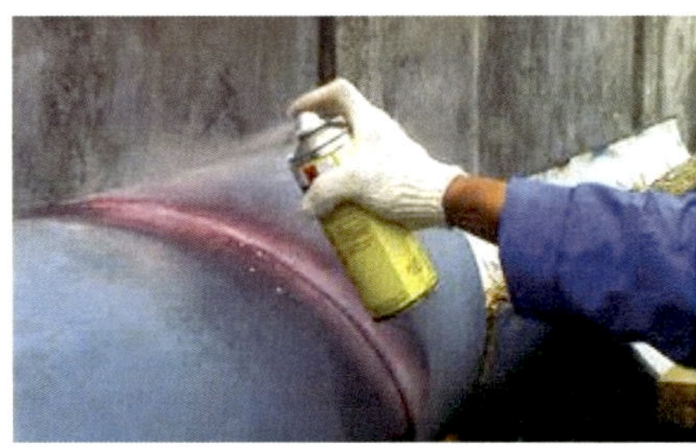

Developer Application

After excess penetrant has been removed, a white developer is applied to the sample. Several developer types are available, including: non-aqueous wet developer, dry powder, water-suspendable, and water-soluble

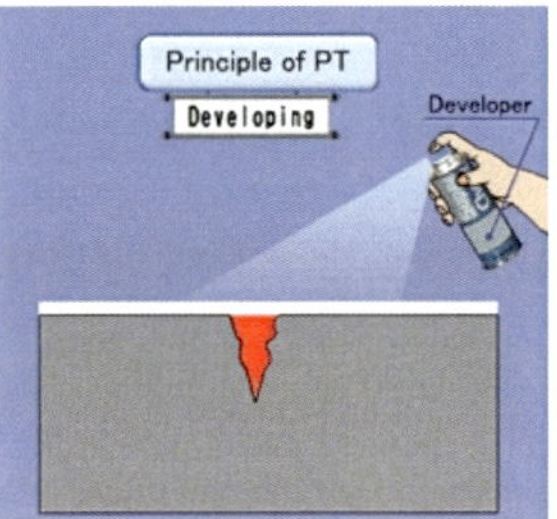

Developer Application

The developer draws penetrant from defects out onto the surface to form a visible indication, commonly known as bleed-out. Any areas that bleed out can indicate the location, orientation and possible types of defects on the surface

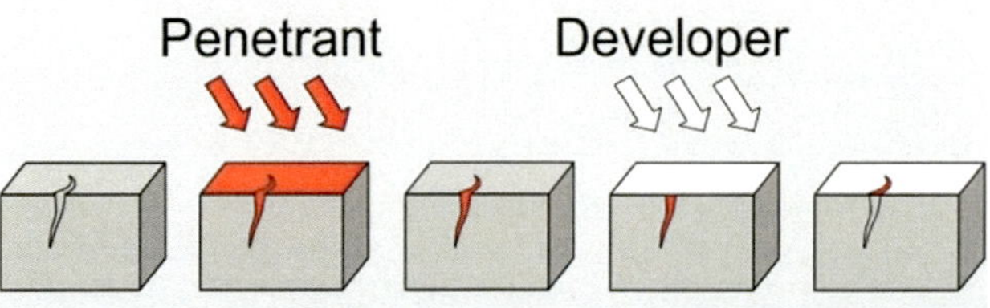

Why Cracks Form

Loss of coolant, severe overheating as well as sudden changes in operating temperature from hot to cold can all create the kind of conditions that cause cracks to form

Thermal Expansion

When metal is heated it expands. Aluminum expands at nearly twice the rate of cast iron, which creates a mismatch in expansion rates on bimetal engines with aluminum heads and cast iron blocks

Engine Overheating

Elevated operating temperatures can push a head beyond its design limits causing the metal to deform. This, in turn, may cause cracks to form as the metal cools and contracts

Engine Overheating

When overhead cam heads get hot, they often swell and bow up in the middle. This may cause the OHC camshaft to seize or break as well as cracks to form in the underside of the head

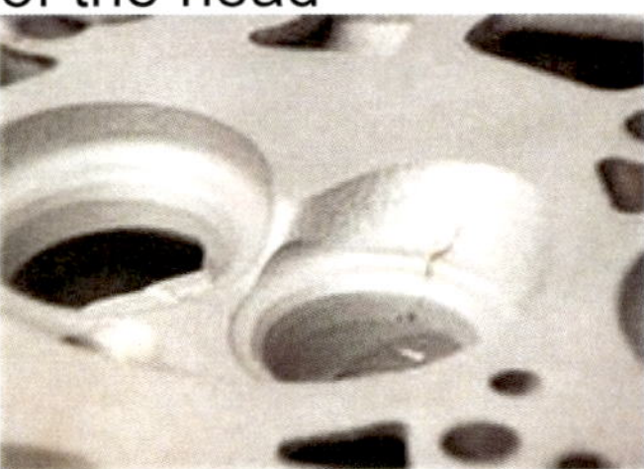

Head and Block Inspection

- All heads should be thoroughly inspected for cracks using a variety of techniques before any machine work is done. Better to find out the head is cracked before you rework the valve guides and seats than afterwards.

- Think of crack detection as your first line of defense against comebacks

Repair or Replace?

- Cracks do not necessarily mean a cylinder head has to be replaced. In fact, many cracked heads that were once thought to be "un-repairable" are now being fixed

- Repairing a cracked cylinder head always involves a certain amount of risk, but when done properly is usually much less expensive than replacing a cracked head with a new or used casting

Repair or Replace?

- Cracks in cast iron heads are most often repaired by "pinning" (installing a series of overlapping threaded pins)
- Most small cracks in cast iron as well as aluminum heads can be repaired by pinning

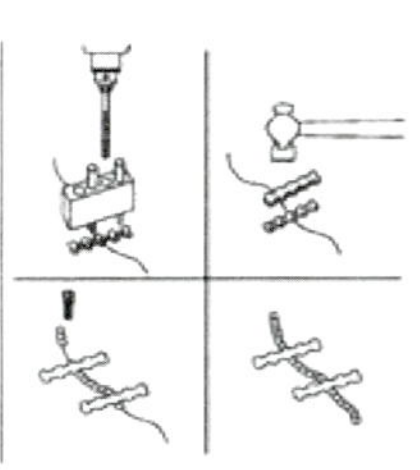  

Aluminum Head Repair

- Larger cracks in aluminum heads typically require TIG (Tungsten Inert Gas) welding. Larger cracks in cast iron heads can often be repaired by furnace welding or flame spray welding
- On aluminum heads, cracks are welded 77 percent of the time. For diesel heads, welding is performed 41 percent of the time

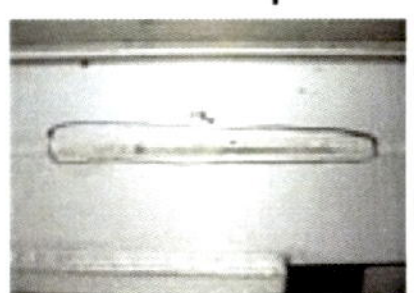

Pinning Cracks

The technique involves drilling holes in both ends of the crack to keep it from spreading, then drilling holes at various intervals along the length of the crack, installing overlapping pins to fill the crack, then peening over the pins with an air hammer to seal and blend the surface

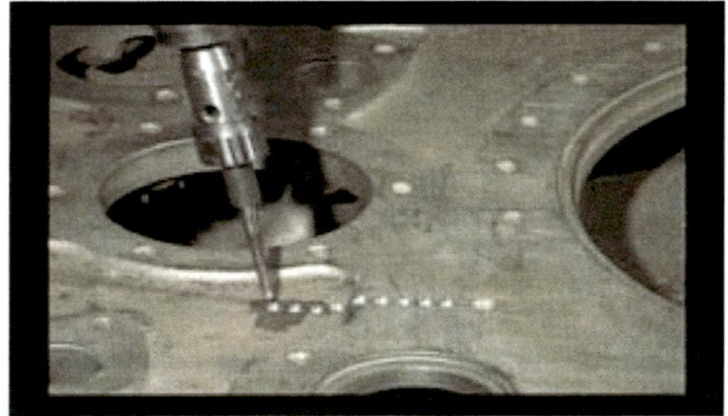

Pinning Cracks

The technique involves drilling holes in both ends of the crack to keep it from spreading, then drilling holes at various intervals along the length of the crack, installing overlapping pins to fill the crack, then peening over the pins with an air hammer to seal and blend the surface

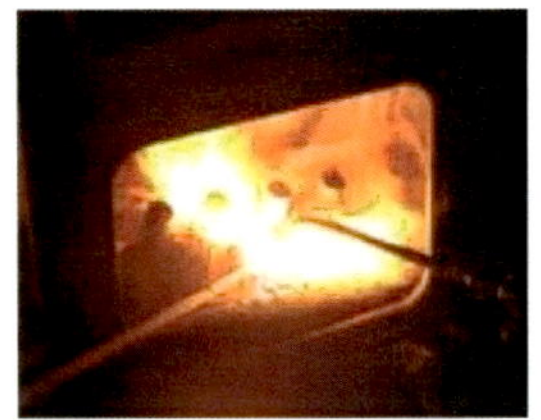

Welding Techniques

- Furnace welding cast iron is often called the "black art" of crack repair because it requires lots of heat and operator skill.
- Learning how to furnace weld cast iron is not something an inexperienced welder can pick up quickly

Welding Techniques

- To furnace weld a cast iron head, the head is first preheated to 1,300 degrees F (cherry red) in an oven
- This step is absolutely essential to minimize thermal shock and to relax the metal so it won't distort when the torch is applied to the casting

Cast Iron Heads

- Because of the high temperatures involved in furnace welding, the induction hardening of integral valve seats is usually destroyed
- For this reason, some shops prefer to braze weld cracks rather than furnace weld. With braze welding, the job can be done at 800 to 900 degrees so the head doesn't get as hot

Aluminum Heads Welding

- Cracks in aluminum heads are most often repaired by TIG welding (though pinning also works with small accessible cracks).
- The head must be clean, grease-free and dry before grinding the crack all the way out

Aluminum Heads Welding

- Just grinding the surface and welding over a crack will likely be a short-lived temporary fix because the underlying crack is still there and will continue to grow
- When exposed to air, aluminum forms an oxide coating that contaminates the weld and interferes with fusing. A TIG welder prevents the formation of the oxide layer by bathing the weld with a steady supply of inert gas (usually argon)

Aluminum Heads Welding

- Because aluminum can crack when subjected to too much heat in a concentrated area, the head must be preheated in an oven to 450 to 550 degrees F to eliminate thermal stress
- The added heat also makes the head easier to weld because aluminum conducts heat away from the weld area rapidly

Check for flatness

- The surface of the head must be flat to seal the head gasket against the block.
- Excessive warpage, roughness or any damage can cause the head gasket to fail.
- If the head exceeds the maximum allowable out-of-flatness specs, it must be resurfaced or replaced

The Head Gasket

It is located between the block and the head and its function is to keep the engine compression

Head Gasket

Its purpose is to seal the cylinders to ensure maximum **compression** and avoid leakage of **coolant** or engine oil into the cylinders; as such, it is the most critical sealing application in any engine, and, as part of the combustion chamber, it shares the same strength requirements as other combustion chamber components

Gasket Materials

- Copper and brass were originally used as head gaskets
- Today are designed with **multiple** thin **layers** of cold-rolled **steel** (MLS) ,coated with a very thin layer of elastomeric material

Copper Head Gaskets

- Copper head gaskets provide the strongest combustion seal, they're commonly used in high-performance applications where extreme cylinder pressures will be encountered .
- Copper head gaskets are re-usable, and they do not have to be re-annealed

Modern copper head gaskets

- Copper gaskets come in many thicknesses, and in most cases are reusable
- The most common is the ICS (Integral Combustion Seal) copper head gasket with coolant and oil passage seals bonded to the sealing surfaces of the head gasket. These built-in seals eliminate the need for additional sealants

Modern Copper Head gaskets

These built-in seals eliminate the need for additional sealants. These head gaskets are installed dry and there are two versions available: one for engines with machined O-ring combustion seals and one for those without O-rings

Graphite Head Gaskets

- Graphite head gaskets can be used on aluminum heads with an iron block.
- Graphite is excellent in handling high temperatures and is anisotropic (draws heat away from hot spots). It also seals very well too.

Teflon Head Gasket

- Though aluminum is lighter than cast iron, it expands twice as fast.

- This quicker cycle of expansion and contraction stresses head gaskets. To handle this, a non-stick coating is often applied to the gasket, like **Teflon** or **molybdenum**.

Symptoms of a Blown Head Gasket

- As the coolant heats up, and pressure rises, it will be forced into the cylinder during the intake stroke.

- This coolant will flash into steam during combustion and will be visible as a white cloud coming from the tailpipe

Symptoms of blown gasket

- To determine the exact cylinder, there two tests can be performed. Visually inspecting the spark plugs (or Injector) will likely turn up one that is visibly wet or distinctively cleaner than the rest

- A compression gauge will show lower pressure in the affected cylinder.

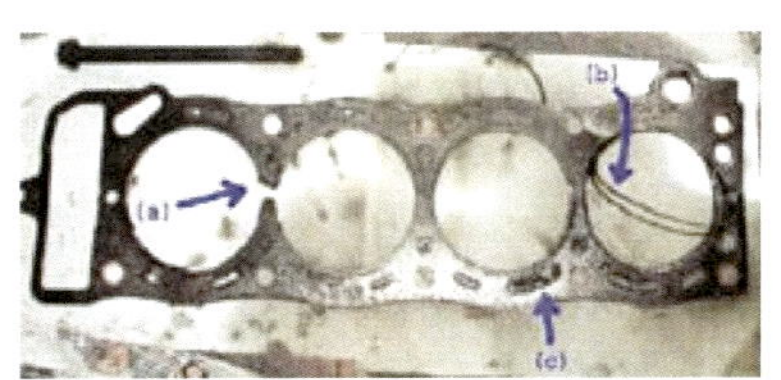

Coolant in the Oil

- Coolant entering the oiling system is one of the more common head gasket problems, and potentially one of the most damaging to your engine
- Symptoms occur much sooner at the rocker arms. Look at the bottom side of the oil cap and down into the oil filler neck for the same thick and tan oil mixture

Oil in the coolant

- Sometimes a blown head gasket will occur in ways that allow engine oil to enter into the cooling system
- Regular inspection of the coolant is the best way to catch this problem before further engine damage occurs. If any oil has mixed with the coolant, it will have changed in color to a light tan, close to the color of coffee with a lot of milk added

Coolant Leaks

- You may also see a blown head gasket that causes only a coolant leak
- Early symptoms are overheating and visible puddles of coolant on the bilge
- There are many other areas on an engine that more commonly leak coolant, and it can be difficult to determine the source

Compression Loss

- The first indication of the blown head gasket is an audible hissing or tapping sound, which can be difficult to locate, as it sounds like a bad lifter or rocker arm
- Depending on the location, you may see the air leak on the engine. The most accurate test for this problem is a compression gauge

Compression Tester

Engine Compression Testing

- An engine is essentially a self-powered air pump , so it needs good compression to run efficiently, cleanly and to start easily
- As a rule, most Gas engines should have 140 to 160 lbs. Of cranking compression with no more than 10% difference between any of the cylinders
- For Diesel engines the cranking compression are between 400 and 1200 PSI

Low Compression Evaluation

- Low compression in one cylinder usually indicates a bad exhaust valve
- Low compression in two adjacent cylinders typically means you have a bad head gasket

Low Compression??

Low compression in all cylinders would tell you That the rings, valves, gaskets and cylinders are worn and the engine needs to be overhauled

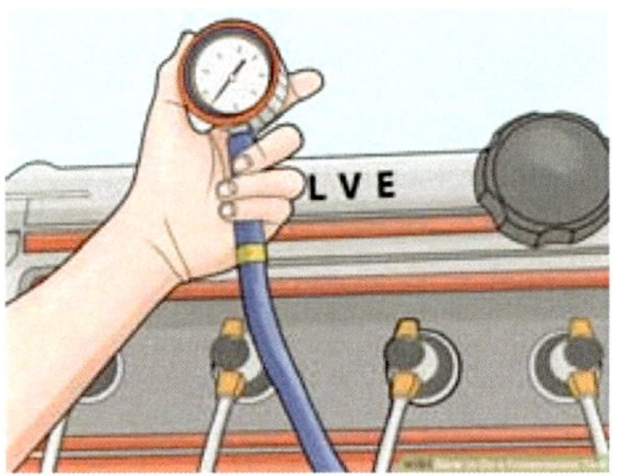

IS IT THE RINGS OR THE VALVES?

- If compression is low in one or more cylinders, you can isolate the problem to the valves or rings by squirting a little 30 weight motor oil into the cylinder through the spark plug hole and repeating the compression test. The oil temporarily seals the rings.
- If the compression readings are higher the second time around, it means the rings and/or cylinder is worn
- No change in the compression readings would tell you the cylinder has a bad valve

Head Bolts

- Head bolts are engineered to stretch within a controlled "yield zone". Once they reach this zone, they spring back to provide a precise level of clamping force over the entire area of the gasket
- Head bolts should never be reused. Damage to engine and cylinder heads can occur.

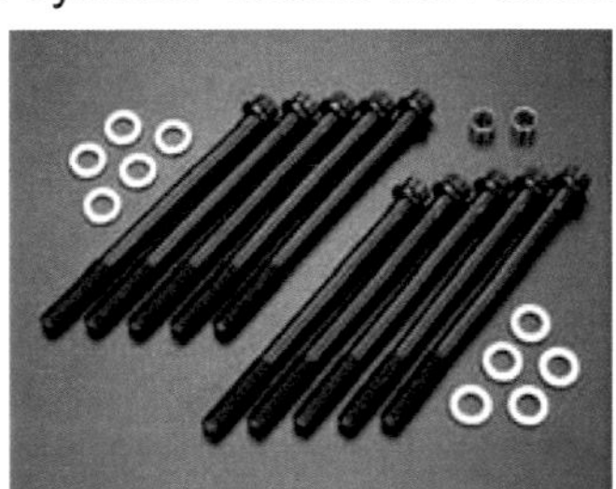

Head Bolt Torque

- Run the engine for the next week but keep the boost psi under 30psi during this time. (crucial) By doing this , the different heat cycles will get the head to 'settle' better.
- After a weeks time has passed, do the Line Torque Procedure again

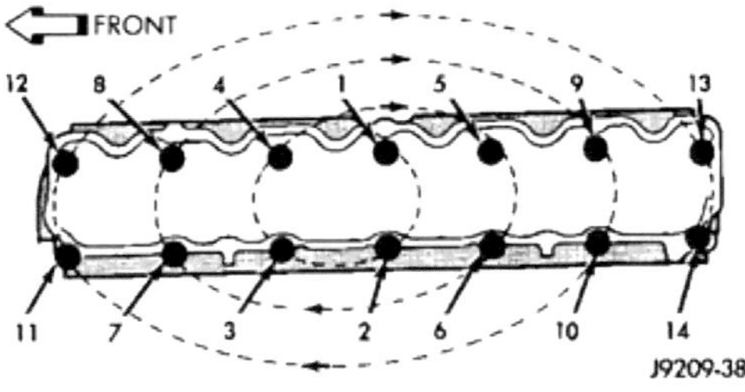

Bolt Torque Sequence

Each manufacturer has its own sequence of torque. I generally recommend starting the sequence from the center (With half the maximum torque) of the head moving in a spiral down to the ends

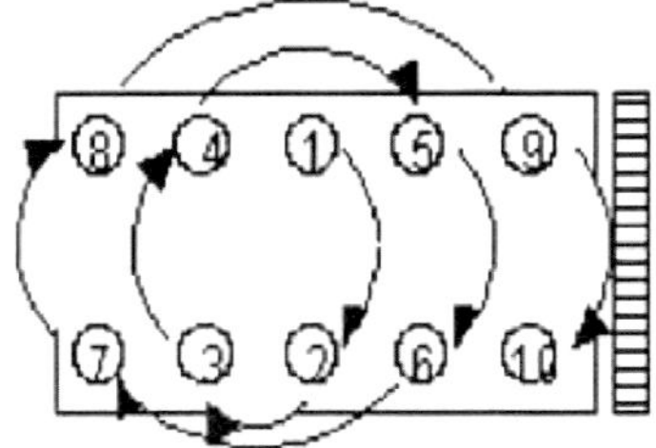

Camshaft description

The **camshaft** is an apparatus often used in piston engines to operate poppet valves

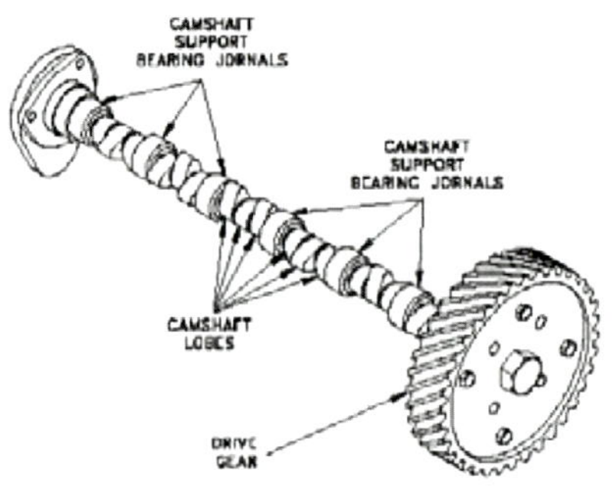

Camshaft Materials

- Camshafts can be made out of several different types of material. These include , Chilled iron castings

- A chilled iron camshaft has a resistance against wear because the camshaft lobes have been chilled, generally making them harder

- Billet Steel : When a high quality camshaft is required, engine builders and camshaft manufacturers choose to make the camshaft from steel billet. This method is also used for low volume production

Camshaft Description

It consists of a cylindrical rod running the length of the cylinder bank with a number of oblong *lobes* or cams protruding from it, one for each valve. The cams force the valves open by pressing on the valve, or on some intermediate mechanism, as they rotate

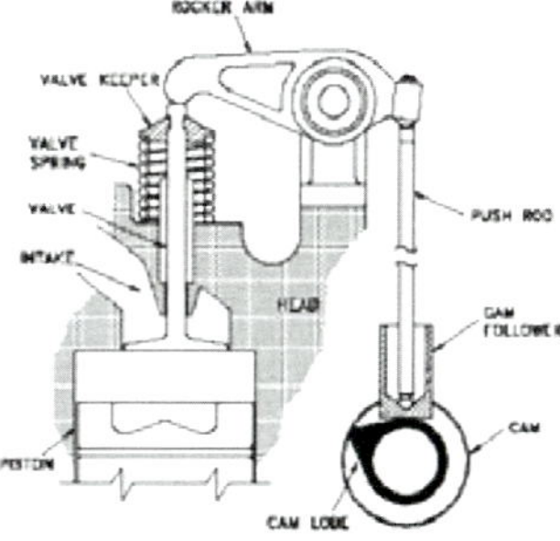

Camshaft

The **Camshaft** is driven by a gear train coming from the crankshaft

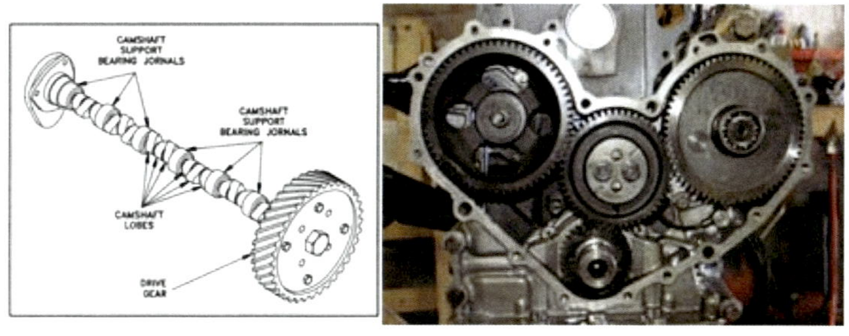

Camshaft Lobe

Every **Camshaft Lobe** is made up of:
1.   Base circle
2.   Ramps
3.   Nose.
The distance from the base circle diameter to the top of the nose is called lift and it determines how far the valves are opened.

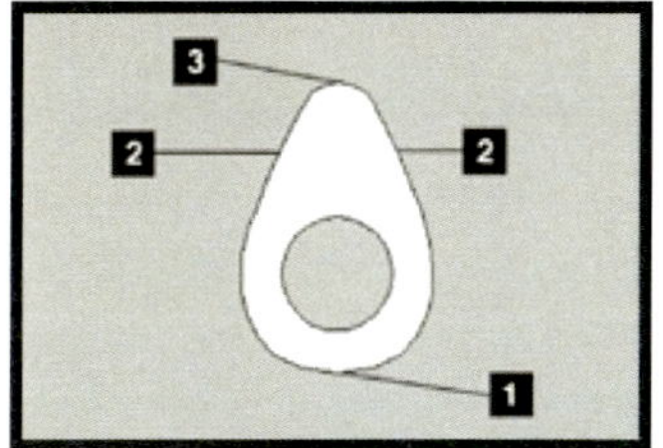

1. Base Circle

2. Ramps

3. Nose

Opening and Closing Ramps

The shape of the opening and closing ramps determine how fast valves open and close and the shape of the nose determines how long the valve remains open.

1. Fast opening
2. Long open period
3. Fast close
4. Slow close

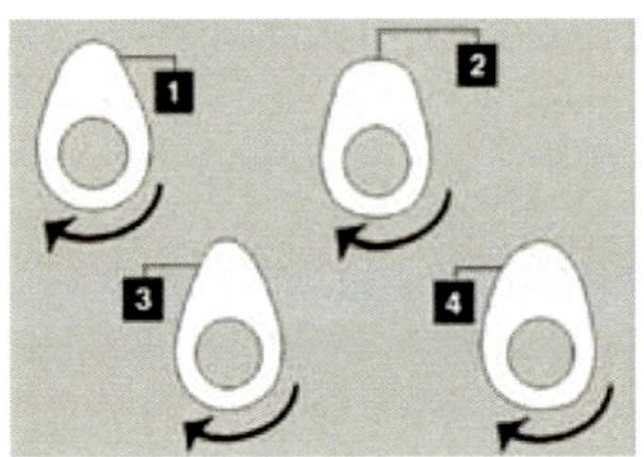

1. Fast Opening
2. Long Open Period
3. Fast Close
4. Slow Close

Camshaft /Oil Pump/Fuel Pump

In some designs the camshaft also drives the distributor (gas) or the Fuel Injection Pump (Diesel) and the oil pump

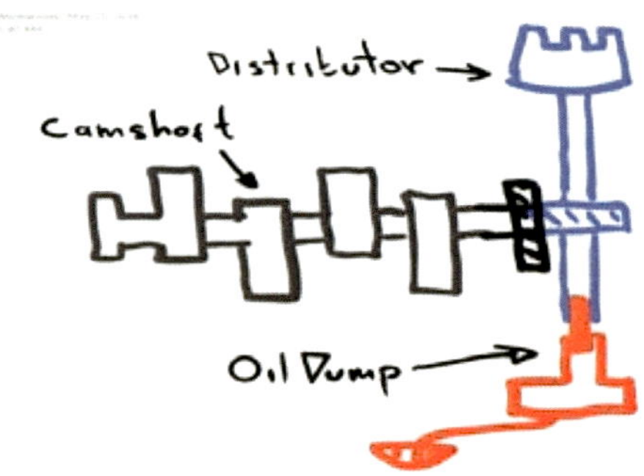

Timing

The relationship between the rotation of the camshaft and the rotation of the crankshaft is of critical importance

Gear to Gear connection (Oposite Rotation)

Timing Chain Connection (Equal Rotation)

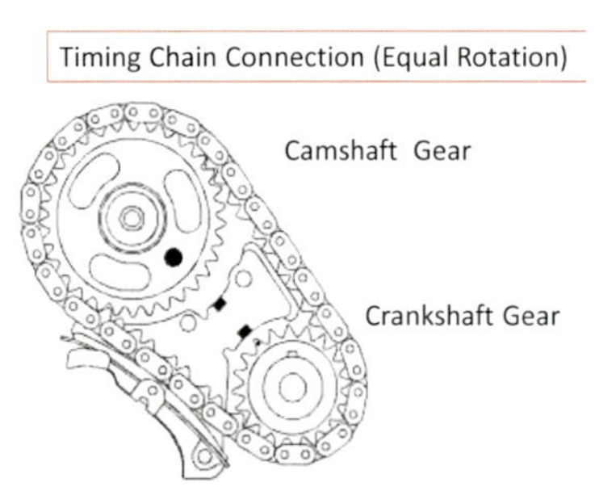

90

Timing

The valves control the flow of air intake and exhaust gases (in Diesel Engines), they must be opened and closed at the appropriate time during the stroke of the piston

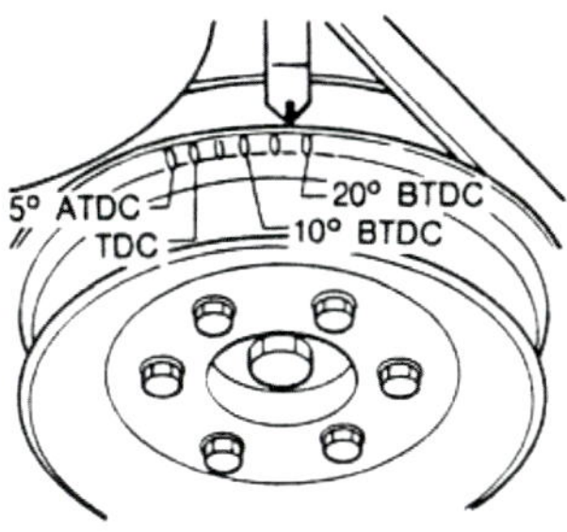

Timing

The timing of the camshaft can be <u>advanced</u> to produce better <u>low end torque</u> or it can be <u>retarded</u> to produce better <u>high end torque</u>

Timing Mark

A **timing mark** is an indicator used for setting the timing of the ignition system of an engine, typically found on the crankshaft pulley (as pictured) or the flywheel, being the largest radius rotating at crankshaft speed and therefore the place where marks at one degree intervals will be furthest apart

Timing Marks

The picture below show a Timing mark on pulley at 6° before TDC.

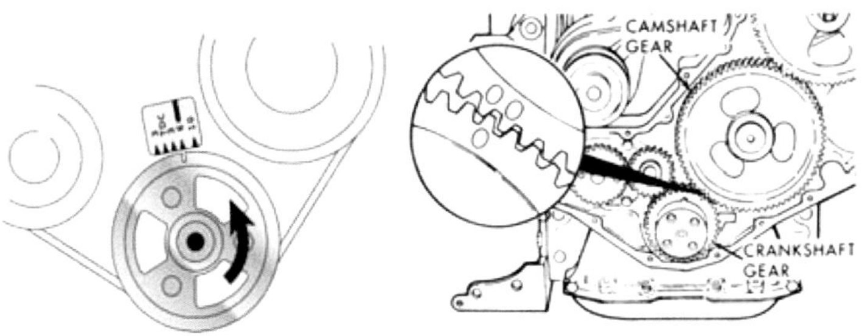

Timing Mark (Gas)

On older engines it is common to set the timing using a timing light, which flashes in time with the ignition system (and hence engine rotation), so when shone on the timing marks makes them appear stationary due to the stroboscopic effect

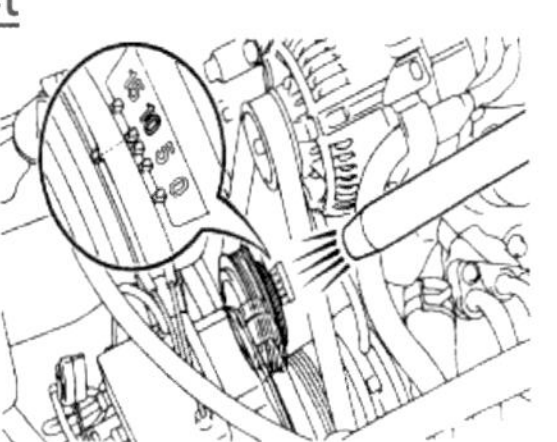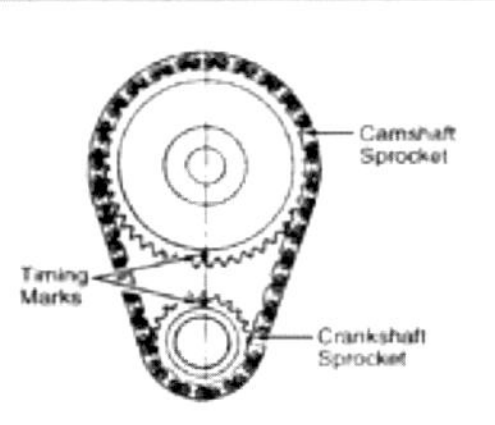

Timing Mark (Gas)

The ignition timing can then be adjusted to fire at the correct point in the engine's rotation. The timing can be adjusted by loosening and slightly rotating the distributor in its seat

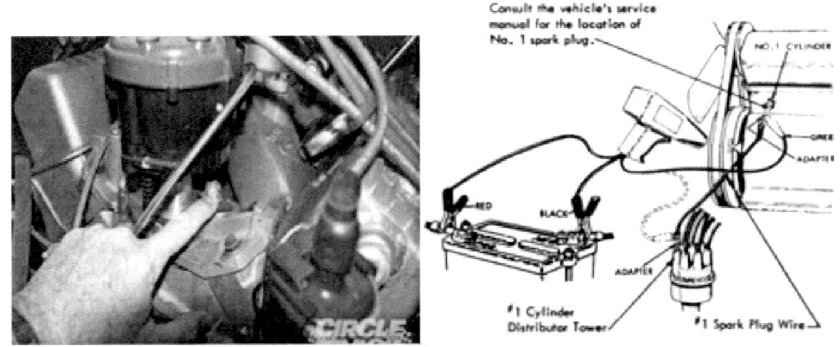

Timing light

The timing light is connected to the ignition circuit and used to illuminate the timing marks with the engine running

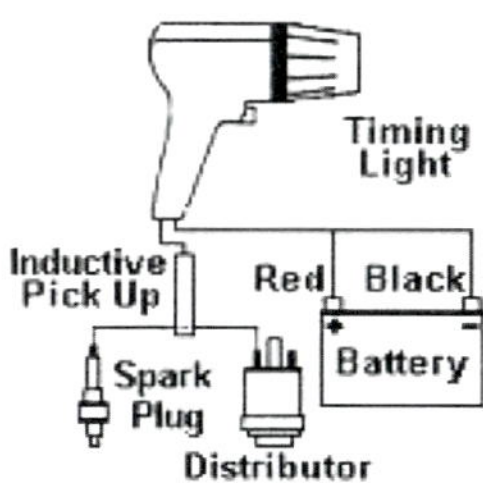

Timing light

The apparent position of the marks, frozen by the stroboscopic effect, indicates the current timing of the spark in relation to piston position. On most automotive engines, the timing is set based on the #1 cylinder

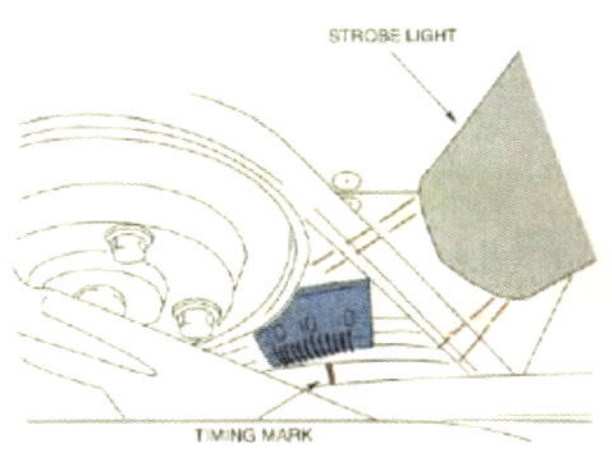

Timing Mark (Gas)

Modern engines usually use a crank sensor directly connected to the engine management system

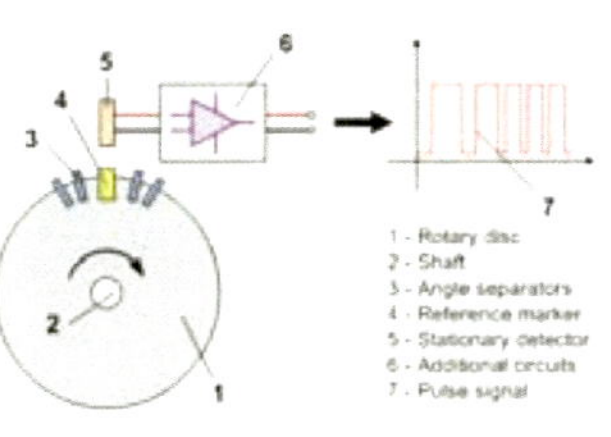

Diesel Time Adjustment

On Diesel Engines Can gear and Fuel Pump gear shall be synchronized

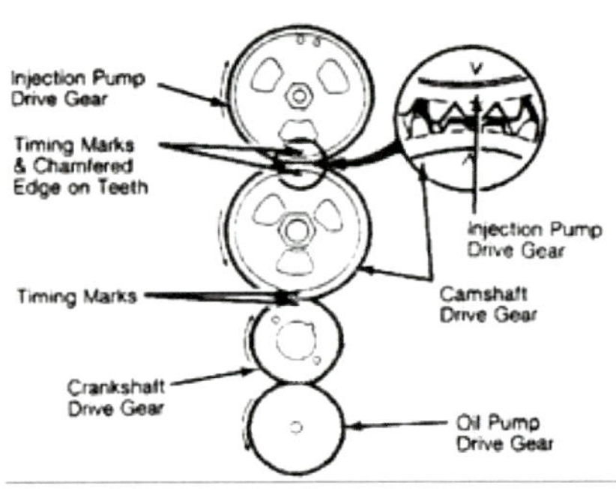

Diesel Timing

The Timing can be adjusted with the fuel injection pump

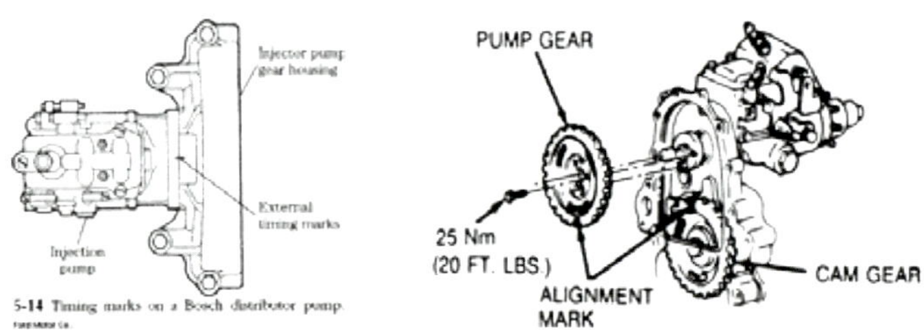

Timing Light (AC / DC)

The Sun timing light draws power from a wall outlet or, if equipped with an inverter, from the engine's 12-V or 24-V batteries

Diesel Timing Light

A transducer clamps over No. 1 fuel line to trigger the strobe when the injector opens and the sudden drop in fuel pressure contracts the line. The instrument also tracks how many crankshaft degrees the timing advances as engine speed increases

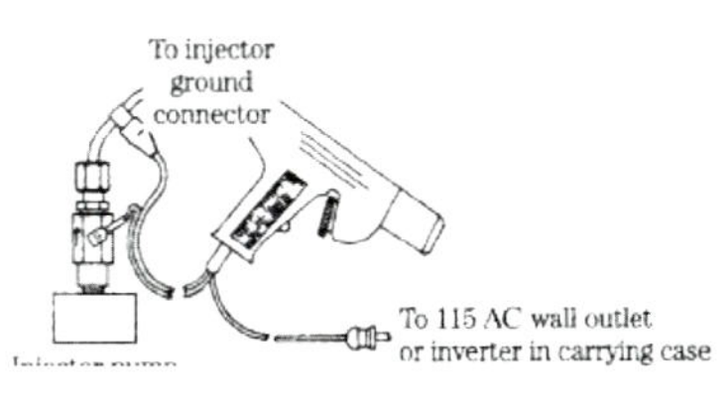

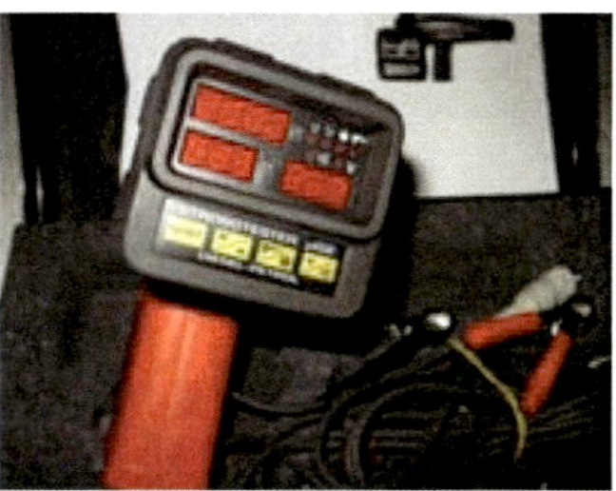

Timing on Diesel Engines

Since there are no spark plugs on a diesel engine, a spark-triggered timing meter or light cannot be used to read the ignition timing. A few diesel timing meters have been developed that employ a luminosity probe to detect ignition

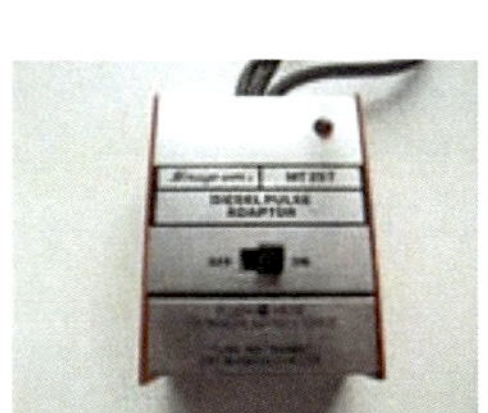

Timing Belt

Timing belt is a toothed belt that connects the engine crankshaft to the camshaft or camshafts

Timing Belt

The timing belt connects the crankshaft to the camshaft(s), which in turn controls the opening and closing of the engine's valves

Timing Belt

It has teeth to turn the camshaft(s) synchronized with the crankshaft, and is specifically designed for a particular engine. In some engine designs, the timing belt may also be used to drive other engine components such as the water pump and oil pump

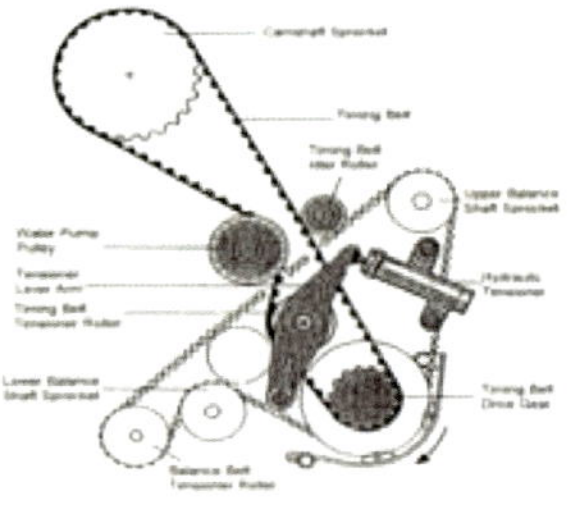

Timing Belt Replacement Intervals

- Your timing belt should be replaced every 50-70,000 miles (800 Hrs Marine Diesel Engines)
- In an engine, the valves and piston share the same air space. They never touch, unless your timing belt breaks or skips, and this is a catastrophic failure that requires removing the head and replacing bent valves

Belt Tensioner (Hydraulic)

Belt Tensioner (Mechanical-Spring)

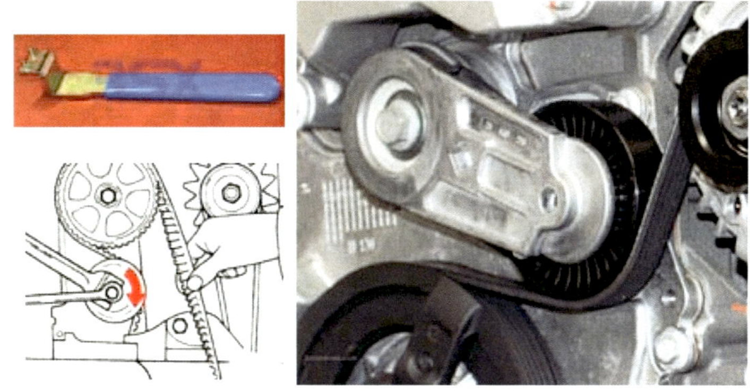

Timing Chain

The timing chain keeps internal engine components in sync with each other by causing valves to open and close at the proper time

Timing Chain

- It is durable and generally needs no maintenance
- If the timing belt is loose or improperly adjusted, it may "jump time" (skipping a tooth or more, usually on the cam gear). This results in a loss of synchronization and engine performance

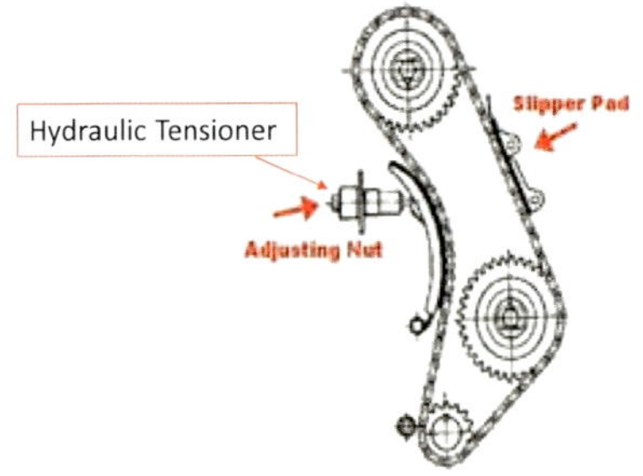

Timing belt or Timing Chain

- A **timing belt/Chain** controls the timing of the engine's valves, connects the crankshaft to the camshaft(s), at the correct timing.
- Camshaft drives, whether gears, belts or chains are also able to even out wear, since the chain or belt can be made such that the number of teeth on the belt is coprime to the number of teeth on the crankshaft and camshaft sprockets

Timing Belts   Vs   Timing Chain

- Belts are quieter
- Are less expensive
- Are mechanically more efficient
- Considerably lighter
- Belts do not require lubrication
- a broken timing belt can cause expensive damages

- Chains are More durable
- A timing chain is noisier
- Less efficient, and more expensive

- A sloppy timing chain can result in poor running, valve clatter, and loss of power

Cylinder Numbering

- In a straight engine the injectors (and cylinders) are numbered, starting with #1, usually from the front of the engine to the rear

- In a V engine, cylinder numbering varies among manufacturers. Generally speaking, the most forward cylinder is numbered 1

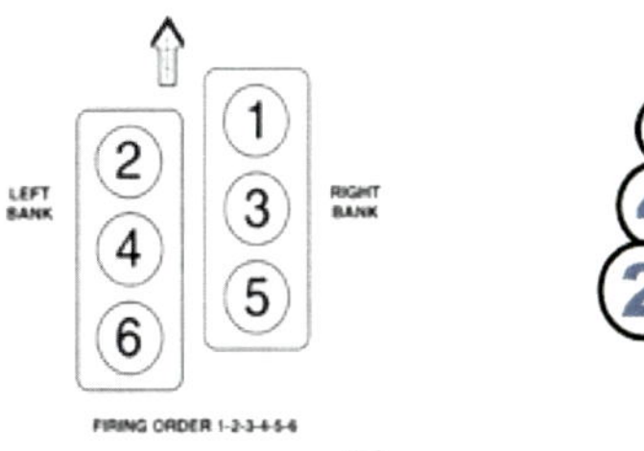

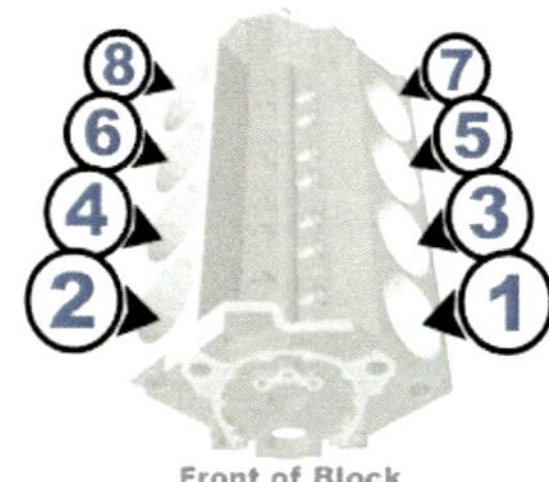

Block Firing Order

The **firing order** is the sequence of power delivery of each cylinder in a multi-cylinder reciprocating engine

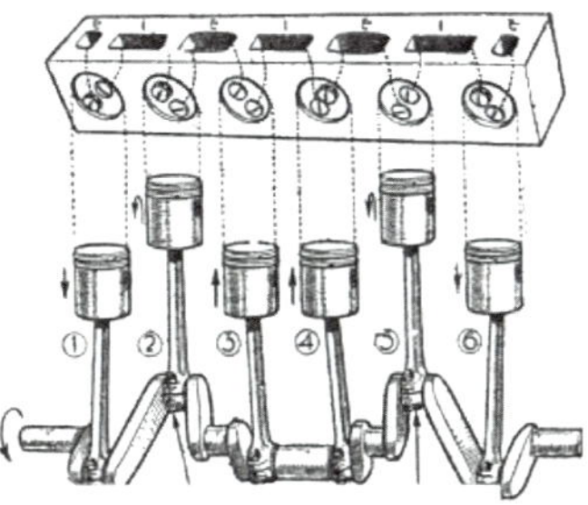

Firing Order

This is achieved by sparking of the spark plugs in a gasoline engine in the correct order, or by the sequence of fuel injection in a Diesel engine

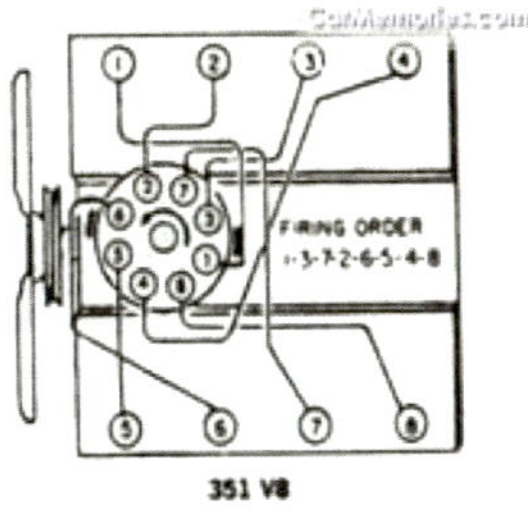

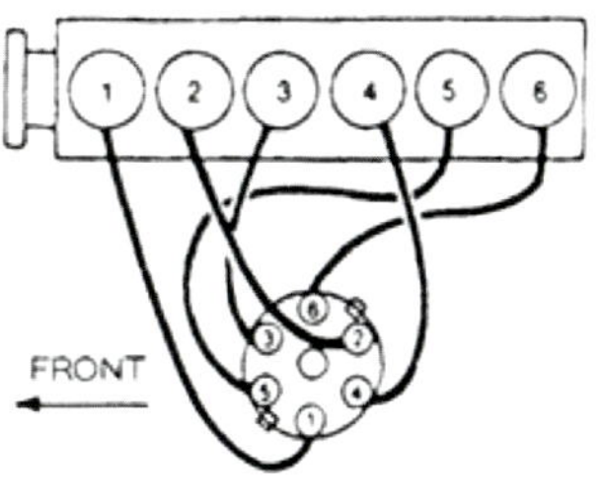

Firing Order (Gasoline)

In a gasoline engine, the correct firing order is obtained by the correct placement of the spark plug wires on the distributor

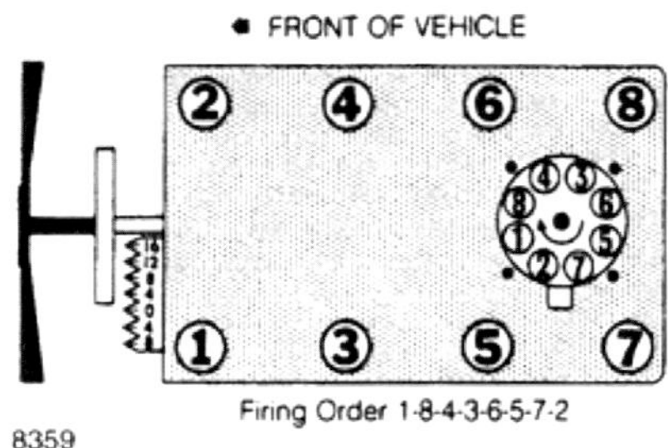

Clockwise Rotation (Gas)

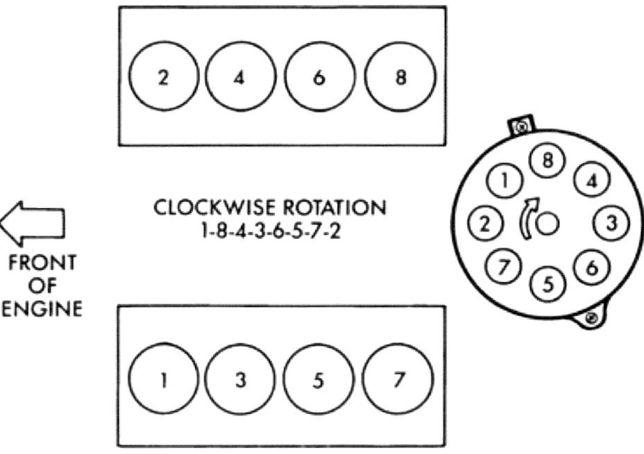

Firing order

- The order in which combustion occurs in the cylinders of an engine. Also the order in which spark is distributed to the plugs by the distributor
- When designing an engine, choosing an appropriate firing order is critical to minimizing vibration and achieving smooth running, for long engine fatigue life and user comfort, and heavily influences crankshaft design

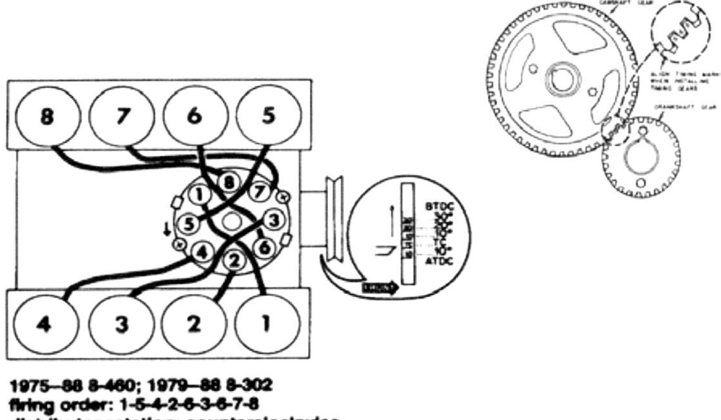

Firing Order (Gas)

In a gasoline engine, the correct firing order is obtained by the correct placement of the spark plug wires on the distributor

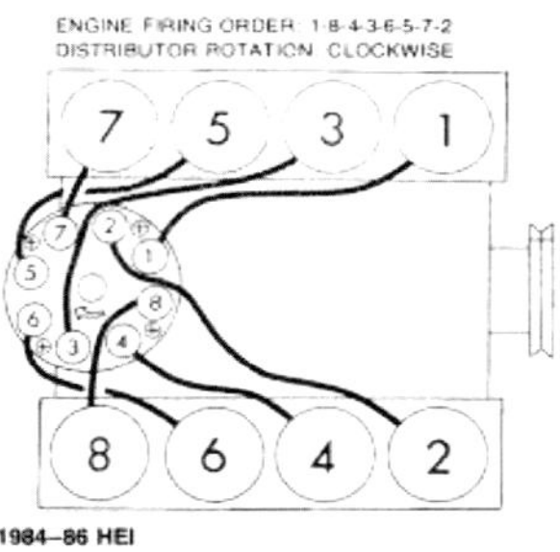

Firing Order (Diesel)

In a Diesel engine, the correct firing order is obtained by the correct placement of the Fuel injection pump engaged with the camshaft

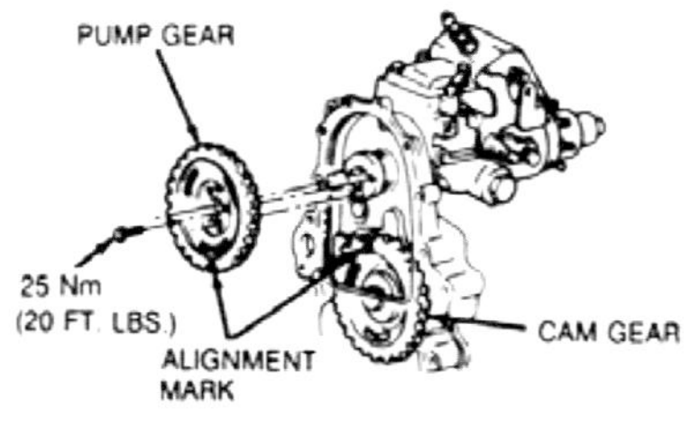

Diesel Engine Firing Order

The ships engine cylinders are numbered from the flywheel end of the engine, whereas car engine cylinders are numbered from the free end of the engine

Timing Marks

A **timing mark** is an indicator used for setting the timing of the ignition system of an engine, typically found on the crankshaft pulley (as pictured) or the flywheel, being the largest radius rotating at crankshaft speed and therefore the place where marks at one degree intervals will be farthest apart

Diesel Pump Timing

Diesel pump timing involves the timing of the fuel that is injected into the cylinders of the diesel engine

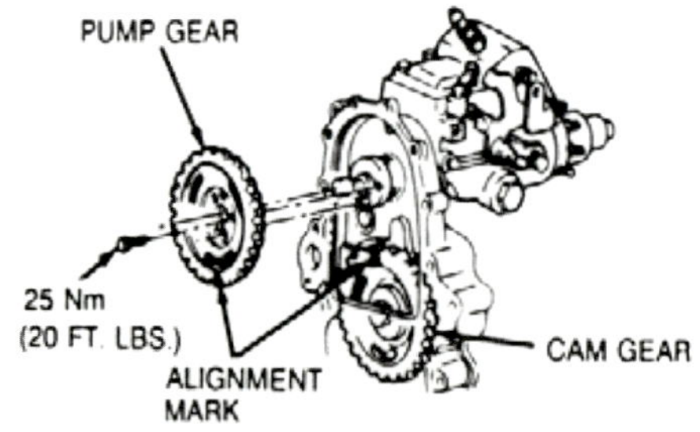

Diesel Pump Timing

The pump is usually driven by a timing belt or by a Cam gear, and the diesel fuel is slightly injected right before the engine pistons are at their farthest point from the crankshaft

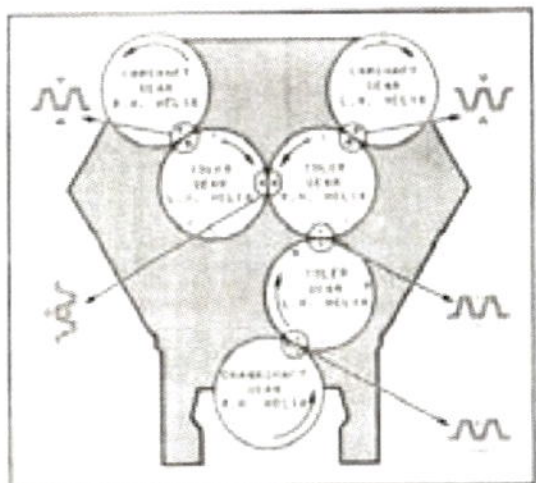

Diesel Pump Timing

A general rule to follow, and a good starting point for tuning, is that the more fuel you're running, the more conservative the ignition timing

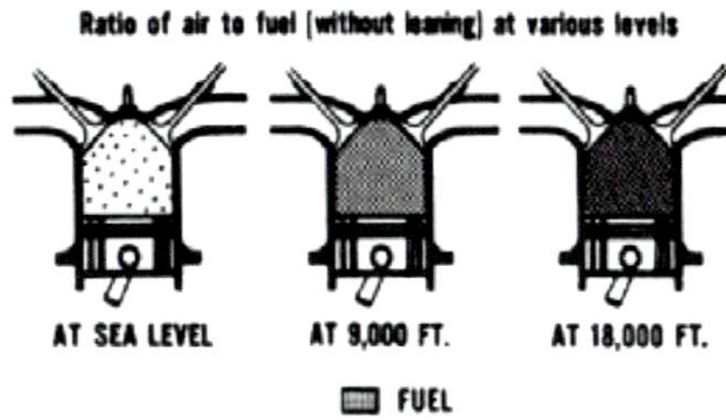

Figure 2–12   Changes in Fuel-Air Mixture with Increased Altitude

Diesel Pump Timing

The timing of the fuel injection can be increased or decreased by rotating the diesel pump to position it towards or away from the crankshaft. Timing can also increase or decrease injection pressures

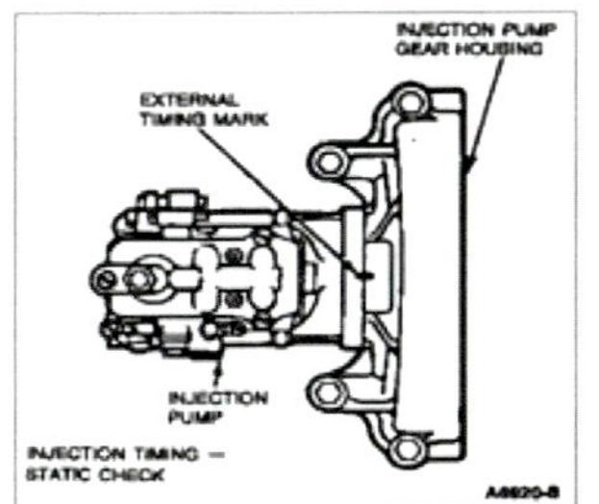

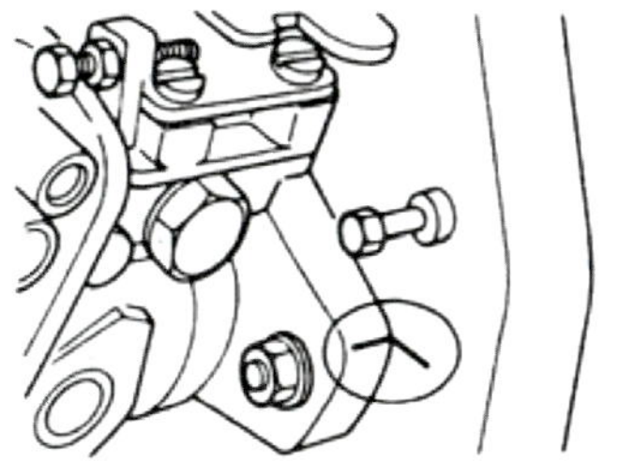

Diesel Pump Timing

Reference marks stamped on the pump body and mounting flange enable the adjustment to be replicated

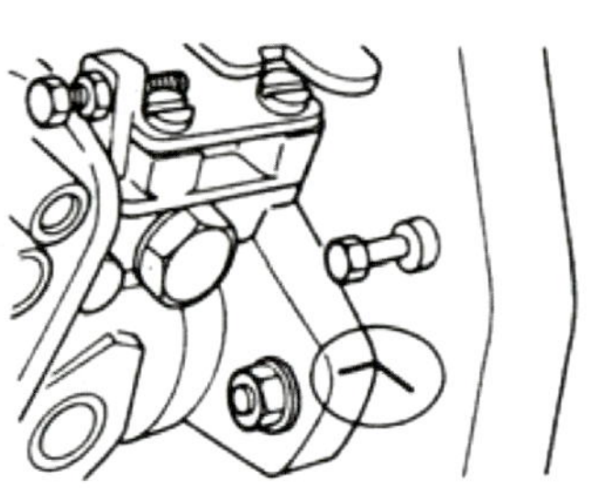

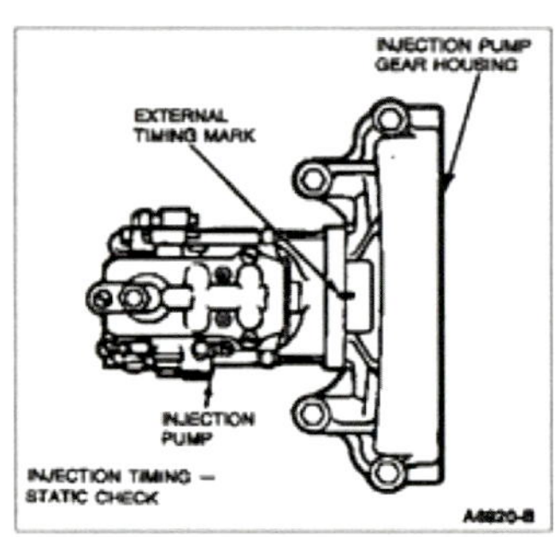

Removing the Fuel Injection Pump

Before removing a pump, bar the engine over until both valves on No. 1 cylinder close and the timing mark on the harmonic balancer or flywheel aligns with its pointer

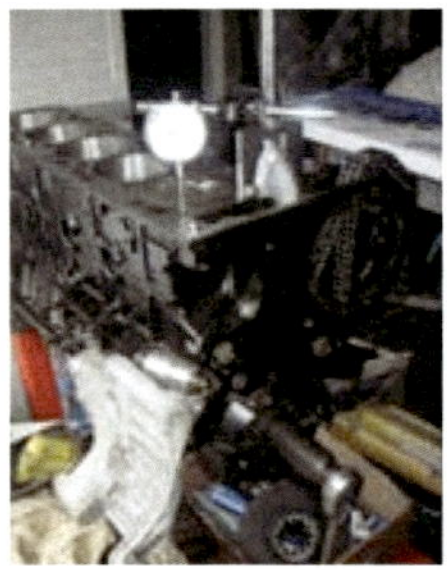

Timing Marks

The **Timing Marks** are used to align the gears and help ensure proper valve and injection timing

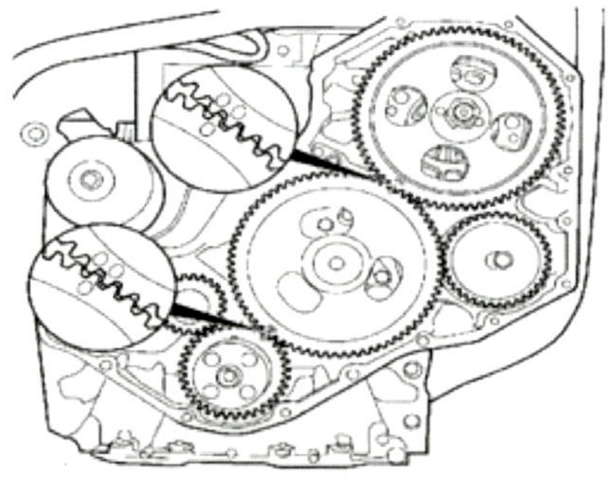

Removing Fuel Injection Pump

This procedure indexes pump-gear timing marks for easy assembly. If the same pump is reinstalled, the reference marks on the pump body and flange should be valid

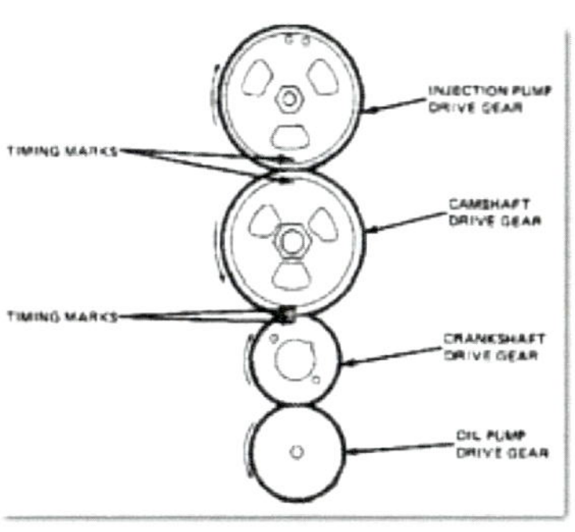

Fuel Injection Pump Gear

The camshaft gear drives the **Fuel Injection Pump Gear**. Since they are the same size, they turn at the same speed

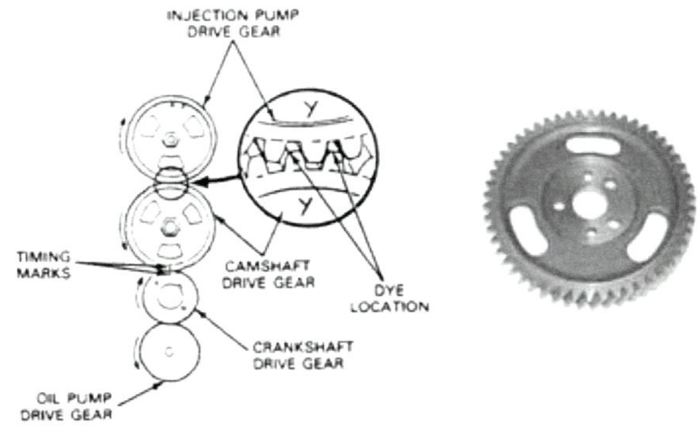

Fuel Injection Timing

Substituting another pump puts the marks into question and the engine should be retimed, either statically or dynamically

Fuel Injection Timing

On the injection pump gear are timing marks for 3,4 and 6 cylinder engines as the same gear is used in all the engines. To time the pump align the mark with the correct number for

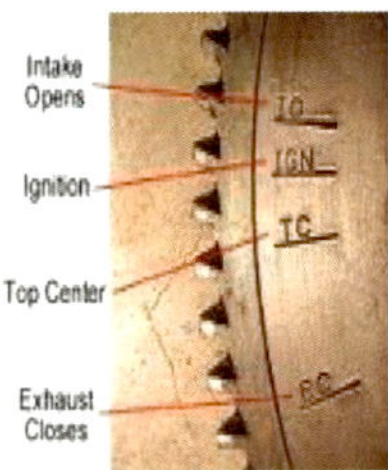

Flywheel Timing Marks

Flywheels for small Diesel engines generally have two marks inscribed on their rims: one representing TDC and the other, always in advance of the first, indicating when fuel should begin to flow from No. 1

Timing Gear Housing

The **Timing Gear Housing** protects all the timing gears and seals the front of the engine block

Gear Train Assembly

The **Gear train** assembly is a series of gears that transfers power from the crankshaft to other major components of the engine

Crankshaft Gear

The **Crankshaft Gear** is mounted on the crankshaft and drives the gear train

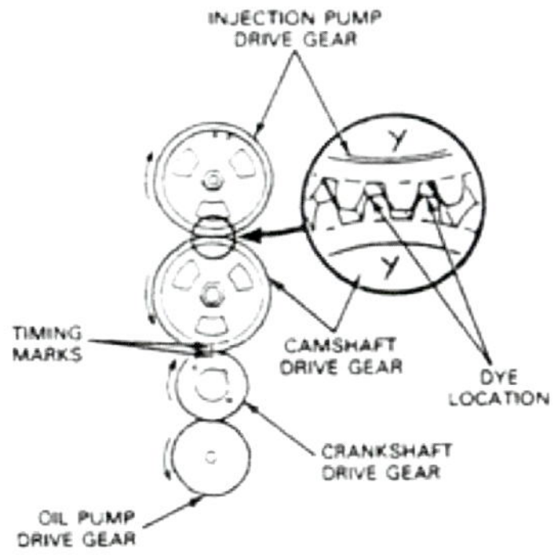

Camshaft Gear

The **Camshaft Gear** meshes with the idler gear. It turns at half the speed of the crankshaft to make sure the intake and exhaust valves open and close at the correct time

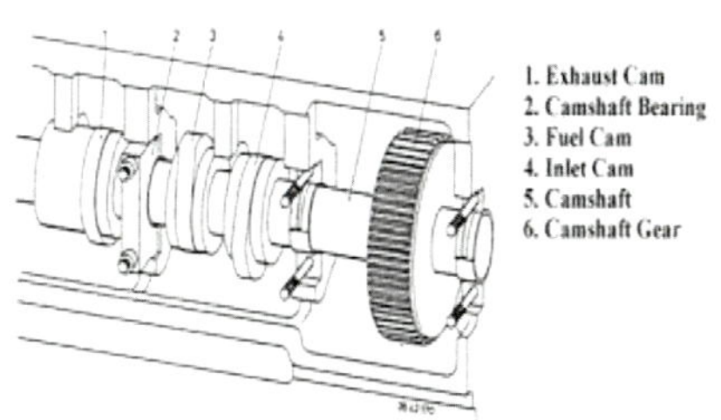

Oil Pump Gear

The **Oil Pump Gear** is driven by the crankshaft gear. The oil pump creates engine oil flow to enable oil circulation throughout the engine

Water Pump Gear

The **Water Pump Gear** is usually driven at the same speed as the crankshaft. The water pump circulates coolant throughout the engine

Air Compressor Gear

The air compressor (Option available on truck engines) is driven from idler gears and turns at the speed recommended by the manufacturer

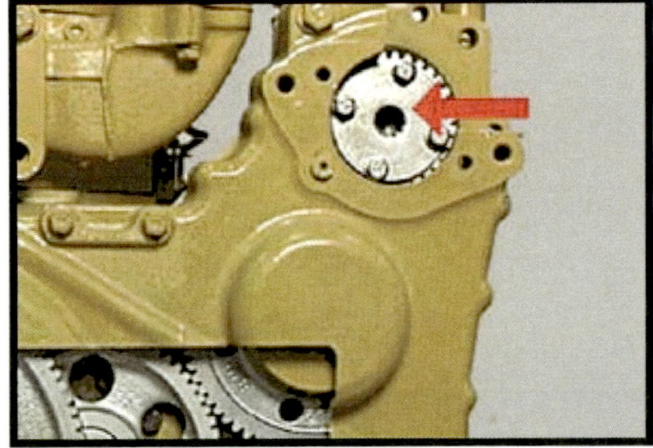

Chapter 7
Engine Block

Engine Block Parts

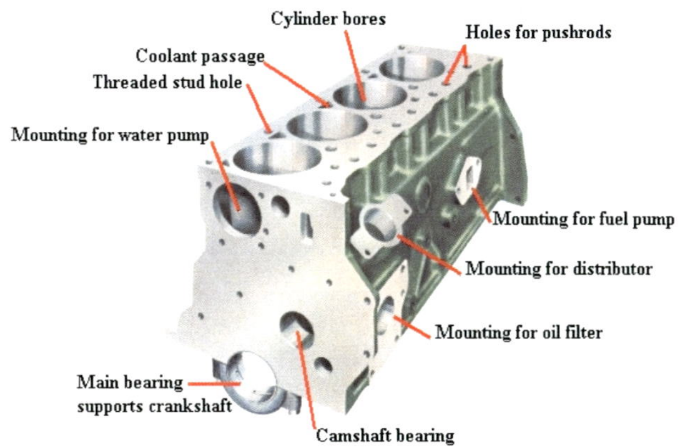

Block Configuration

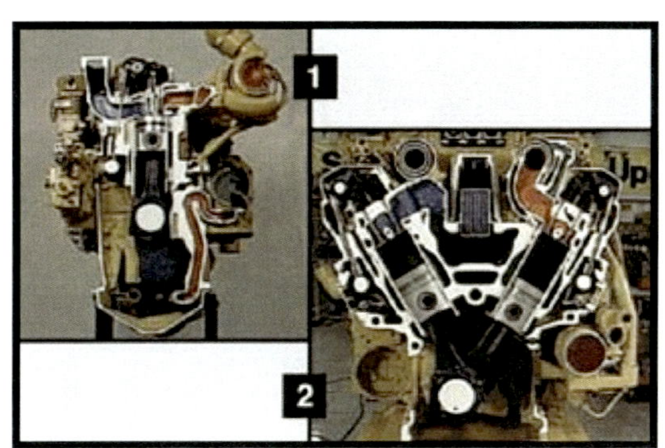

Blocks Design

Blocks are normally cast-iron and contain casts or drilling

1.   to allow for the passage of air, oil and coolants

2.   lubrication

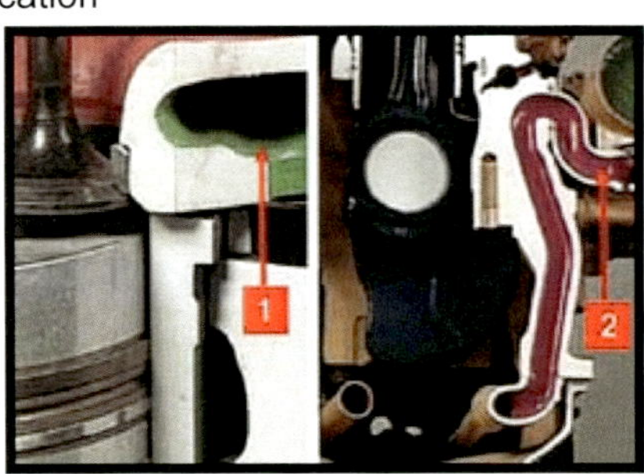

Cylinder Liners

Cylinder Liners (or the cylinders) are made of a cast molybdenum alloy iron for extra hardness. The internal surface of each liner is induction hardened, then honed in a cross-hatched pattern to aid oil control.

Water Jacket

Cylinder liners form the **water jacket** wall between coolant and pistons.

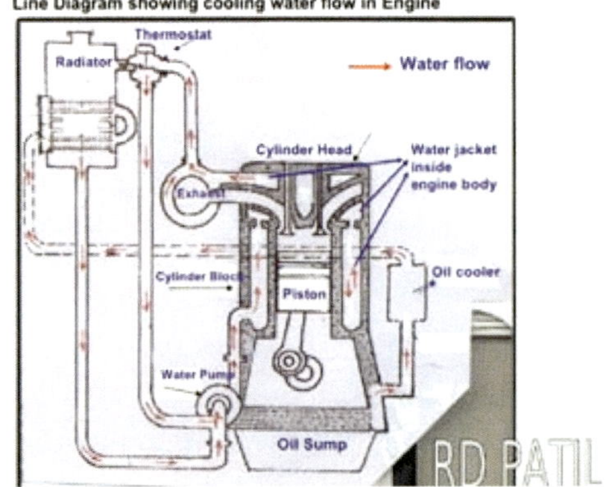

110

Wet and Dry Liners

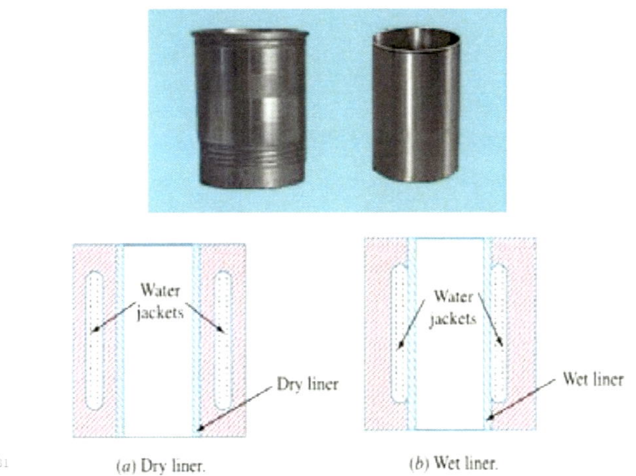

What does it mean to have a "Dry sleeve" engine

Dry sleeves are pressed into the block and do not come in direct contact with the coolant. much harder to replace usually requiring to heat up the block and ice down the sleeve

Dry Liners

What does it mean to have a "wet sleeve" engine

- **Wet sleeves** are the cylinder wall that can be removed and replaced at rebuild time.
- They are surrounded by coolant and they are o-ringed on the bottom. Usually they are easy to R&R without special tools

Wet Sleeve Installation

- Clean the o ring grooves in the block thoroughly

Wet Sleeve Installation

Measure the liner protrusion before you install the o rings

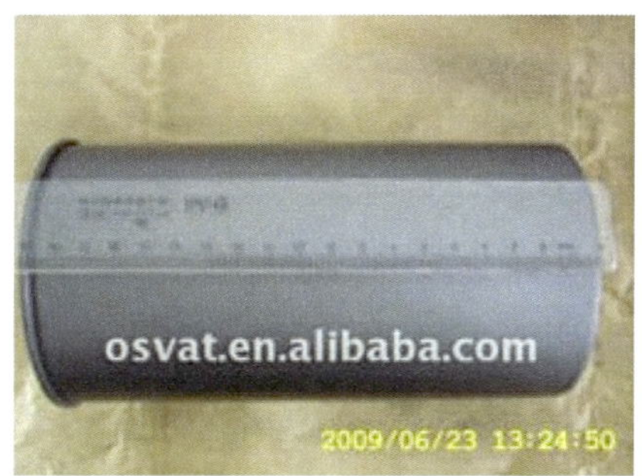

Wet Liner Installation

Wet Sleeve Installation

Install the o rings paying special attention to the location specified

Wet Sleeve Installation

Install the pistons and rods in the sleeves

Wet Sleeve Installation

- Position the sleeves and pistons so that rod and piston are aligned properly. Front markings toward the front of the engine.
- Carefully lower the sleeve piston assembly into the bore

Wet Sleeve Installation

- Push down to start the liner through the first o ring
- Gently tap sleeves completely in

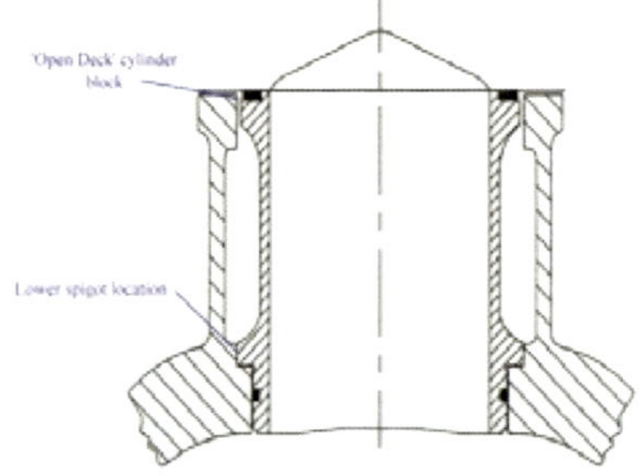

Wet Sleeve Installation

Hold the liner down with a bolt and washer
to prevent it from moving

Wet Sleeve

- Only one sleeve can be replaced

Engine Block Honing

- Engine honing is the process in which a mechanic
  uses fine-grade stones to deglaze the cylinder walls
  of an engine block before the pistons are put back
  into the engine
- This assures a tight, leak-free fit between the piston
  rings and the cylinder walls

Engine Block Honing

Larger engine blocks must be bolted to an engine stand. Make sure your stand has the appropriate weight capacity for your application. When using power tools such as electric drill, always use eye protection

Engine Block Honing

Attach the engine hone to an electric drill using the long protruding shank that extends away from the cutting stones. Then coat both the stones and the cylinder wall with light household oil

Engine Block Honing

Run the drill at low speed, between 300-700 rpm, moving the hone up and down inside the cylinder around once a second. Be careful not to push or pull the stones more than 3/4 inch out of the cylinder

Engine Block Honing

- Stop the drill after 10 seconds and check your progress. Look for a 45-degree crosshatch pattern on the cylinder walls

- If the pattern is circular with the lines running horizontally along the cylinder wall, move the drill more quickly

- If the lines are running vertically along the cylinder wall, move the drill more slowly

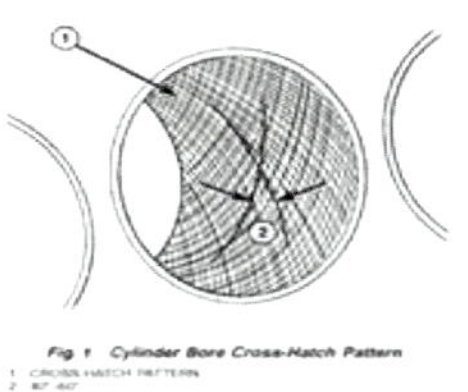

Fig. 1 Cylinder Bore Cross-Hatch Pattern

Engine Block Honing

- Do not run the motor hone for more than three minutes. Excessive deglazing could damage the engine block by wearing away too much of the cylinder wall

- Stop as soon as there is an even cross hatched pattern across the entire cylinder wall

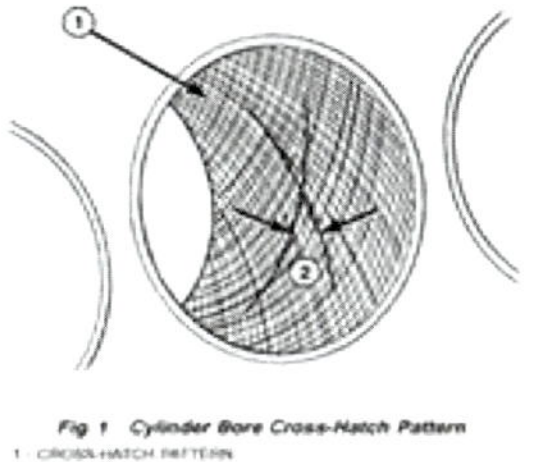

Fig. 1 Cylinder Bore Cross-Hatch Pattern

117

Crankshaft

The crankshaft is a component which converts the up and down movement of the pistons of the cylinder into continuous rotary motion which is transmitted further to move the propeller shaft

Crankshaft

The Crankshaft of the Marine Diesel Engine acts as the outlet of power from the engine for supply to the huge propeller immersed in the sea which helps to propel the ship forward

Crankshaft

The **Crankshaft** is a carbon steel forging that is totally hardened.
The big end of the connecting rod turns the crankshaft, located at the bottom of the engine block. The crankshaft transmits rotary motion to the flywheel and driveline (clutch, transmission and other driven devices), providing energy suitable for work.

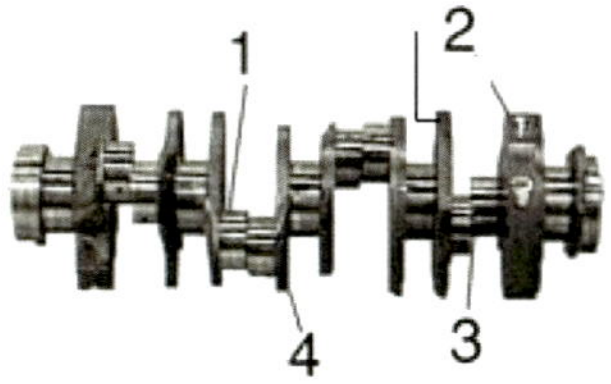

1. Connecting Rod Bearing Journal
2. Counterweights
3. Main Bearing Journal
4. Web

Forces on the Crankshaft

There are several types of forces which come to act upon the crankshaft of engines used in marine propulsion

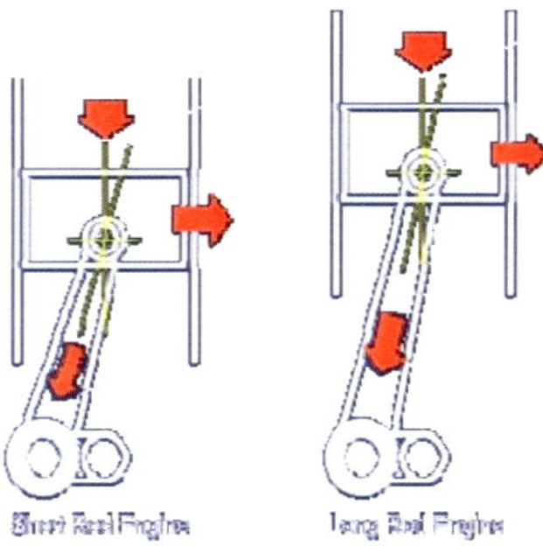

Forces on the Crankshaft

These forces are due to a variety of factors including but not limited to the weight of the pistons, combustion loads, the axial load from the propeller which is immersed in the sea, compressive loads of webs on journals and so forth

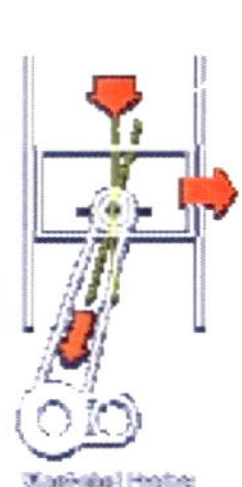
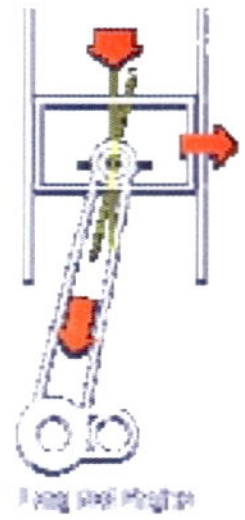
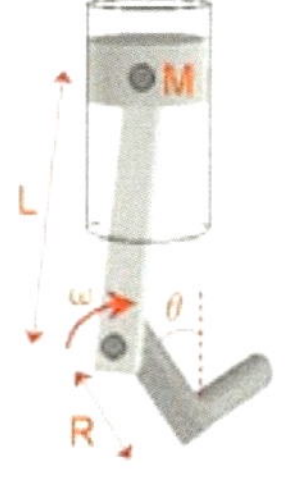

Forces on the Crankshaft

The obvious source of forces applied to a crankshaft is the product of combustion chamber pressure acting on the top of the piston

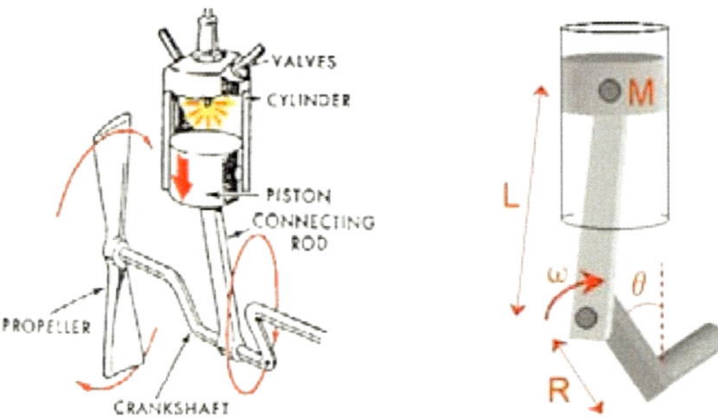

Forces on the Crankshaft

- Most of these forces have alternating patterns which gives rise to fatigue and the materials used for construction need to have substantial Ultimate Tensile Strength.
- Apart from that the other properties required in the material of a crankshaft are wear resistance, tensile strength, and ductility

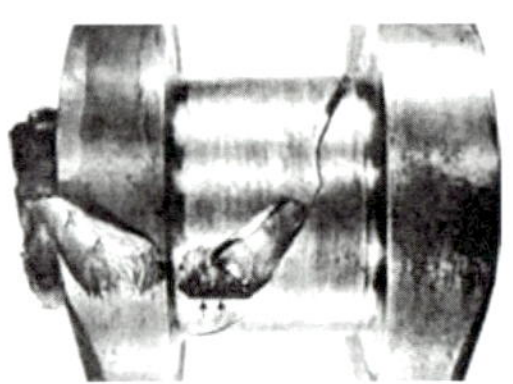

Forces on the Crankshaft

- The obvious source of forces applied to a crankshaft is the product of combustion chamber pressure acting on the top of the piston
- Contemporary diesel high-performance Compression-Ignition (CI) engines can see combustion pressures in excess of 200 bar (2900 psi).

Dynamic Forces

- There is another major source of forces imposed on a crankshaft, namely Piston Acceleration .
- The combined weight of the piston, ring package, wristpin, retainers, the conrod small end and a small amount of oil are being continuously accelerated from rest to very high velocity and back to rest twice each crankshaft revolution
- Since the force it takes to accelerate an object is proportional to the weight of the object times the acceleration

Crankshaft Materials

- The material for construction also depends on the speed on the engine and slow speed marine diesel engines have crankshafts fabricated out of plain carbon steel with a percentage of carbon lying between 0.2 & 0.4%, while the alloy steels are used for engines having a relatively higher speed
- The majority of crankshaft are made of high strength alloy steel 42 CrMo and soft nitride treatment after finish machining

Fabrication Crankshaft

- **Fully-built Crankshafts** are those in which all the various components are shrink-fitted after separate fabrication

- **Welded Crankshafts** are those in which the crank-shaft is made by welding case web crank pins and half journal units.

Crankshaft Balance

- How can it be made to spin properly again?
- One way is to put an identical weight on the other side. The first weight is balanced by the second weight. It acts as a counterweight.

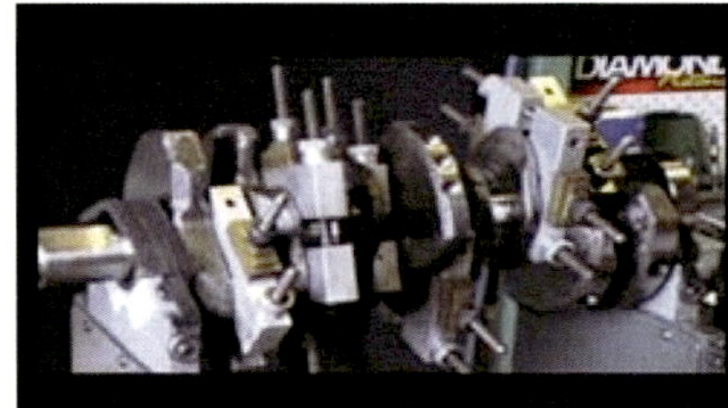

Dynamic Balance

- Crankshaft counterweights keep the rotating components in balance and help the crankshaft turn as smoothly as possible.
- The crankshaft turns because of the forces transmitted through the connecting rods. It must also be held in place. That's done by bearings.

Understanding Crankshaft Balancing

- Since different rods and different pistons are different weights, it is impossible to make a crankshaft that is balanced "right out of the box" for any rod and piston combination
- All crankshafts must be balanced to your specific rod and piston combination

Crankshaft Balancing

- The counterweights are designed to offset the weight of the rod and pistons. You have the weight of the crankshaft and the pistons and rods

- At any point in the assembly's rotation, the sum of all of the forces are roughly equal to zero

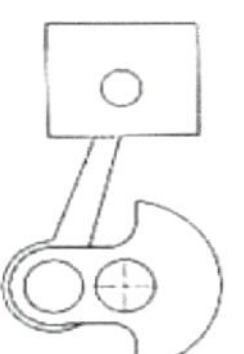 

Vibration Damper

A **Vibration Damper** is a unit that counteracts the twisting or torsional vibration caused by force variations (usually from about 3 to 10 tons [2,724 to 9,080 kg]) on the piston and subsequently the crank.

The Harmonic Balancer

- Since the harmonic dampener (front) or flywheel (rear) play a part in the balancing of the assembly, they must be installed on the crankshaft when it is balanced

- Both the harmonic balancer and the fly wheel should be integrated during the crankshaft balance

Rubber Element Vibration Damper

The Rubber Element Vibration Damper is made up of an inner iron-alloy steel flange or has a mounting facility, and an outer cast-iron weight assembly (inertial mass). The inner flange is bonded to the outer weight by a rubber compound.

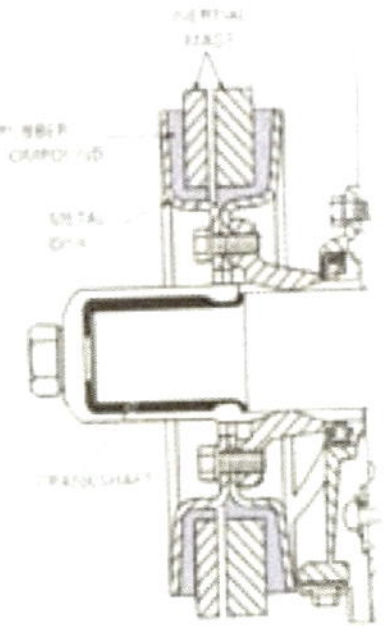

Viscous Vibration Damper

The **Viscous Vibration Damper** is a two-piece housing. A weight (inertial mass) and a special grease-like fluid is placed between the two pieces which are then welded together. The clearance between the housing and the weight is about 0.010 in.

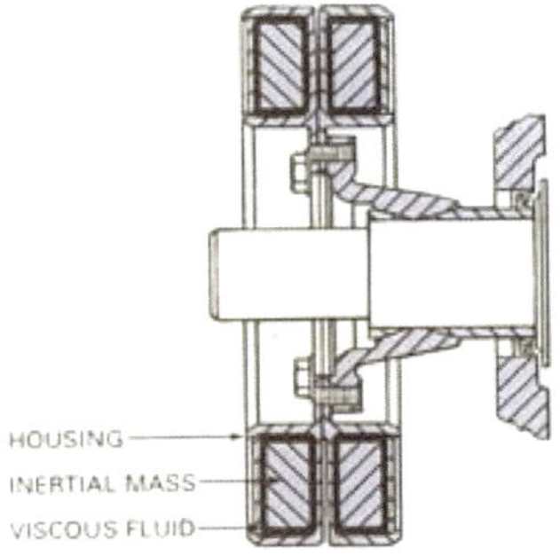

Flywheel Assembly

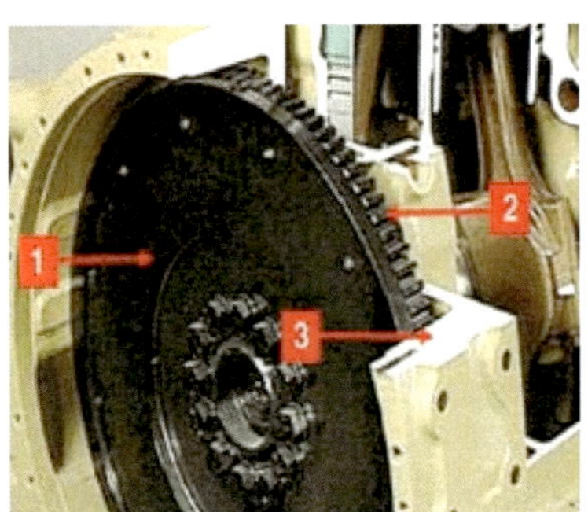

1. Flywheel
2. Ring Gear
3. Flywheel Housing

Crankshaft Lubrication

- The engine lubrication system includes the lubricating oil, oil pump, oil filter and the oil passages.
- Oil lubrication provides a barrier between rotating engine parts to prevent damage by friction, damage which can result in huge auto repair bills

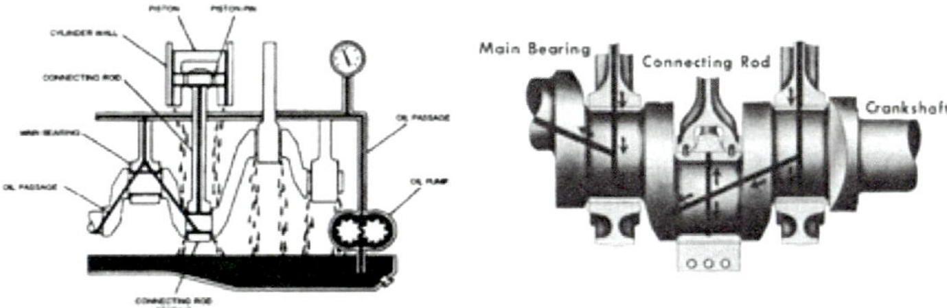

Drilled Lightening Holes and Drilled Oil Holes

The crankshaft has drilled oil holes to get oil from the main bearing journals to the connecting rod journals and bearings. These passages are plugged at one end by a cup plug or set screw.

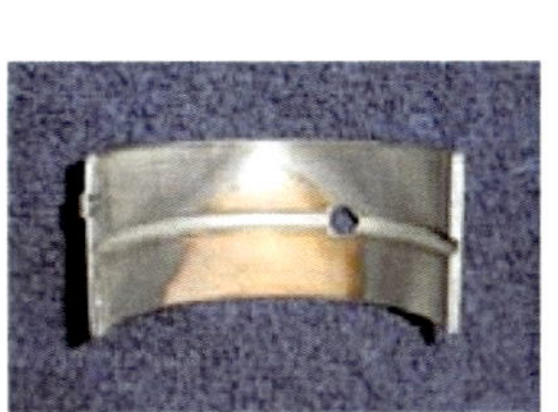

Thrust Bearing Surface

A main **Thrust Bearing Surface** is located at one of the main bearing journals. The crankshaft webs on either side of this main bearing journal have wide ground surfaces that limit the crankshafts back and forth movement, called endplay

Crankshaft Bearings

- To protect moving parts and reduce friction, Diesel engine oil provides a barrier between the rotating or moving engine components
- Ideally, a film of oil should exist between moving components. This is called **full film lubrication**

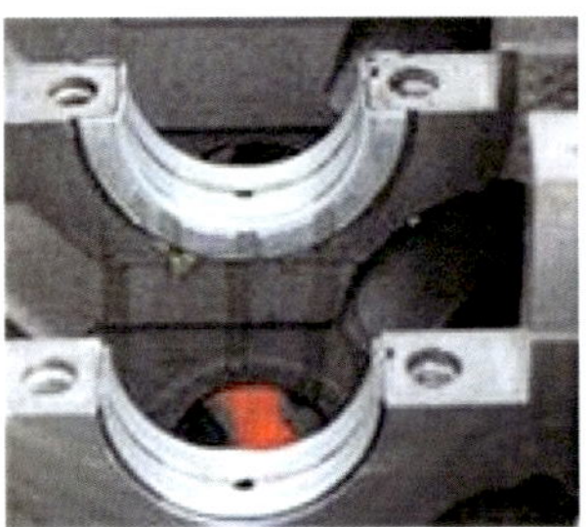
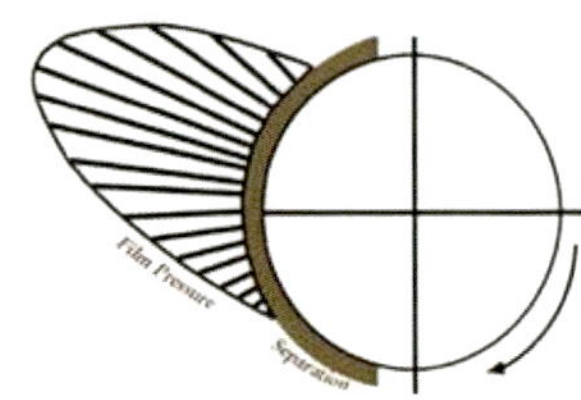

Crankshaft bearings

- The crankshaft has passages bored into it that allows oil to travel to all the bearing surfaces.
- Crankshaft bearings are made of a softer material than the crankshaft journal or cap they sit in

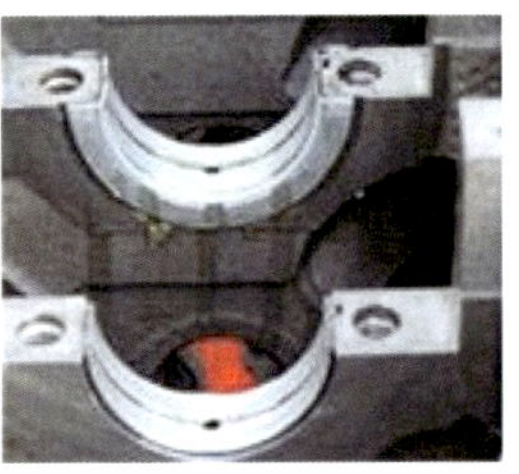

Crankshaft Bearings

If the crankshaft journals are worn, scored, or out of round the gap between the two is affected. Whenever there is contact between the journal and the bearing there will be excessive wear

Crankshaft Bearing

- Check the bearing for metal particles embedded into its surface. This would indicate two metal parts in contact with each other creating these metal shavings
- Whenever you find metal shavings in the bearings the oil pump must be inspected

Checking Clearance Bearings

- Before installing the crankshaft permanently it is always wise to check the main bearing and rod bearing clearances with plastigage
- Plastigage is packaged in calibrated envelopes. These envelopes not only protect the plastic threads but also serve directly as scales to measure the bearing clearance

Checking Bearings

Warpage can be corrected when it is not too far out of specifications. The block can be line bored and oversized bearings installed

Installing New Bearings

- Bearings come **standard, oversized, or undersized**. They have the information stamped on the back on the insert

- An **oversized** bearing has the same material thickness on the inside of the bearing facing the crankshaft journal with more material on the outside facing the bore. The oversized bearing is used after a block has been line bored

How to use plastigage

- Remove the bearing cap and wipe the oil from the bearing insert and crankshaft journal

- Tear off a piece of Plastigage as long as the full bearing width. Lay the piece of Plastigage across the full width of the lower bearing shell about 1/4" off center

How to Use Plastigage

Install and tighten the bearing cap to the proper torque specifications. Do not rotate the crankshaft while making this check.

How to Use Plastigage

Remove the loosen main bolts and remove the main bearing cap. The flattened Plastigage will adhere to either the bearing shell or the crankshaft

How to Use Plastigage

Compare the width of the flattened Plastigage at the WIDEST point with the graduations on the piece of envelope torn off in Step 2.

How to Use Plastigage

- The number within the graduations, which matches the flattened Plastigage width, indicates the total bearing clearance in thousandths of an inch or in millimeters.

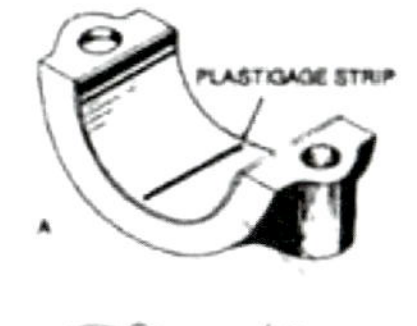

End-play Crankshaft

- The **rear thrust bearing** controls the crankshaft end-play.
- By eliminating the majority of the forward and backward movement, close tolerances can be preserved within the total rotating assembly.

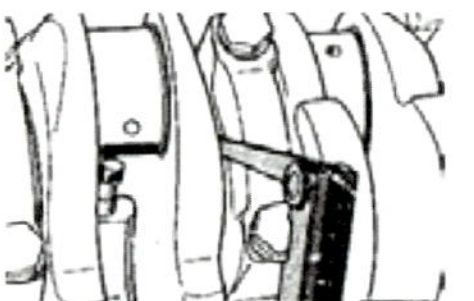  

End Play Crankshaft

Though it may be possible that a newly rebuilt engine could have a misaligned main cap and associated bearing, the much more likely cause is a bad torque converter

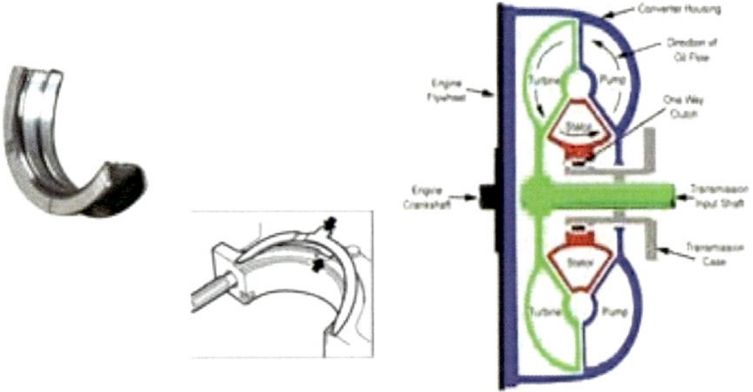

End play Crankshaft

- The torque converter naturally pushes up against the crankshaft.
- Under normal circumstances the amount of forward torque converter force is limited. However, when transmission oil is higher than normal working temperatures, the internal pressure in the torque converter drastically increases

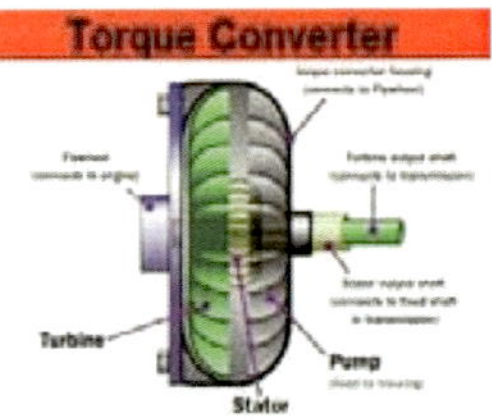 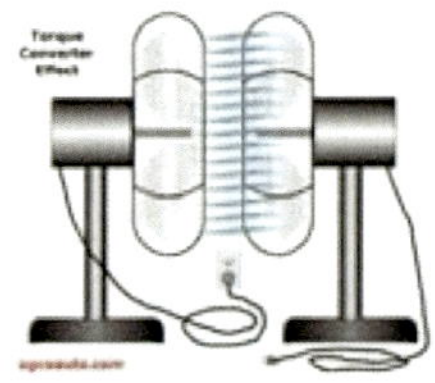

Connecting Rods

There are several parts to a connecting rod:
1. Rod-eye, gudgeon-end or small-end, that holds the piston pin bushing

2. Piston pin bushing. Bushings are a type of bearing that distribute load and can be replaced when worn.

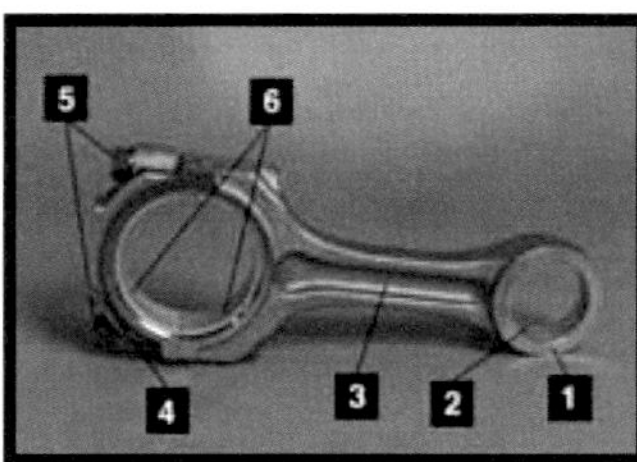

1. Rod End
2. Piston Pin Bushing
3. Shank
4. Rod End Cap
5. Rod Bolts and Nuts
6. Big-end Bearing

Connecting Rods

3. Shank between small and big ends. It has an I-beam shape for strength and rigidity.
4. The crankshaft journal bore and cap are at the big end of the connecting rod. These surround the crankshaft   bearing journal and attach the connecting rod to the crankshaft.

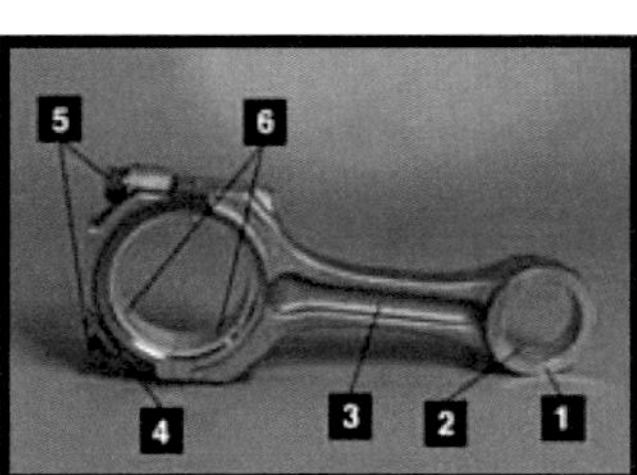

1. Rod End
2. Piston Pin Bushing
3. Shank
4. Rod End Cap
5. Rod Bolts and Nuts
6. Big-end Bearing

Connecting Rods

5. Rod-bolts and nuts secure the rod and cap to the crankshaft. This is called the crank end or big end of the connecting rod.
6. Connecting rod in big-end bearings are in the crank-end. The crankshaft turns inside the connecting rod bearings, which carry the load.

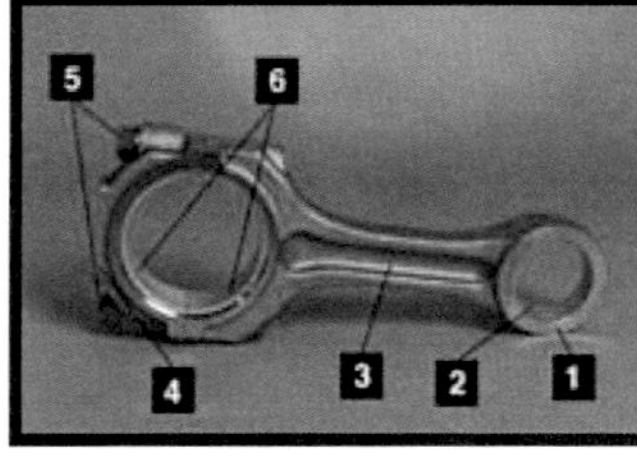

1. Rod End
2. Piston Pin Bushing
3. Shank
4. Rod End Cap
5. Rod Bolts and Nuts
6. Big-end Bearing

Chapter 9
Pistons & Rings

Diesel Engine Piston

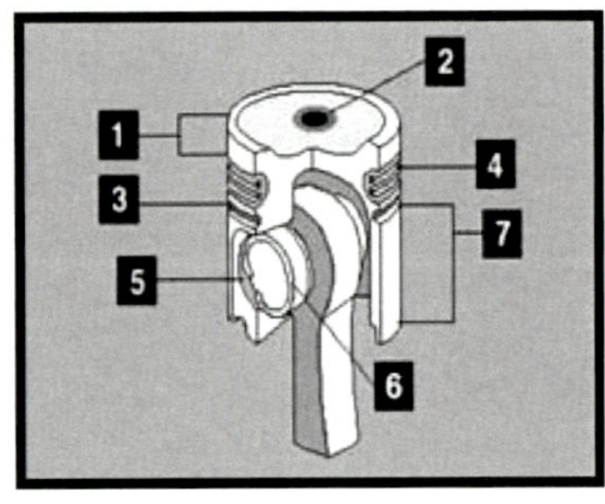

1. Crown
2. Ring Grooves
3. Piston Pin
4. Retaining Ring
5. Thrust Skirt

Piston Construction

Pistons can be of various construction:

1. Cast aluminium crown with iron band for compression rings and a forged aluminium skirt, electron
    beam welded (right).

2. The single piece cast aluminium piston with an iron band that carries the piston rings.

Piston Construction

Two-piece Articulated, comprising forged steel crown with pin bores and bushes, and a separate cast aluminium skirt, held together by the wrist pin.

Forged Piston Advantages

- Forged pistons generally run 18 to 20 percent cooler than cast pistons.

- Forging also increases cracking resistance, and may allow a piston to survive a close encounter with a valve without shattering (thus saving the engine).

Diesel Engine Pistons

- Diesel Engines have double the torque of gasoline engines and can achieve up to 40 percent better fuel economy.

- Pistons in 2 & 4-stroke diesel engines can change direction hundreds of times a second and are exposed to extremes of heat and pressure.

Piston Head

- The combustion chamber in most diesel engines is actually in the top of the piston rather than the cylinder head
- There is a bowl-shaped recess that swirls and compresses the air as the piston comes up

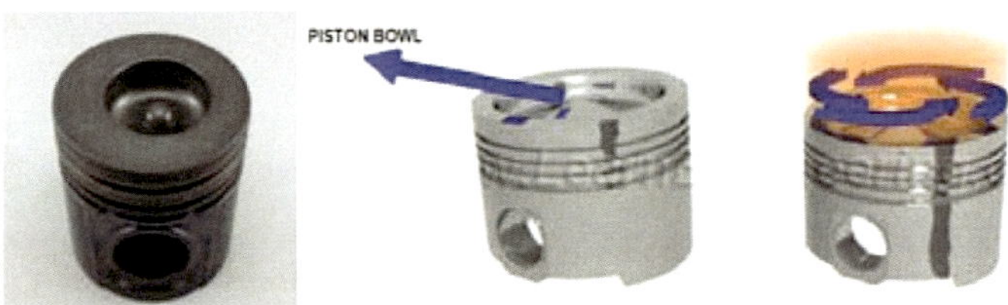

Piston Dimensions

- This design means the piston will be relatively long and heavy compared to a piston in a gasoline engine .
- But the extra weight doesn't matter much because a stock diesel engine typically operates at relatively low rpm (under 4,000 rpm)

Piston Weight & Momentum

At higher speeds, the weight of the piston on the upstroke creates a lot of momentum and load on the small end of the rod, the wrist pin and the wrist pin eyelets in the piston

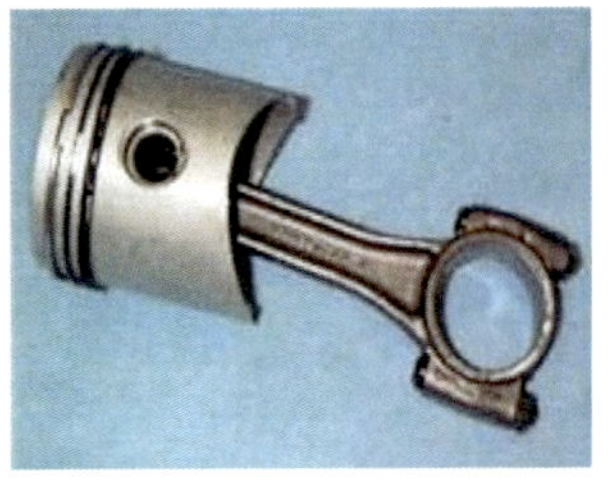

Compression Pressures

- The pistons in a diesel engine must also withstand significantly higher compression pressures and operating temperatures than those in a gasoline engine
- Because of this, the top piston ring runs hot. To reduce top ring pound out, most stock diesel pistons have a steel or iron insert for the top ring groove .

Boost Pressures

As long as engine speed remains below the design limit of the stock piston, most pistons will hold up fairly well to increased turbocharger boost pressures or even a brief shot of nitrous. But if the engine revs too high, the wrist pin eyelets can be literally pulled out of the piston

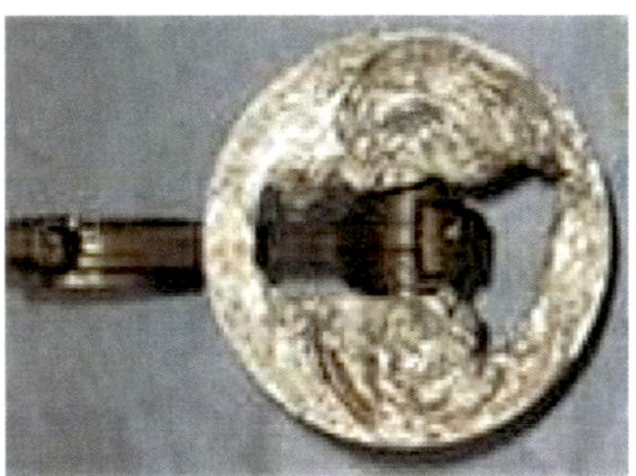

Inside of Piston

1. Under-crown Area
2. Oil Cooling Galleries

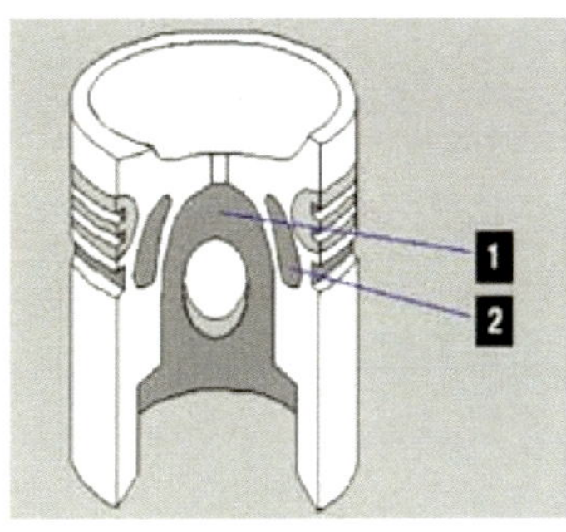

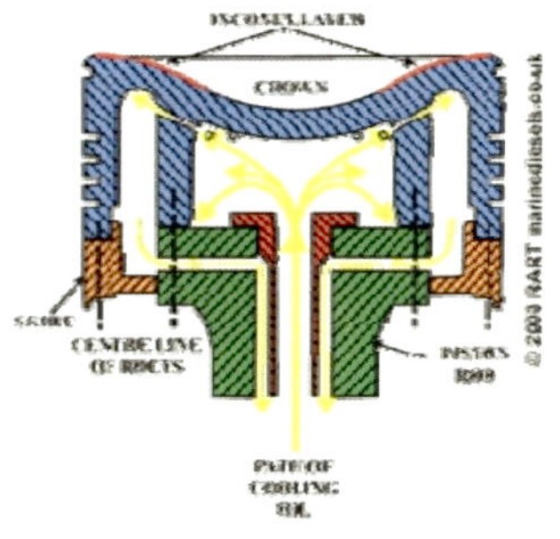

Piston Rings

1. Compression Rings
2. Oil Control Rings

Piston Rings

The efficiency of the engine depends upon the effective sealing between the piston and liners

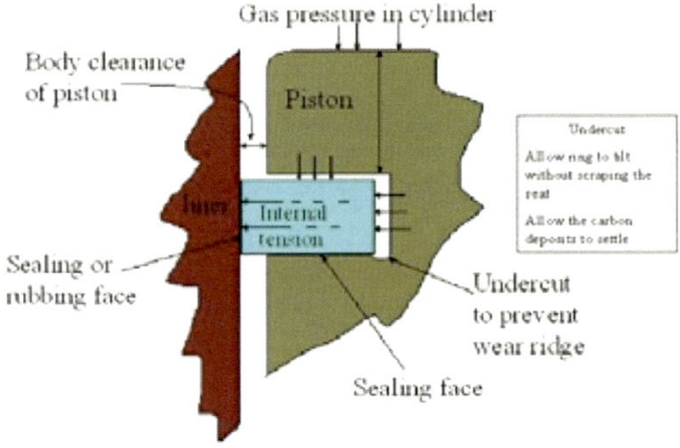

Piston Rings

Piston rings seal the gas space by expanding outwards due to the gas pressure acting behind them

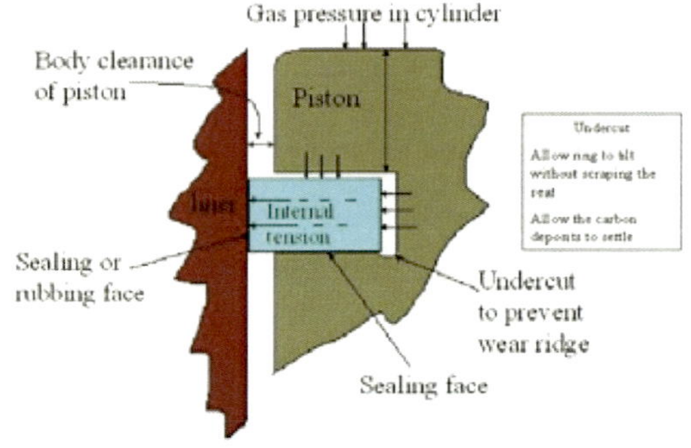

Piston Rings

- The top ring position is controlled by the operating temperature and exhaust or scavenge port trimming.
- The lower act as a scrapper or oil distribution around liner wall which could result if lubricant is allowed to leak past the compression rings and be burnt with fuel oil.

Piston Rigs Diameter

The diameter of the piston rings, when free, is slightly larger than the cylinder bore. Consequently, when the ring is squeezed into the liner, it presses against the cylinder wall and tend to seal it

Piston Ring Clearance

There are three clearances which are important

i) Side or axial clearances
ii) Butt or Gap clearances
iii) Back clearances

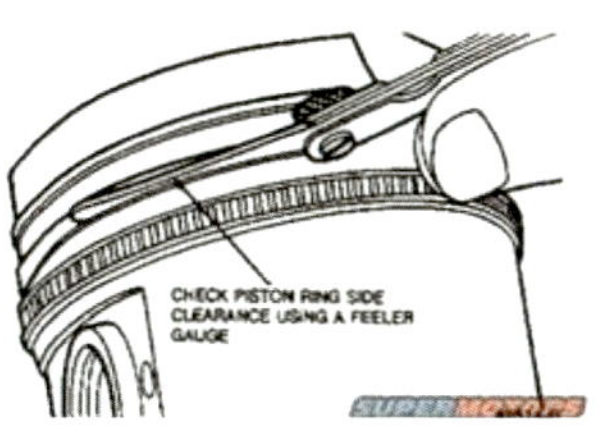

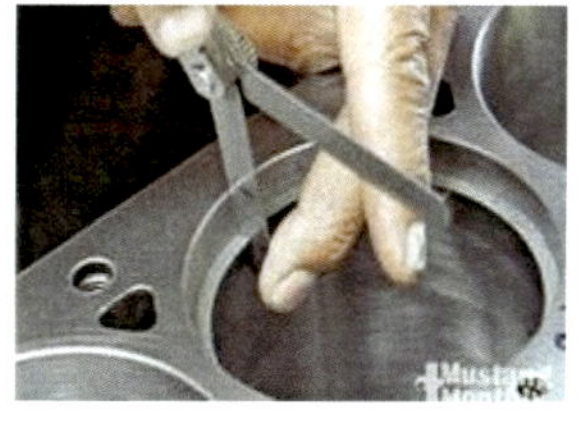

Oil Control Ring

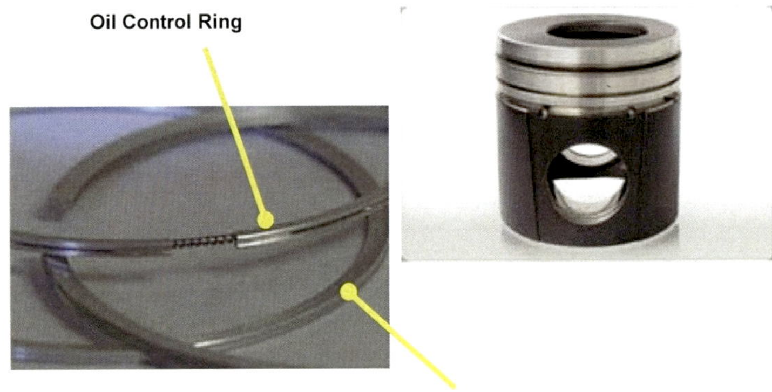

Rectangular Ring

- This ring with its geometrically simple shape performs the necessary sealing functions under normal operating conditions.

- With a peripheral coating and appropriate barrel face the rectangular ring is today used mainly in the top groove in diesel engines

 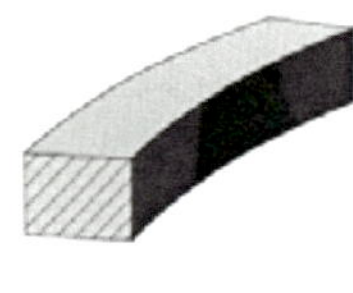

Taper Faced Ring

- Taper faced rings are chiefly installed in the second groove in passenger car gasoline and diesel engines

- **Keystone Ring**. With its tapered sides, radial movement of the ring in engine operation will cause the axial clearance in the groove to increase and decrease

Reusing Old Pistons

After a professional inspection, old pistons can be reused. A good cleaning of each ring groove is recommended before the rings installation

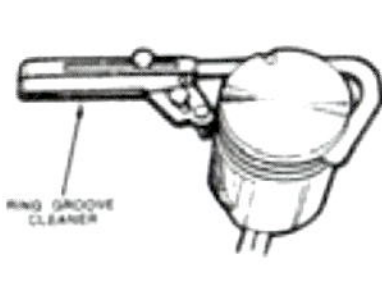

Cutting Piston Rings

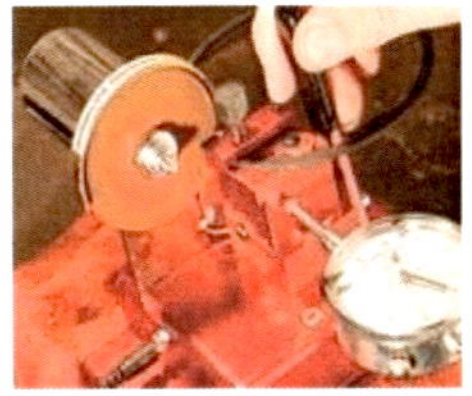

Ring End Gap

- Should not line up when installed

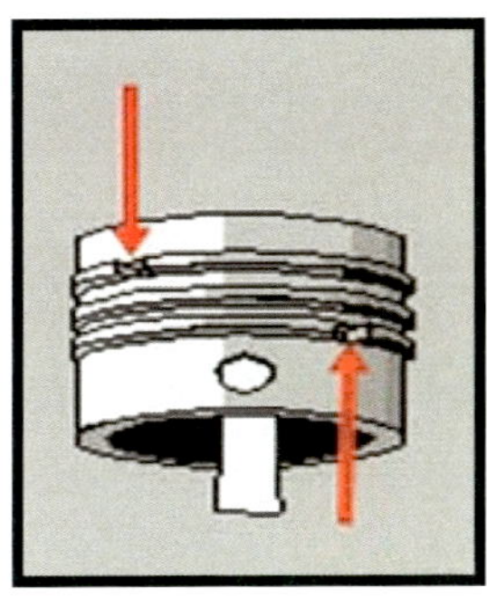

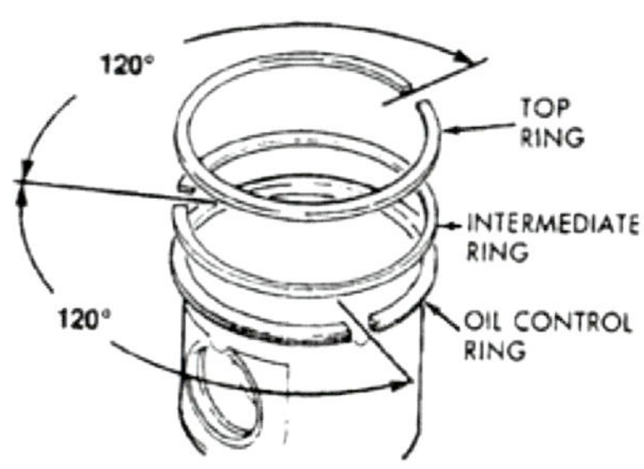

Piston Materials

- Because the pistons need a top ring groove insert for durability, the pistons must be cast rather than forged
- Most use an alloy that contains 11 to 13 percent silicon so the pistons will have some ductility

Piston Materials

The steel pistons are extremely durable and can handle the highest loads, but also tend to be heavy and expensive, costing up to three times as much as conventional cast aluminum pistons

Ceramic Fiber reinforcement

- In recent years, diesel pistons with ceramic fiber reinforcement in the bowl rim have also been developed for high load applications.
- The cast-in fibers increase the load-bearing capacity at the bowl rim, and allow the pistons to withstand extremely high thermal loads without cracking.

Oil Jets

- Another trick that's done with many diesel pistons to use oil jets to help cool the pistons
- When an oil jet is directed at the underside of the piston, oil is deflected into the gallery so it circulates behind the rings to carry away heat

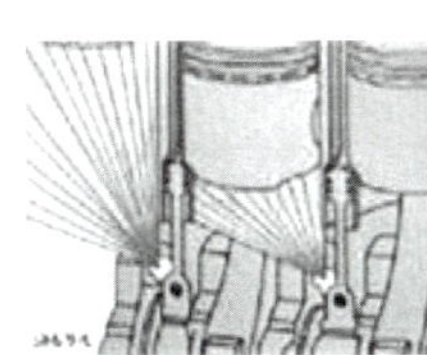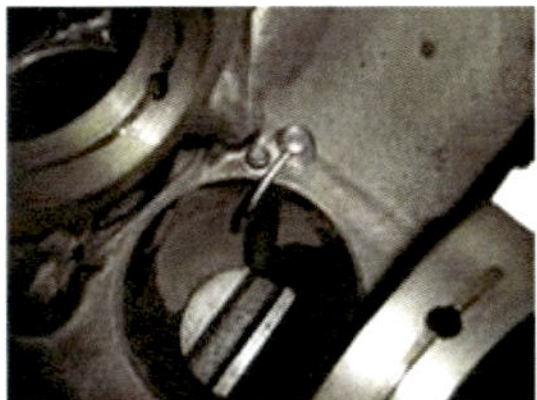

Piston Oil Jets

This technique lowers the temperature of the top ring up to 100 degrees F or more to improve sealing, reduce blowby and emissions, and to extend the life of the piston and rings

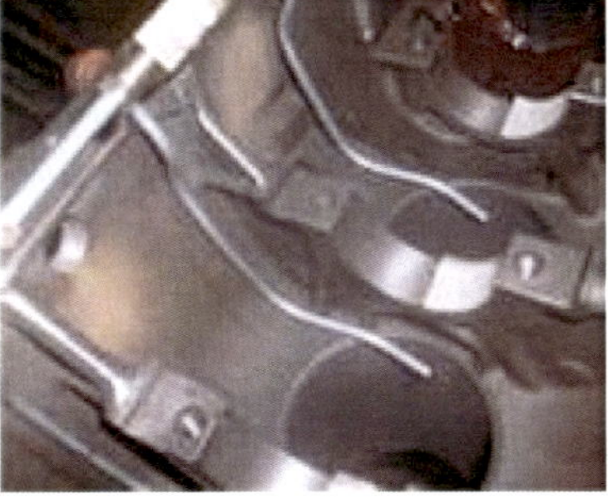

Cooling the Piston

Cooling is improved even more by casting an oil cooling gallery into the top portion of the piston behind the upper ring land to dissipate heat

Anti-scuff Coating

- Many stock diesel pistons come with anti-scuff coatings on the side skirts
- Aftermarket performance pistons are also available with or without these side coatings, as well as oil-shedding undercoatings to improve cooling, and heat-reflective top coatings to help the pistons run cooler

Anti-scuff Coating

- The anti-scuff side coatings are a good idea for break-in and to protect the pistons and cylinders in the event the engine's oil supply is disrupted for any reason
- An insulating top coating on a diesel piston can slow heat transfer into the piston to reduce thermal expansion and stress on the piston and top ring

Piston Clearance

If the stock cast pistons in a diesel engine are being replaced with stronger forged pistons, increased piston-to-cylinder clearance may be required

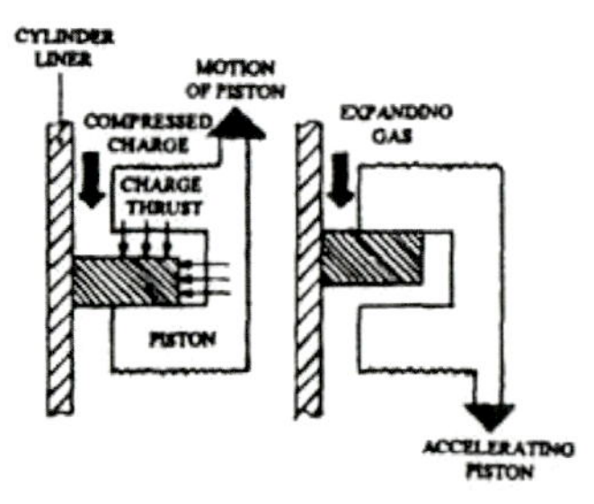

Piston Clearance

For Cast pistons the factory recommends about .002″ of clearance. With forged pistons, the clearance may have to be opened up to .006″ or more

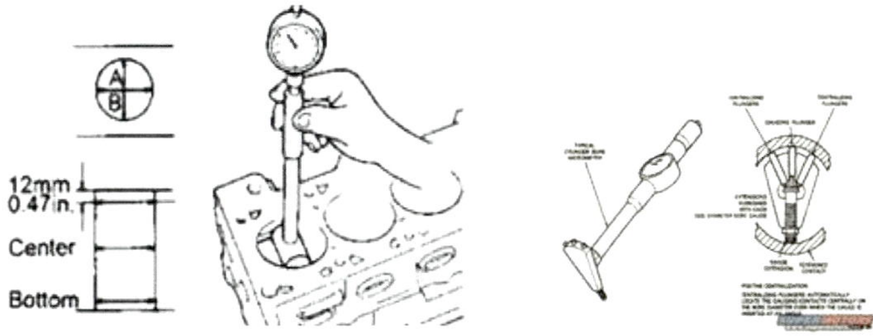

Piston Upgrades

- The forged pistons offer greater strength than the cast pistons, and are a good upgrade for pulling and racing applications
- Forged pistons and stronger rods usually become necessary in a light truck diesel engine once horsepower is boosted beyond 550 to 600 horsepower
- The forging process increases the density of the metal and significantly improves its strength (up to 40 percent or more over conventional cast pistons) .

Piston Failure Analysis

- When you tear down a diesel engine and find debris embedded in the tops of the pistons, or dents or dings in the piston crowns, it means the engine ingested some kind of contaminants.
- It may be the fault of the turbocharger, the intercooler, or a missing or damaged air filter or intake ductwork

Piston Failure Analysis

- If you find a piston with the impression of a valve stamped into the top of it, the engine may have had a valve sticking problem
- This can be caused by a buildup of fuel deposits on the valve stems from prolonged idling

Piston Failure Analysis

- The buildup of fuel deposits is called "wet stacking" and it can be prevented by adding a fuel conditioner to the fuel tank and/or be installing an idle speed controller that kicks up the idle speed if the engine is left idling for a period of time

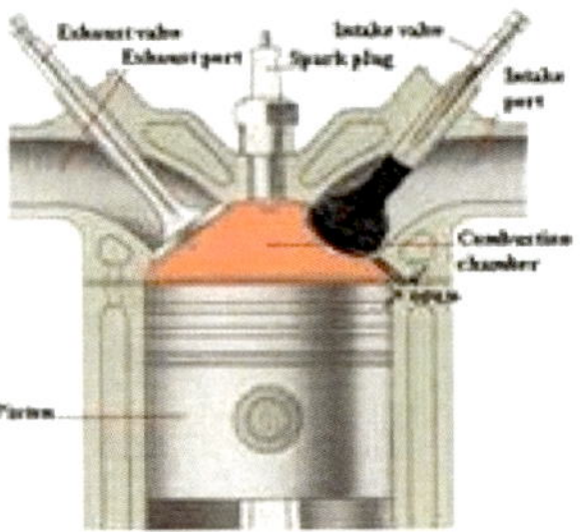

Piston Failure Analysis

- Blue smoke coming out of the tailpipe is a classic indication of oil entering the combustion chamber past worn valve guides, or worn, cracked, broken or improperly installed piston rings.
- .

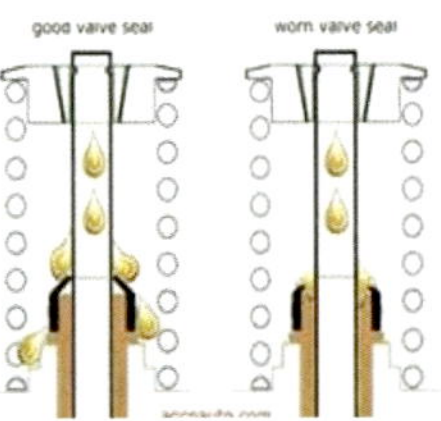

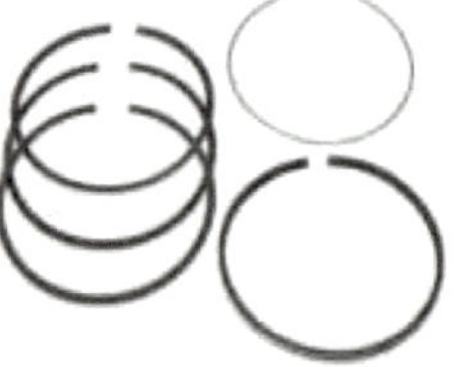

Low Oil Level & Overheating

Lean Mixture

Chapter 10
Lubrication System

Diesel Engine oil Pump

To avoid the need for priming, the pump is always mounted low-down, either submerged or around the level of the oil in the sump

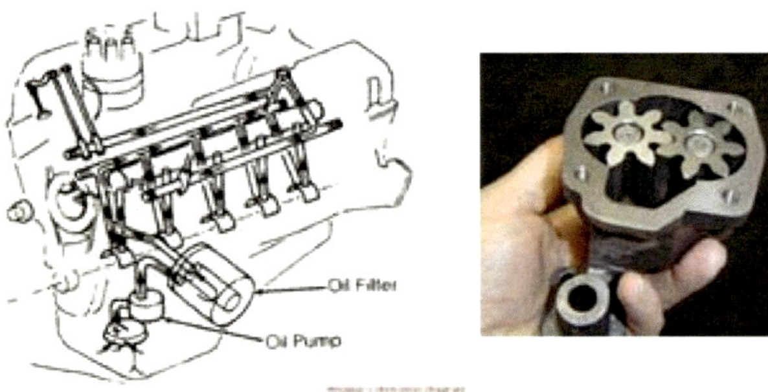

Diesel Engine oil Pump

Placing the oil pump low-down uses a near-vertical drive shaft, driven by helical skew gears from the camshaft

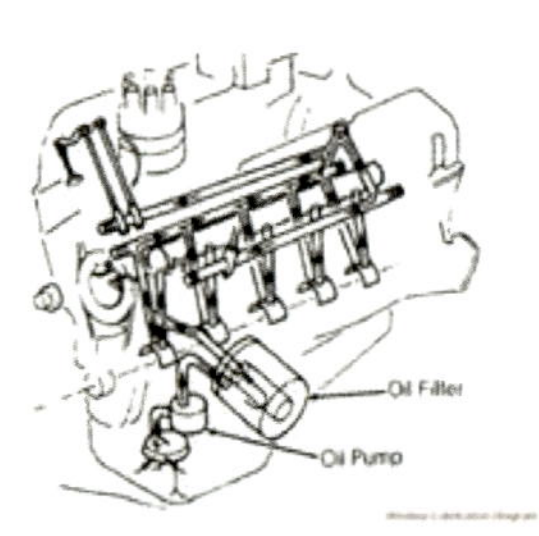

Engine Oil Circulation

- The **oil pump** in a Diesel engine circulate the engine oil under pressure to the rotating bearings, the sliding pistons and the camshaft of the engine
- This lubricates the bearings, allows the use of higher-capacity fluid bearings and also assists in cooling the engine

Oil Pressure

One of the first notable uses in this way was for hydraulic tappets in camshaft and valve actuation. Increasingly common recent uses may include the tensioner for a timing belt/chain or variators for variable valve timing systems

Crescent-Type Oil Pump

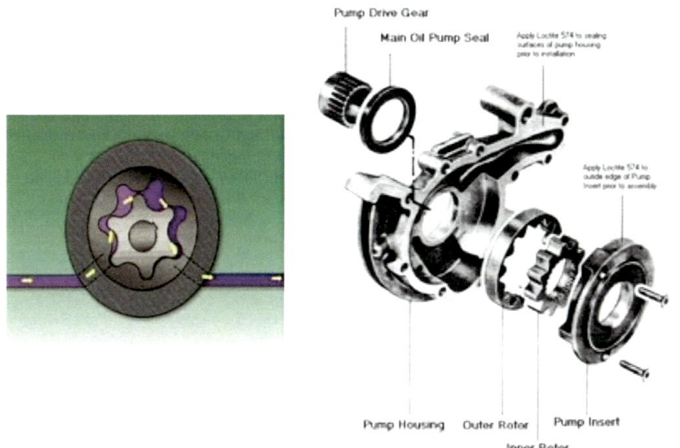

Hydraulic Timing Chain Tensor

Timing Chain Tensor

Oil Pressure Sensor & Switch

- The oil pressure generated in most Diesel engines should be about 10 psi per every 300 revolutions per minute (rpm)

- Local pressure (at the crankshaft journal and bearing) is far higher than the 50, 60 psi . set by the pump's relief valve

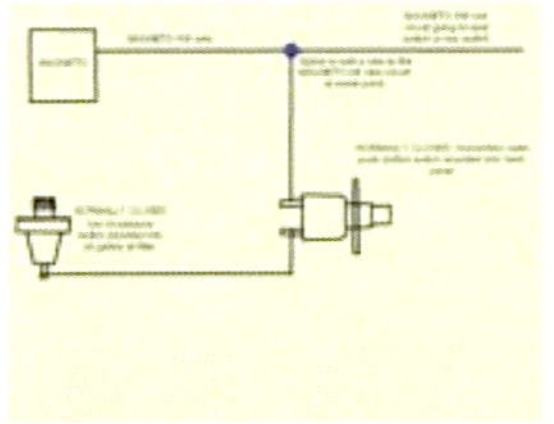

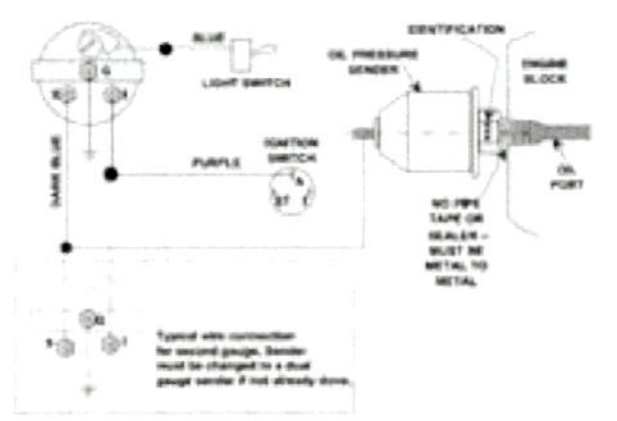

Oil Pressure gauge

- The oil pressure gauge, or warning lamp, gives only the pressure at the point where its sender enters that part of the pressurized system
- this is not a "closed system" in which oil pressure is balanced and identical everywhere. All engines are "open systems", because the oil returns to the pan by a series of controlled leaks

Mineral Oil

- The objective of the refining process is to isolate the desired base oils, also known as mineral oils
- The problem is that after refining operations are performed, a wide variety of chemical components remain that can affect the size and structural arrangement of the molecules. As a result, there may be weak links that break down and degrade the ability of the oil film to perform all of the critical tasks within an engine when operating conditions run to extremes.

Mineral Oil or Petroleum-based oils

- Deep within the earths crust are vast reserves of petroleum crude oil
- Crude oil is far from useful in automotive applications.
- It contains many impurities that must be removed through a distillation process that separates the crude into gases, fuel liquids, lubricant fractions, and heavier components such as asphalt

Synthetic oil

- Is a lubricant consisting of chemical compounds which are artificially made (synthesized) from compounds other than crude oil (petroleum)
- Is used as a substitute for lubricant refined from petroleum, because it generally provides superior mechanical and chemical properties than those found in traditional mineral oils.

Synthetic Oil Vs Mineral Oil

- Measurably better low and high temperature viscosity performance
- Better chemical & shear stability
- Decreased evaporative loss
- Resistance to oxidation, thermal breakdown and oil sludge problems
- Extended drain intervals with the environmental benefit of less oil waste.
- Improved fuel economy in certain engine configurations.
- Better lubrication on cold starts

Oil Grades

- The Society of Automotive Engineers (SAE) has established a numerical code system for grading motor oils according to their viscosity characteristics
- SAE viscosity gradings include the following, from low to high viscosity: 0, 5, 10, 15, 20, 25, 30, 40, 50 or 60.
- The numbers 0, 5…25 are suffixed with the letter W, designating their "winter" (not "weight") or cold-start viscosity

Total Base Number

- An oil's total base number **(TBN)** is a measure of its reserve alkalinity
- Diesel fuel contains some degree of sulfur that produce sulfuric acid that acts as a corrosive to engine components
- The higher the oil's TBN, the greater its neutralizing ability
- Most current Diesel engine oils have a TBN between 10 and 12

Total Acid Number

- The **TAN** is a companion to the TBN it measures the acidity of an oil
- The higher an oil's **TAN** number, the more acidic it is
- A high TAN can be caused by acid contamination from the environment
- Using the wrong lubricant or burning-high sulfur fuel may also result in high TAN values

OIL GRADES

Single-Grade

- Constant Viscosity to different temperatures
- The 11 viscosity grades are 0W, 5W, 10W, 15W, 20W, 25W, 20, 30, 40, 50, and 60

Multi-Grade

- Variable Viscosity. To low Temperature , low viscosity and high viscosity to high Temp.
- for example, **10W-30** designates a common multi-grade oil

Chapter 11
Cooling Systems

Engine Temp <u>Vs</u> Efficient

At high engine torque output and ambient temperature, the system operates at its maximum to maintain a temperature around 180 F

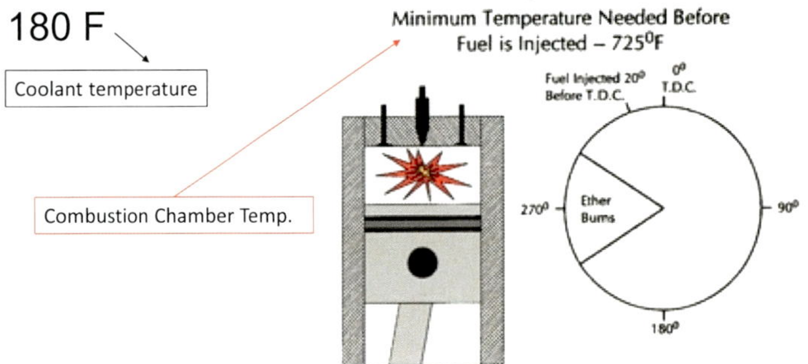

Combustion Chamber Temp.

This high compression heats the air to 550 °C (1,022 °F - Operational Temp)

Engine Temp <u>Vs</u> Efficiency

In a Diesel Eng. Fuel ignites when it is injected into the hot compressed air in a cylinder. Correct engine temperature is critical for efficient Diesel operation

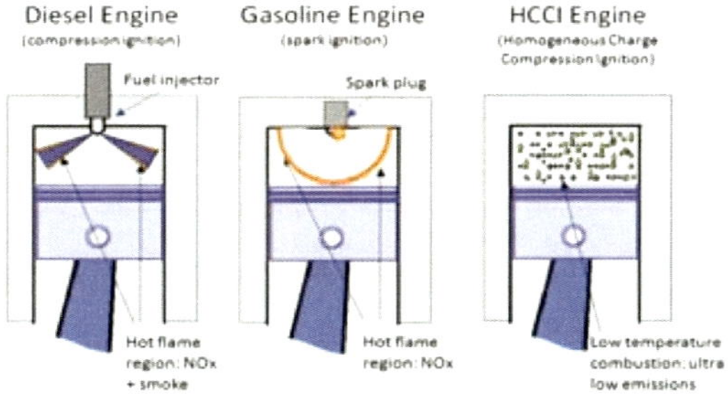

Thermal Efficiency

Diesel engines (as used in ships and other applications where overall engine weight is relatively unimportant) often have a thermal efficiency which exceeds 50 percent

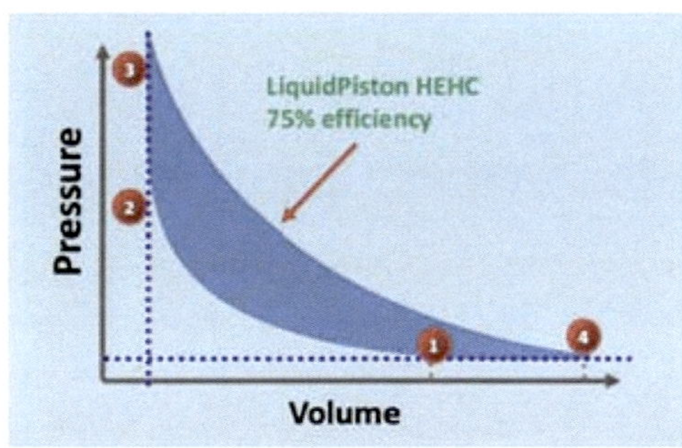

Diesel Engine Operation

In the true diesel engine, only air is initially introduced into the combustion chamber

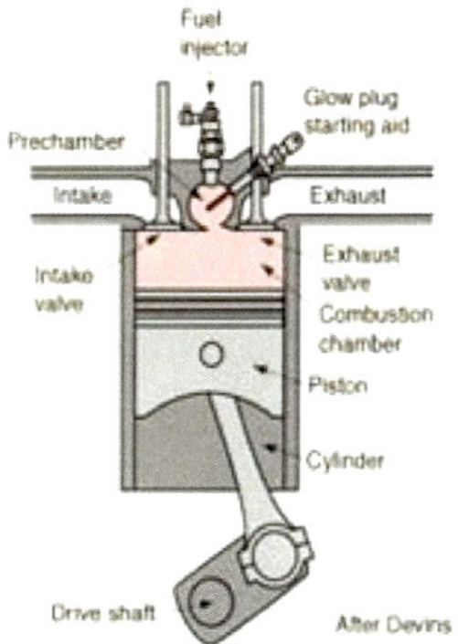

Compression Ratio

The air is then compressed with a compression ratio typically between 15:1 and 22:1 resulting in 40-bar (4.0 MPa; 580 psi) for Diesel compared to 8 to 14 bars (0.80 to 1.4 MPa) (about 200 psi) in the gasoline engine

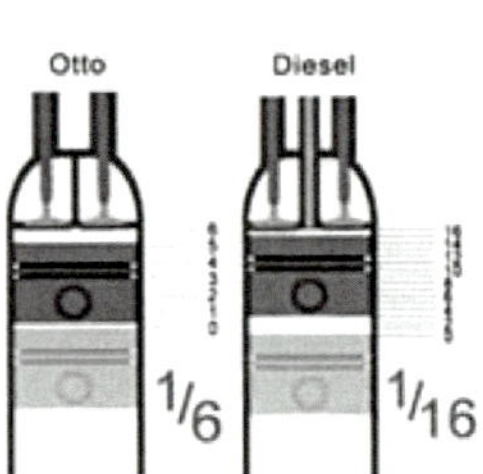

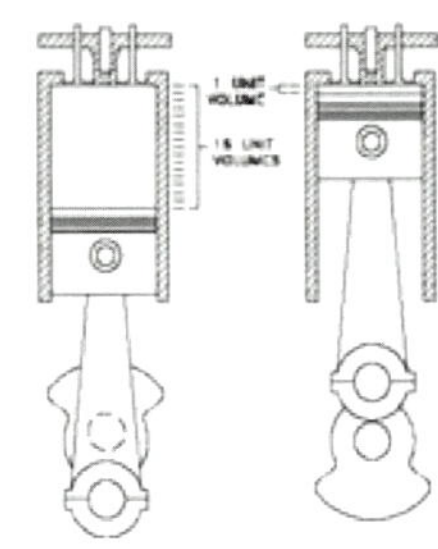

Fuel/Air Injection

At about the top of the compression stroke, fuel is injected directly into the compressed air in the combustion chamber

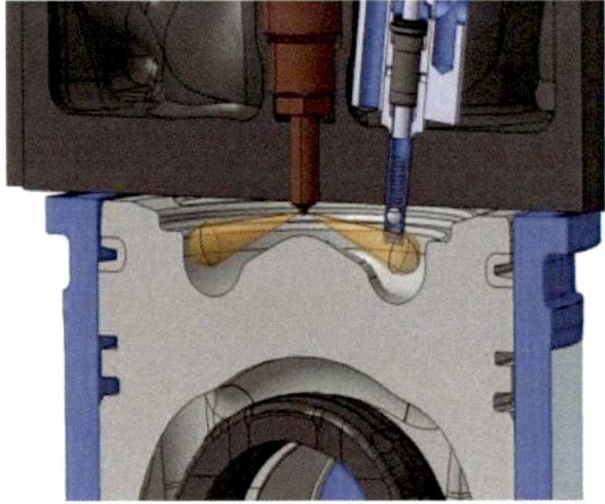

Fuel Vaporization

The heat of the compressed air vaporizes fuel from the surface of the droplets. The vapor is then ignited by the heat from the compressed air in the combustion chamber

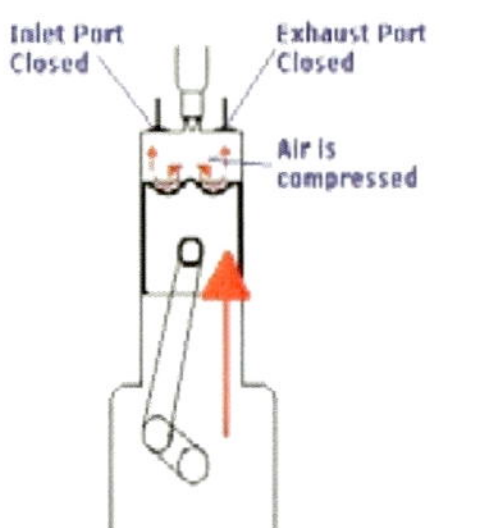

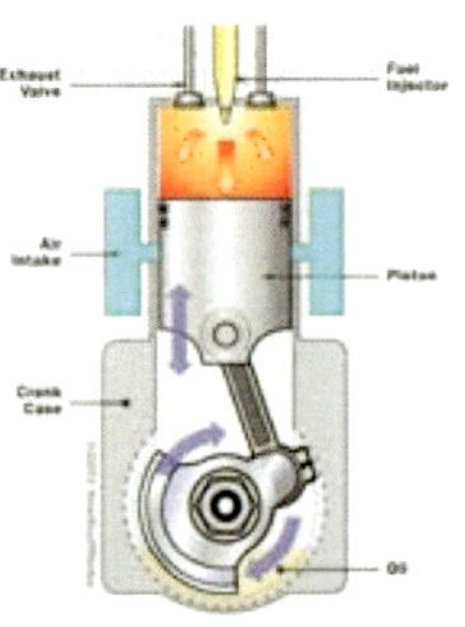

Diesel Knocking

The start of vaporization causes a delay period during ignition and the characteristic **diesel knocking** sound as the vapor reaches ignition temperature and causes an abrupt increase in pressure above the piston

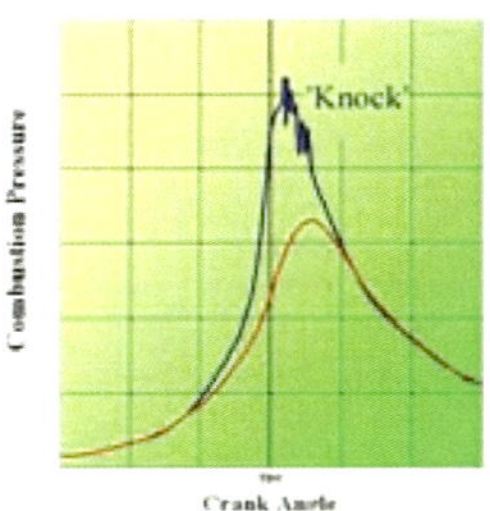

Temperature Management recommendations

Running an engine outside the proper temperature window can result in greatly increased emissions.

Temperature Management recommendations

- When engines are run cool, hydrocarbon emissions increase, fuel efficiency drops, and engine wear rates increase dramatically.
- When engines are run above , oxides of nitrogen emissions increase and engine wear rates greatly increase

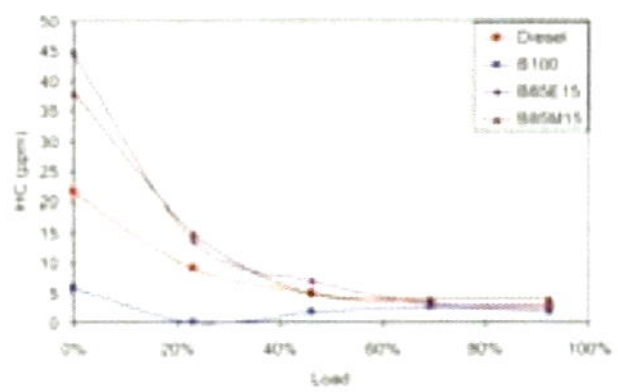

Temperature Management recommendations

- The following are the operating temperatures range of operation.
- Caterpillar :   From 160 F to 210 F
- Cummins: From 170 F to 190 F
- Detroit Diesel : From 160 F to 210 F
- Volvo: From 190 F to 200 F
- Mack: From 170 F to 210 F.
- Yannmar   From   170 F to 190 F

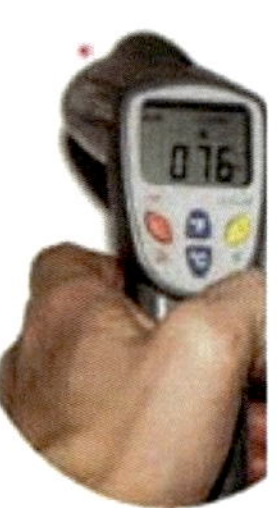

The Most Common Engine Cooling System

Raw Water System.

Close System or Heat Exchanger.

Keel Cooling System.

Raw Water Cooling Systems

Raw water cooling systems draw water from outside the boat (seawater or lake water)

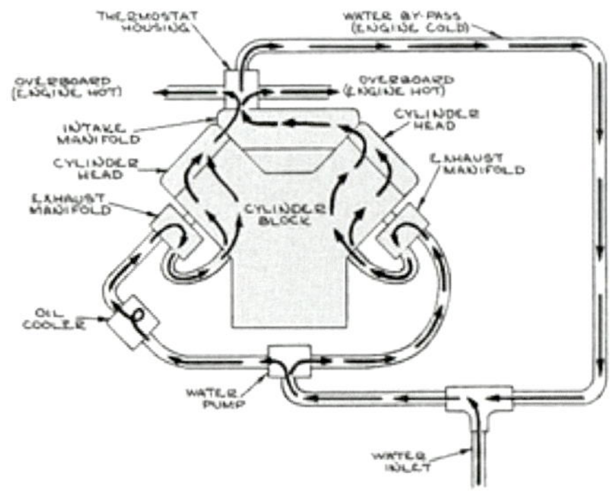

Raw Water Cooling Systems

Water is pumped from the source to the engine block then the engine circulation pump forces the raw water thru the engine block and the water is expelled thru the exhaust

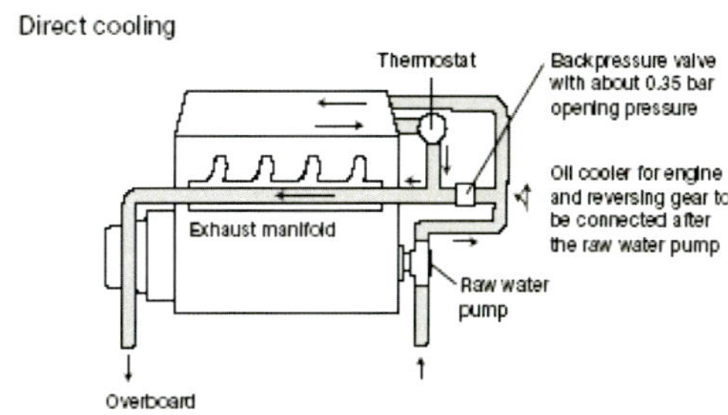

Raw Water Cooling System

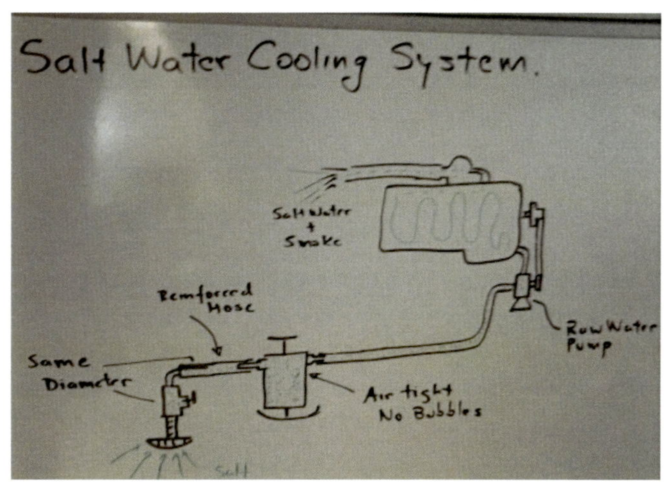

Raw Water Cooling Systems

On larger engines and inboard engines the raw water pump is located inside the boat and is driven by a v-belt or directly off of the crankshaft

Raw Water Cooling Systems

There are hidden dangers that can accumulate over time causing you to spend big Dollars on repairs. The danger is using salt water as a coolant in your engine. Salt water can be highly corrosive

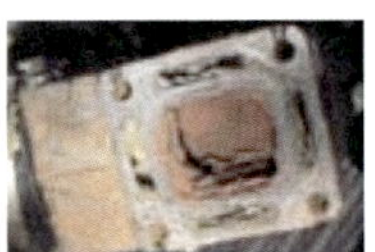

Raw Water Pumps

Salt water Pump and Impeller Kit

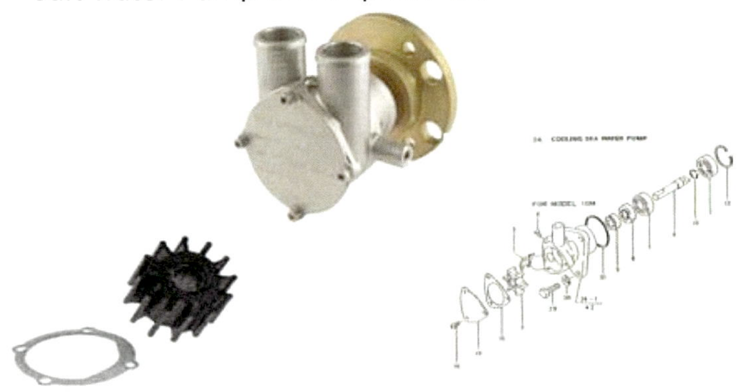

Raw Water Pumps

This pump sucks water from the sea. It typically goes through a strainer as it is sucked towards the pump

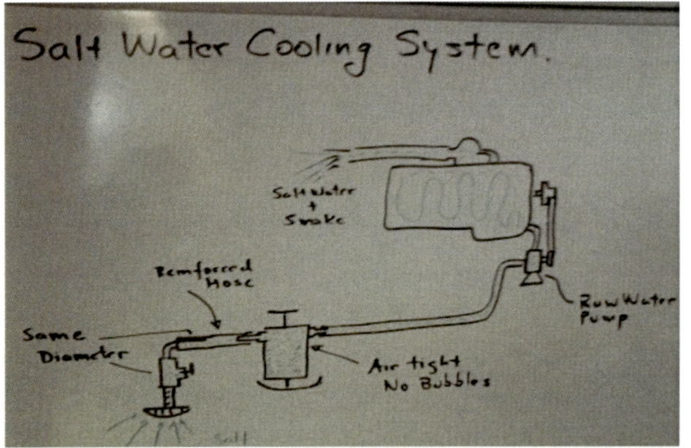

Raw Water Cooling Systems

A flexible impeller pump provides an efficient solution to most raw water pumping needs

Raw Water Cooling Systems

The primary advantage of flexible impeller pumps is that they are self-priming, which means that when the vanes of the impeller are depressed and rebound, they create their own vacuum, drawing fluid into the pump

Raw Water pump Housing

The Cam-housing create a reduction in the volume of the fluid to produce a high output pressure

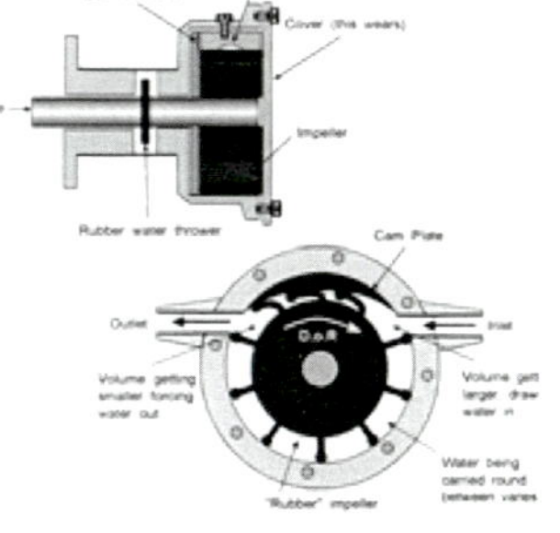

Impeller Removal Tool

Impeller Removal Tool

The impeller housing can be damage using screwdriver or inappropriate tools

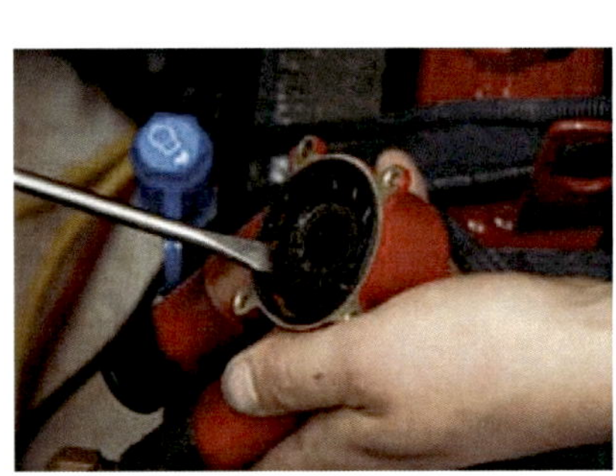

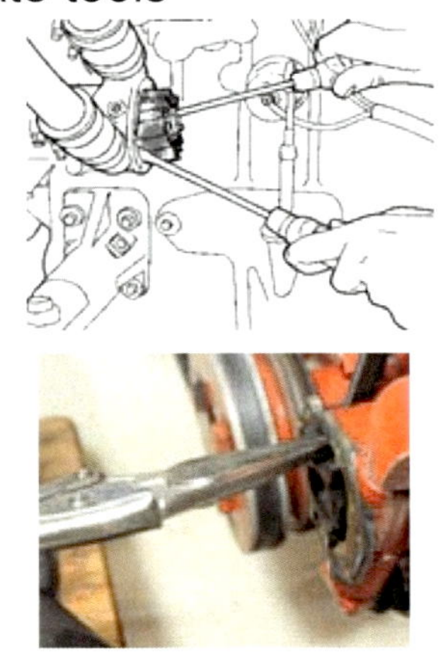

Impeller failure

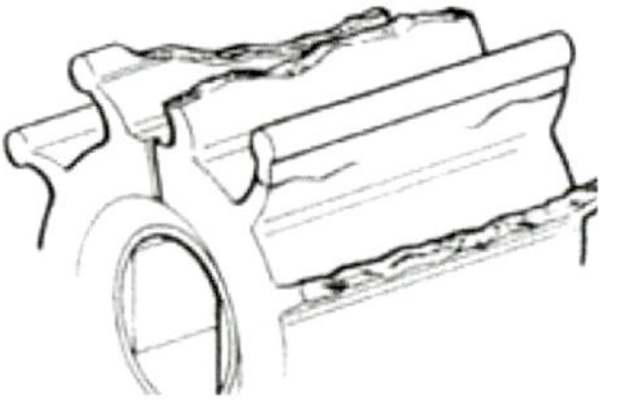

Impeller Plate

Never use screw driver, try to use socket wrench 5/16

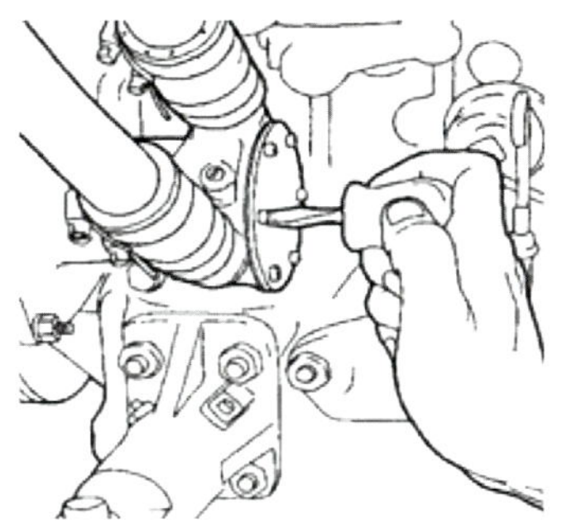

Impeller Plate

Clean the Plate carefully with sand-paper to avoid future leaks

Impeller plate

Don't try to replace the originals bolts by other similar aftermarket because the thread is different and always will have a leak

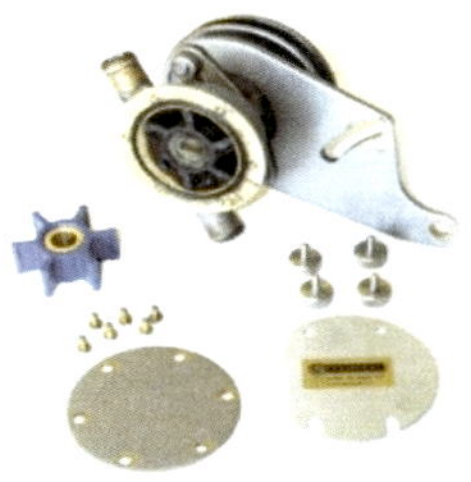

Seals and O-Rings

Always replace the gaskets and never use silicone, because the excess of silicone could clog the heat exchanger

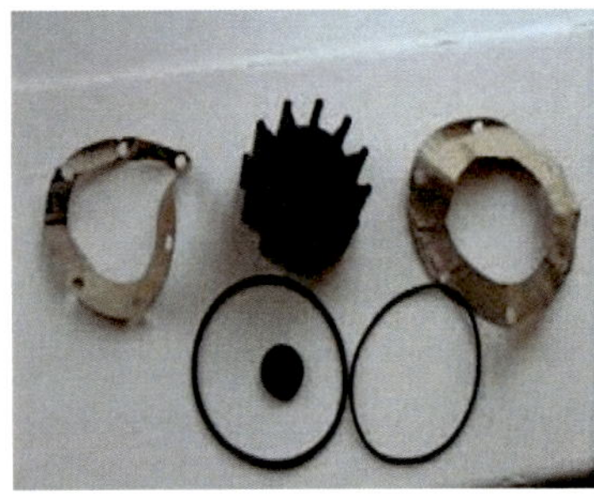

Impeller Plate installation

Before the plate be tight apply grease in between the center of the plate and the impeller , it is help during the cranking and self-priming period

Debris and Particles

After the impeller replacement clean the input of the heat-exchanger to remove impeller blades fractions and debris

Impeller Debris

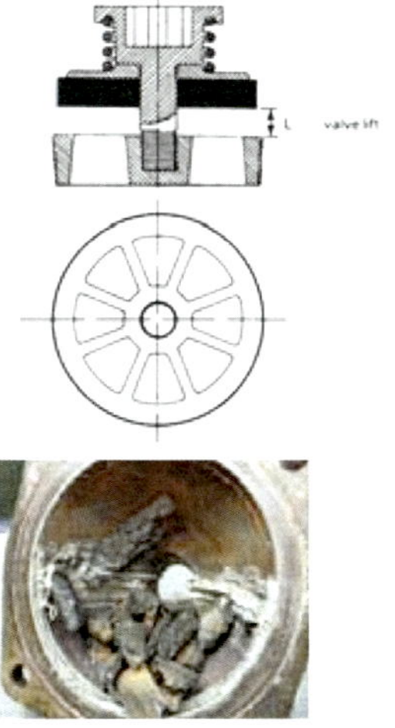

Shaft and Pulley

Remove the belt and verify the shaft by radial play and internal cracking

Sea Water Strainer

Traditional strainer were made of bronze with stainless steel strainer baskets and a removable cover to facilitate pulling out the basket for cleaning

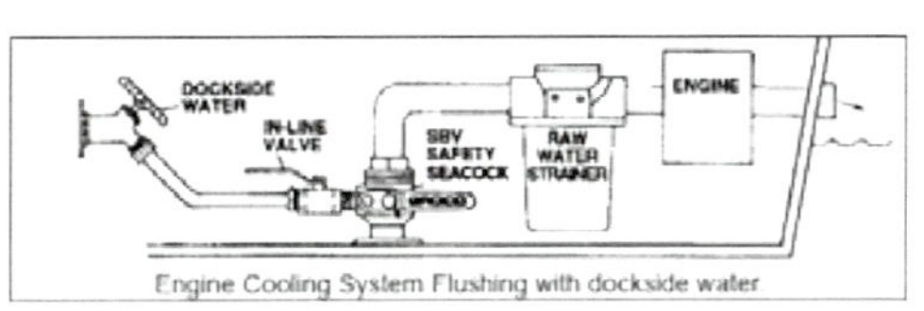

Sea Water Strainer

The cover have a gasket and it is important that upon reassembly the gasket is seated properly

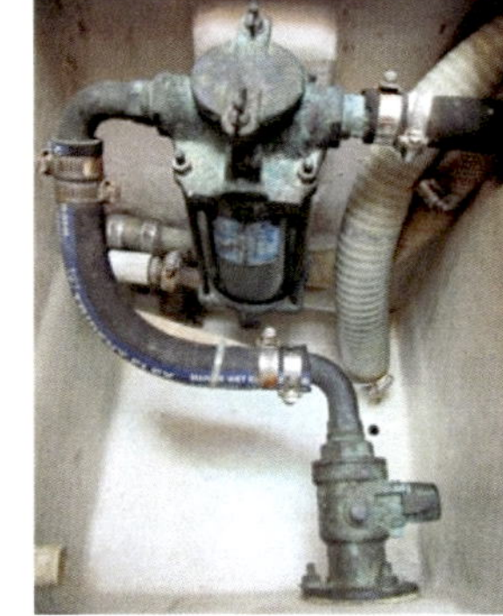

Bronze Strainer

Try to use the same fitting's diameters in both sides to avoid negative suction pressures on the strainer

Raw Water Strainer

Keep the crystal clean to verify easily air bubbles and water contamination

Raw Water Pump head

A dry pump can lift water up to as much as three meters (9 Ft)

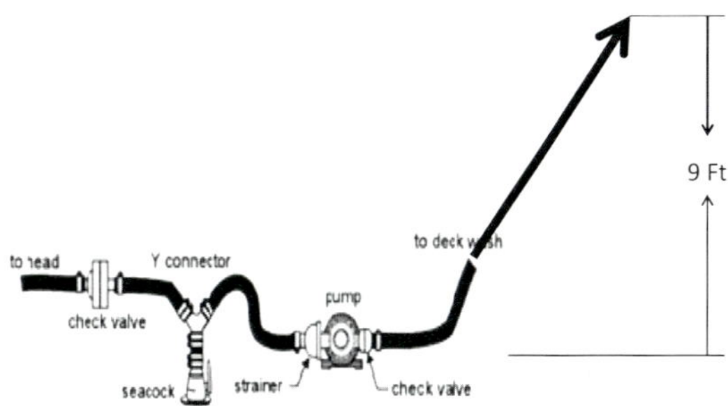

Rubber Impeller / Water line

Thus a flexible impeller pump being used for engine cooling does not need to be manually primed or located below the water line

Rubber Impeller / Water line

- An added feature of a flexible impeller pump is that it can pass fairly large solids without clogging or damaging the pump. This reduces the need for filtration of incoming fluids
- For general or salt water applications, a standard long lasting neoprene rubber impeller is used

Salt Water & Corrosion

- Running salt water through your engine block and exhaust manifolds will lead to destructive corrosion that is unseen until your engine or exhaust manifolds fail.
- Generally speaking, marine engines cooled with raw water, especially ones that use salt water, have a shorter life span than marine engines cooled with a closed cooling system.

Salt Water & Corrosion

- Same water is pumped through engine block, head, exchangers and returned to ocean either by being dumped into exhaust or by a discharge fitting
- Anodes required in raw water system components sometimes
- Cheapest system
- Engine's lifespan very limited due to deposits and corrosion
  - o Definitely a poor choice - even in fresh water it will kill the engine

Raw Water Cooling Systems

- Raw water strainer and pump impeller are regular maintenance items
  - Raw water strainer should have see-through housing
  - Keep in your spare parts , rubber impellers , raw water pumps , hose clamps
  - When an engine overheats, first thing to check is exhaust water and sweater strainer. After that, impeller
- Always install anti-siphon break in raw water at exhaust elbow

Fresh Water Systems with Heat Exchangers

closed cooling systems

Fresh Water Cooling System using a Heat Exchanger

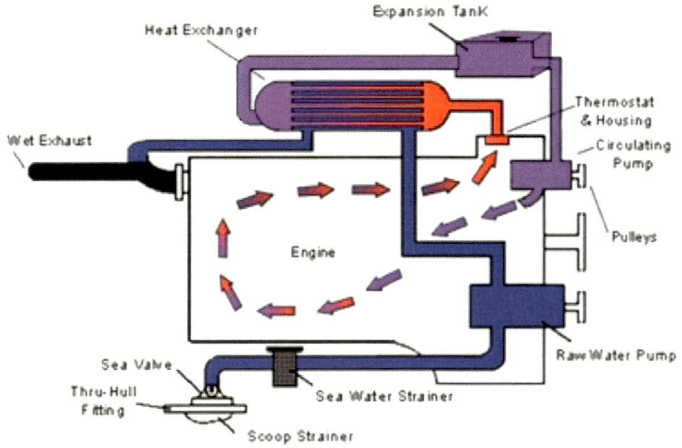

Closed Cooling Systems

The most common type utilizes a Heat Exchanger which functions similarly to the radiator in your car

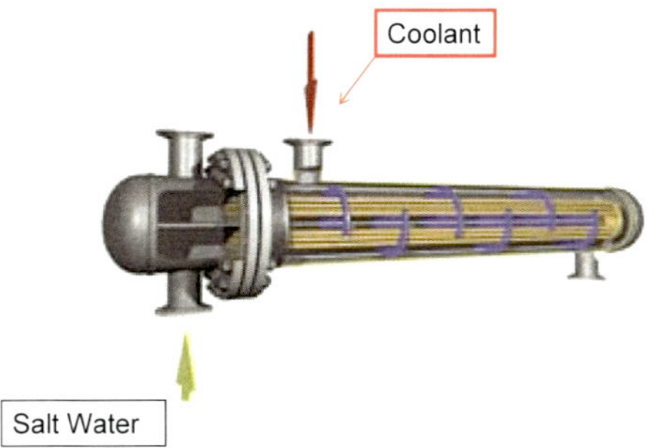

Closed Cooling Systems

- Coolant (antifreeze) is circulated through one side of the heat exchanger where it is cooled by raw water that passes through the other side of the heat exchanger
- The engine coolant is then circulated back into the engine. The raw water is expelled out of the boat thru the exhaust

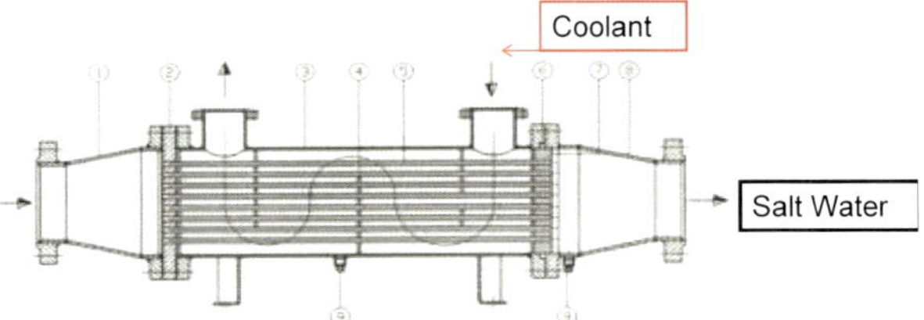

Closed Cooling Systems

The closed circuit normally transfers heat from the engine to the **heat exchanger**. The liquid used is water and anti-freeze (Glycol)

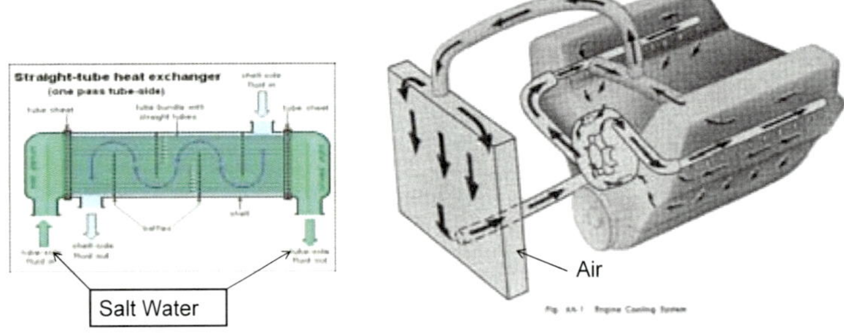

Cooling Pump Capacity

- The required output of the cooling pump is related to engine type and size, **not to** the size of the heat exchanger and exhaust system
- This is true for both raw water as well as closed or heat exchanger systems

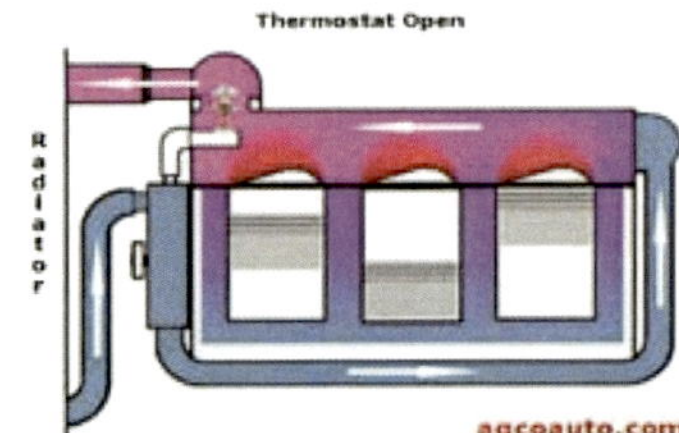

168

Coolant Pump / Raw Water Pump

The coolant is driven by the coolant pump while the salt water is moved by the raw water pump

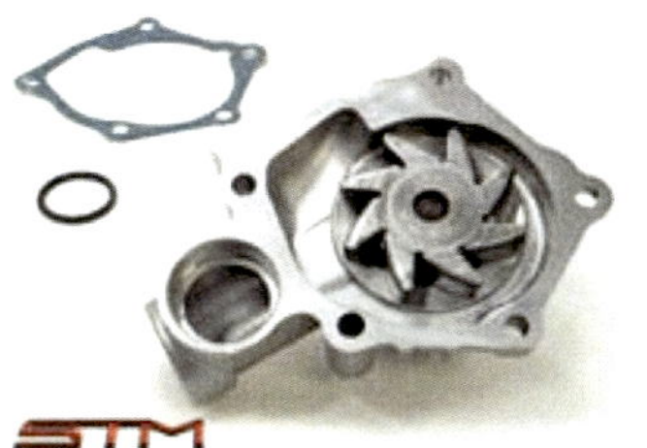 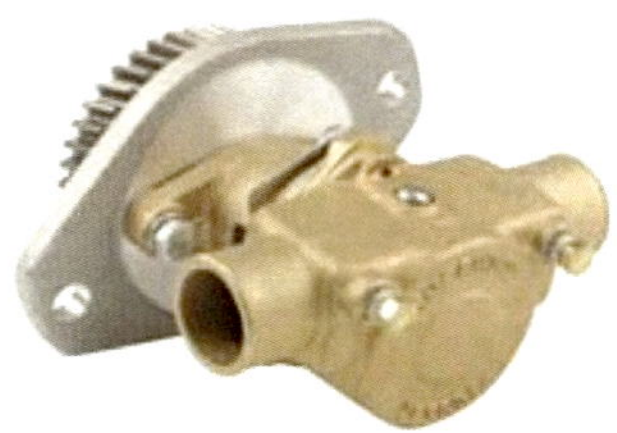

Coolant Pumps or Water pumps

- Are centrifugal Pumps driven directly by a Gear or Belt
- Water pumps are the main reason that engine coolants Should have some lubricating properties

 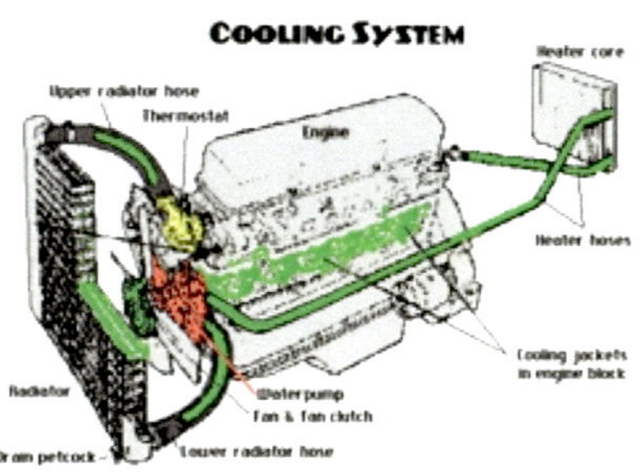

Heat exchangers & Expansion Tanks

Function

Diagnosis

Heat Exchanger

- Heat exchangers may be classified according to their flow arrangement.
- **Parallel Flow**   and   **Cross-Flow**

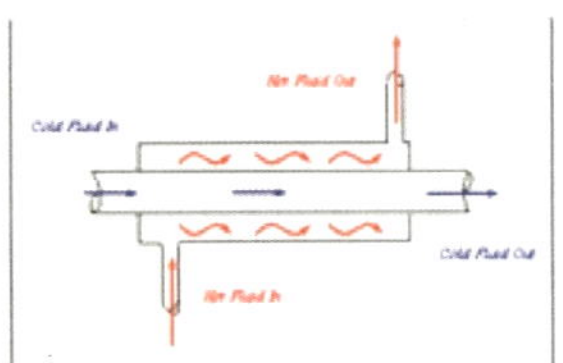

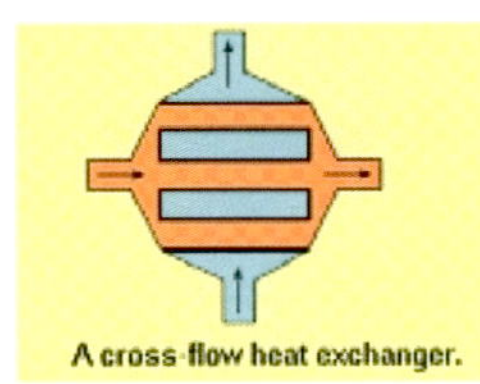

A cross-flow heat exchanger.

Heat Exchanger

- Heat exchangers may be classified according to their flow arrangement.
- In *parallel-flow* heat exchangers, the two fluids enter the exchanger at the same end, and travel in parallel to one another to the other side. In *counter-flow* heat exchangers the fluids enter the exchanger from opposite ends

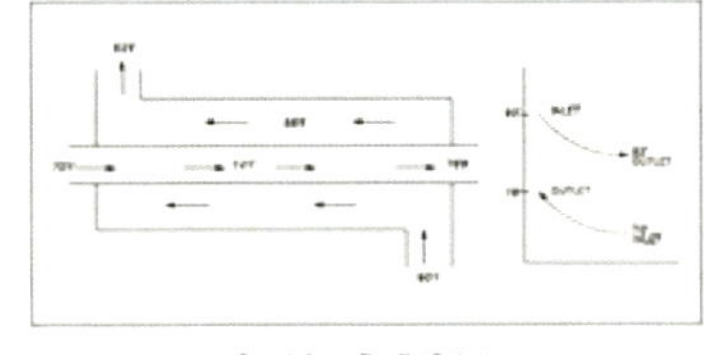

Heat Exchanger

- The counter current design is most efficient, in that it can transfer the most heat from the heat (transfer) medium. In a *cross-flow* heat exchanger, the fluids travel roughly perpendicular to one another through the exchanger.
- In a *cross-flow* heat exchanger, the fluids travel roughly perpendicular to one another through the exchanger

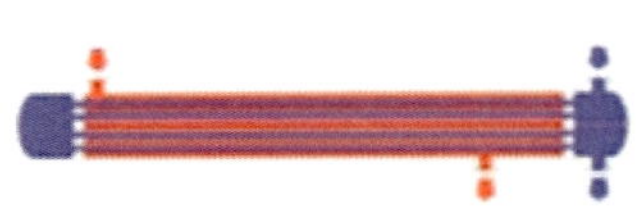

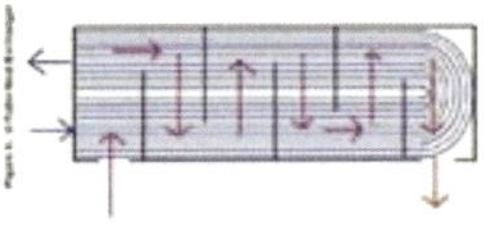

Heat transfer

A Heat Exchanger transfers, or "exchanges," heat from your boat engine's coolant to raw water pumped from the water outside of your boat

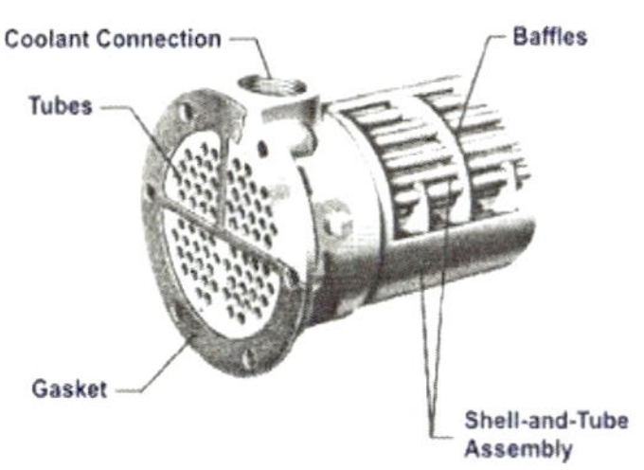

Flow of Water

The raw water is pumped through a bundle of small tubes in a chamber filled with the hot engine coolant. The tubes are cooled by the colder raw water the allowing the tubes to absorb the heat of the engine coolant

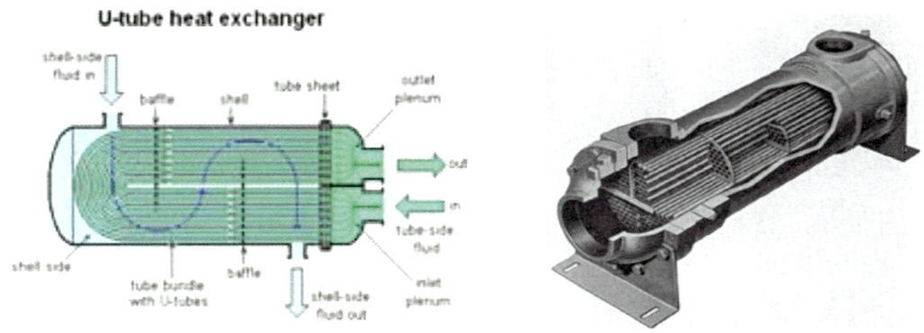

Expansion Tanks

Expansion tanks are an often overlooked but very important part of a closed cooling system. As the engine coolant gets hot it expands, increasing in volume

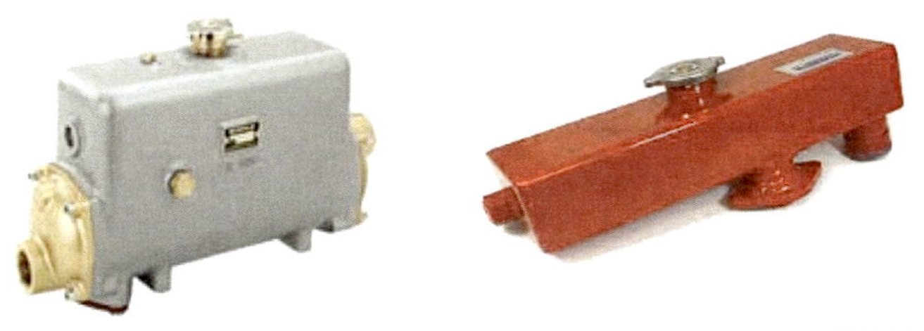

Expansion Tanks

The expansion tank is a small tank that simply provides room for this increase in volume. In some systems, the heat exchanger is a piggy back style. This is when the expansion tank is built on top of the Heat Exchanger

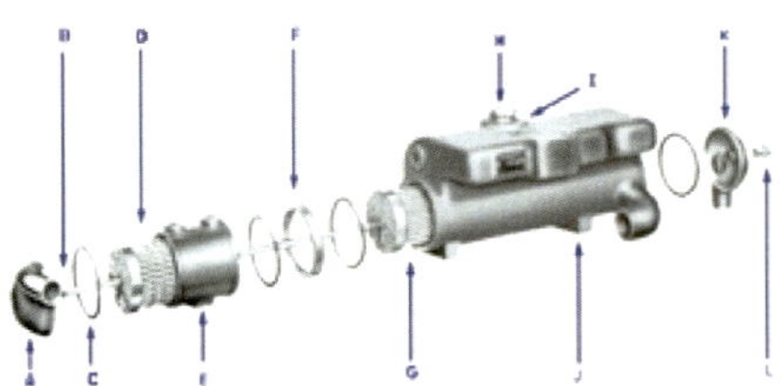

Expansion Tanks

On some Heat Exchanger systems, the expansion tank is a separate tank remotely mounted. Most heat Exchangers have a fitting built into the tank to install a Zink Anode

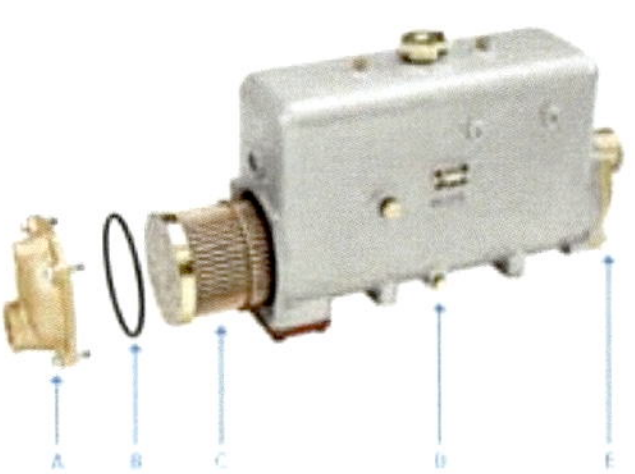

Expansion Tanks

Another important component to the Heat Exchanger is the cap on expansion tank similar to the cap on your cars radiator. It is an important component in, maintaining your cooling system's pressure. It should, be checked regularly for leakage and corrosion

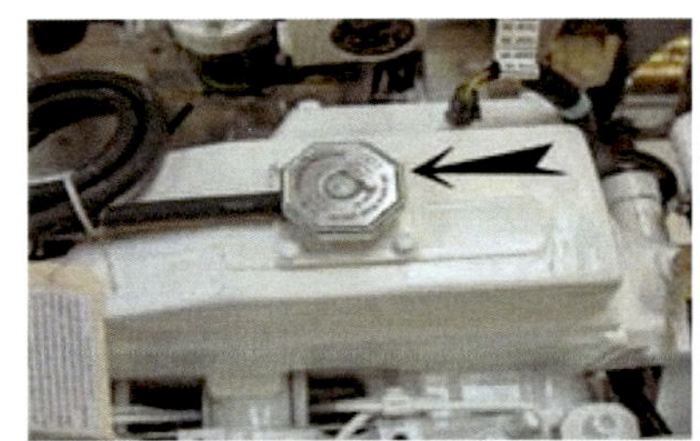

Radiator Cap

Expansion tanks are usually equipped with a pressure cap designed to maintain a fixed operating pressure while the engine is running

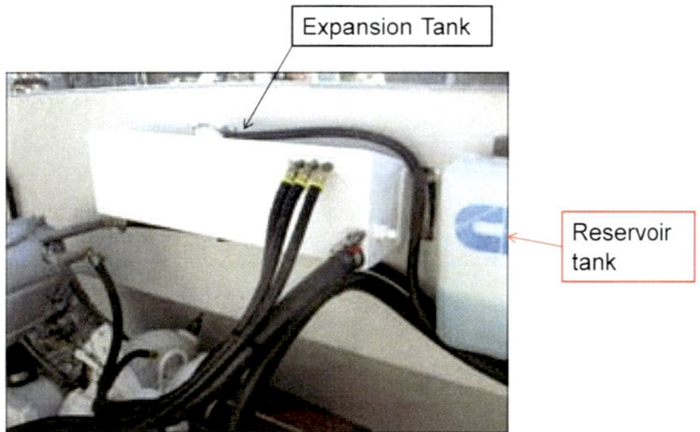

Radiator Cap

- This Cap is also equipped with a vacuum valve to admit surge tank coolant (or air) into the cooling circuit.

- Expansion tanks Caps permit pressurization of a sealed cooling system

Radiator Cap

- For Every 1000 feet of elevation the boil point decreases by 1.25 F
- To compensate for this loss. For each 1psi above atmospheric pressure coolant boil point is raised by 3 F at sea level
- System pressures will seldom be designed to exceed 25 psi; more typically they will range between 7 psi and 15 psi.

The Thermostat Parts

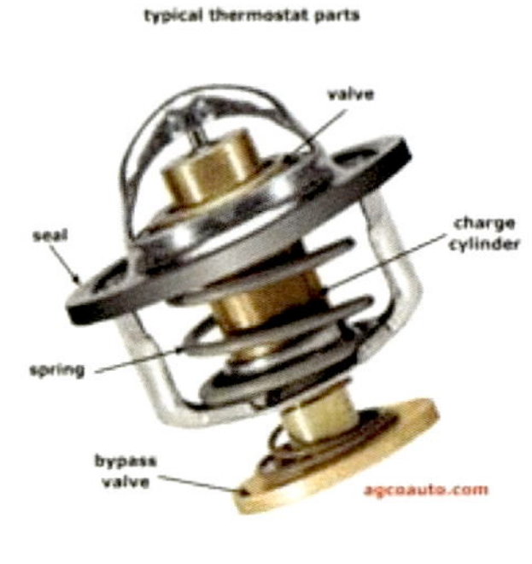

How Thermostat works ?

The thermostat should open and allow coolant in once the engine is adequately heated. Sometimes the thermostat fails and remains closed

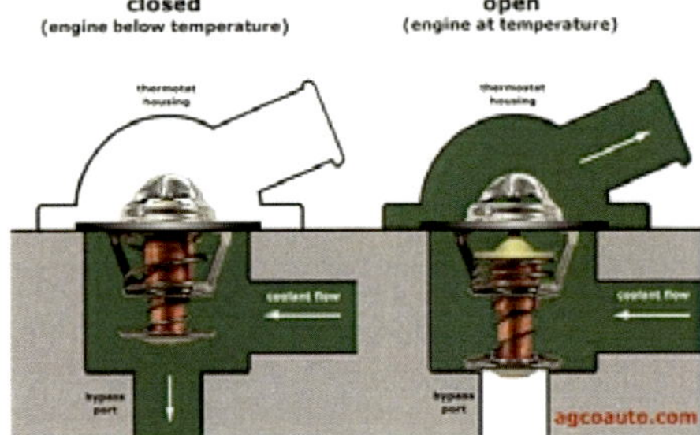

How Thermostat works ?

A stuck thermostat might be to blame if your gauge marker nears the red danger zone within 5 to 15 minutes

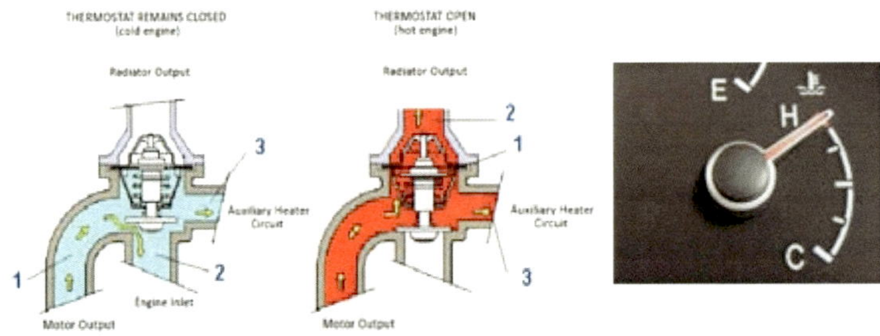

Checking Thermostat

- Thermostat should open at temperature marked on it and be fully open approximately 20 degrees F higher.
- It should close again when immersed in colder water.
- Malfunctioning thermostats cannot be repaired, they must be replaced: Make sure you get the right type

How to Check if the Thermostat Is Stuck closed ?

Unhook your coolant reservoir if you can undo it and pour the antifreeze into a container. Locate your expansion tank and remove the cap

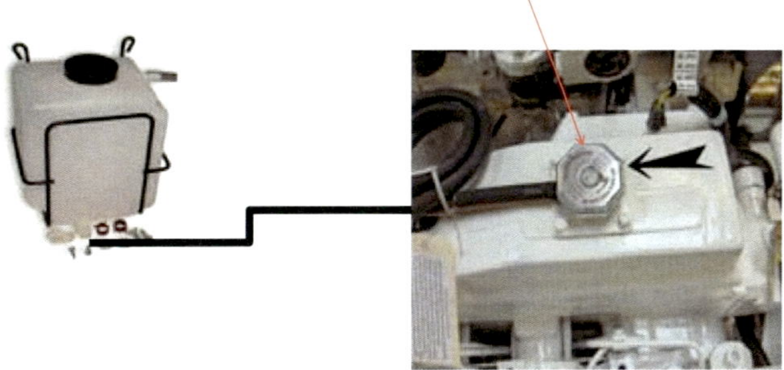

How to Check if the Thermostat Is Stuck closed ?

- Normally you should drain the coolant at least up to the level where the thermostat is located
- You can reuse this coolant if it is brand new

How to Check if the Thermostat Is Stuck closed ?

Remove the thermostat from the thermostat housing taking into account the orientation and gasket position

How to Check if the Thermostat Is Stuck closed ?

Fill a pan with water and place the thermostat into the water until it's completely submerged. Ensure that the part does not touch the bottom of the pan

How to Check if the Thermostat Is Stuck closed ?

The thermostat should remain closed until about 190 ºF (88 ºC). At this temperature, you should see the thermostat begin to open.

The part should be completely open when the water reaches 195 ºF (90.6 ºC). If the thermostat is still closed at this point, it needs to replaced

Method 2 (Testing Thermostat)

- With the engine running at operational temperature . Locate the coolant heat exchanger hoses
- Both hoses should be very warm to the touch. If one hose is very hot and the other one feels cool or even cold to the touch, the thermostat is likely stuck in the closed position

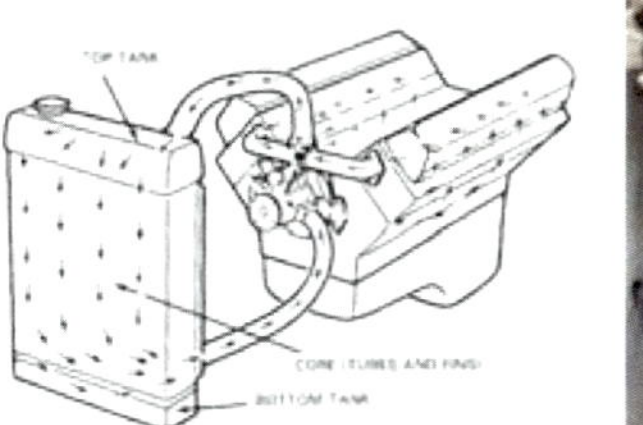 

Temperature Problems

- Temperature gauges, especially the electric type, which are the most common in boats, are often very inaccurate and have temperature scales that are difficult to interpret. Therefore, before worrying about a suspected temperature problem, check your temperature gauge against a thermometer of known accuracy

Engine Coolant

Engine Coolant is a mixture of water , antifreeze (ethylene glycol or Propanediol) and corrosion inhibitors

Chemical	Percentage
1,3-Propanediol 97% to 98% by volume	93 to 95 percent
Sodium Nitrite	0.50 to 1.50 %
2-Ethylhexanoic acid	4.0 to 6.0 %
Sebacic acid	0.0 to 1.5 %
Sodium Tolyltriazole	0.30 to 1.10%
Sodium Molybdate	0.50 to 1.30%
Antifoam	0.05 to 0.10%
Dye	0.00 to 0.02%

Engine Coolant

If the only objective of Diesel Eng. Coolant were to act as a medium to transfer heat, pure water would accomplish this more efficiently than any currently used antifreeze. However water possesses inconvenient boil and freezing points, poor lubricating properties and promotes oxidation

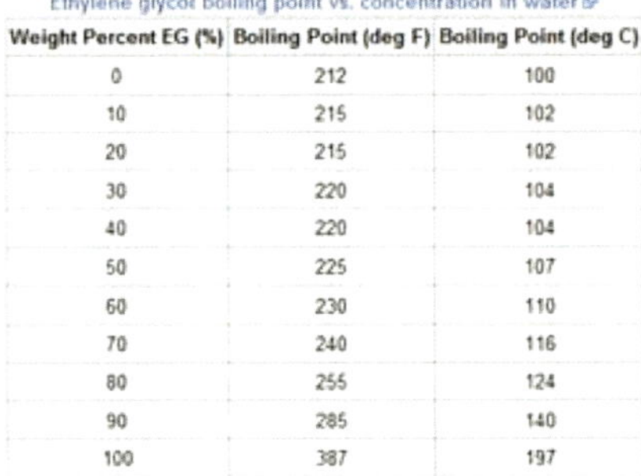

Ethylene glycol boiling point vs. concentration in water

Weight Percent EG (%)	Boiling Point (deg F)	Boiling Point (deg C)
0	212	100
10	215	102
20	215	102
30	220	104
40	220	104
50	225	107
60	230	110
70	240	116
80	255	124
90	285	140
100	387	197

Basic Coolant properties

- The mixture of water, antifreeze and corrosion inhibitors should perform the following:
  - Corrosion Protection
  - Freeze Protection
  - Antiboil Protection
  - Antiscale Protection
  - Acidity Protection
  - Antifoam Protection
  - Ant-dispersant protection

Coolant Analysis

Coolant analysis not only determines coolant condition, also identifies other engine problems that can show up in the cooling system

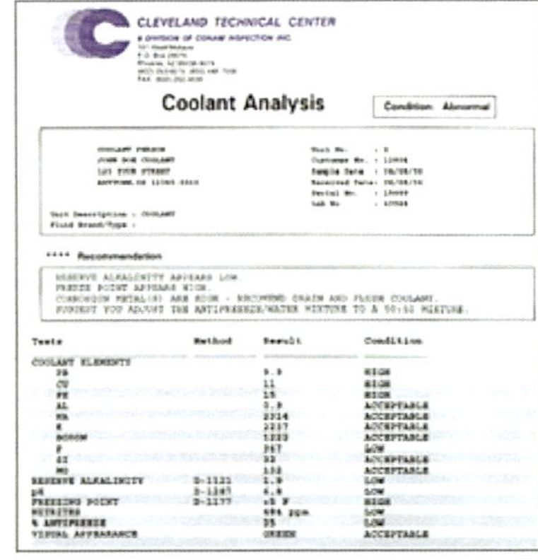

Typical Concentrations (ppm)

The following table shows the normal concentration values appear for each element

Element	Possible Sources	Typical Concentrations (ppm)	
		Clean (no visible debris)	Dirty (visible debris)
Potassium (K)	Additive - pH buffer	no data	no data
Silicon (Si)	Additive - defoamant · corrosion inhibitor for aluminum Dirt	32	81
Boron (B)	Additive - pH buffer · corrosion inhibitor for ferrous metals	552	873
Molybdenum (Mo)	Additive - Anticavitation, silicate	29	130
Phosphorous (P)	Additive - pH buffer, corrosion inhibitor for ferrous metals	553	1061
Magnesium (Mg)	Hard Water - leads to scaling	<1	13
Calcium (Ca)	Hard Water - leads to scaling	6	134
Iron (Fe)	Liners, water pump, cylinder block, cylinder head	4	75
Zinc (Zn)	Brass alloy	1	10
Lead (Pb)	Solder in radiator, oil cooler, after cooler, heater core	4	35
Copper (Cu)	Solder in radiator, oil cooler, after cooler, heater core	3	23
Aluminum (Al)	Radiator tanks, coolant elbows, piping, etc.	7	5
Sodium (Na)	Additive, corrosion inhibitor	3368	4410

Figure 8. Diesel Engine Coolant Analysis

What does it mean to have a "wet sleeve" engine

- **Wet sleeves** are the cylinder wall that can be removed and replaced at rebuild time.
- They are surrounded by coolant and they are o-ringed on the bottom. Usually they are easy to R&R without special tools

What does it mean to have a "Dry sleeve" engine

Dry sleeves are pressed into the block and do not come in direct contact with the coolant. much harder to replace usually requiring to heat up the block and ice down the sleeve

- The most frequent type of diesel-engine failure caused by a poorly maintained cooling system is the perforation of cylinders in wet-sleeve engines.

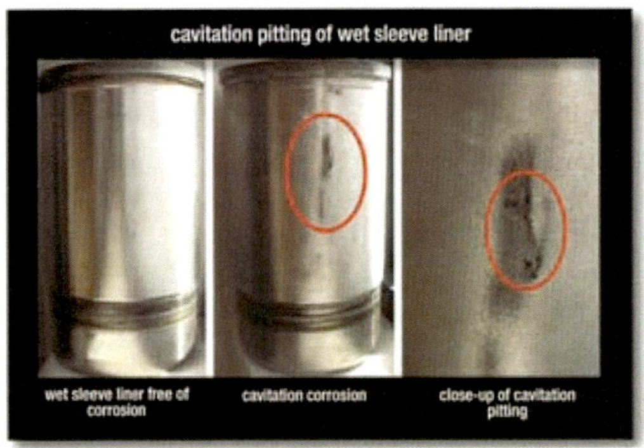

Wet-Sleeve Cavitation

- **Generally speaking sleeves have been attacked by** :
- stray electrical current going to ground through the coolant.

Wet-Sleeve Cavitation

Calcium and magnesium scale that impedes heat transfer and is caused by water with minerals

Wet-Sleeve Cavitation

Chloride (in the water), which "decarbonizes" iron until it is like sand

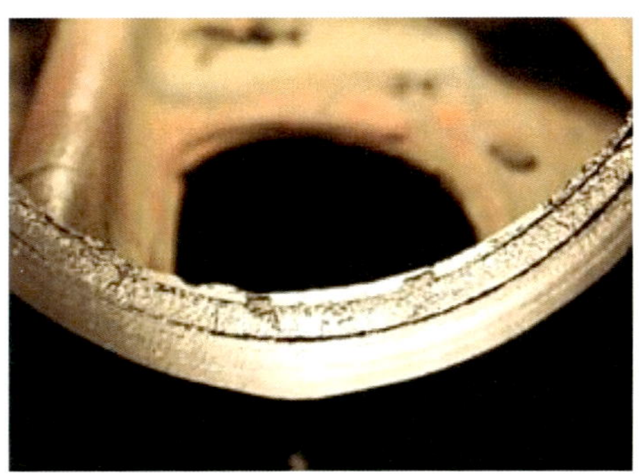

Good Coolant saves Wet-Sleeve Engines

The sleeves are "wet," because they are exposed to the engine's coolant, which is good for efficient heat transfer, but not so good for the sleeve if coolant chemistry is out of whack

Wet- Sleeve Cavitation

In a wet-sleeve cooling system, antifreeze additives create a barrier between the engine's sleeves and the small bubbles that form in the coolant next to the sleeves, the result of pressure differentials

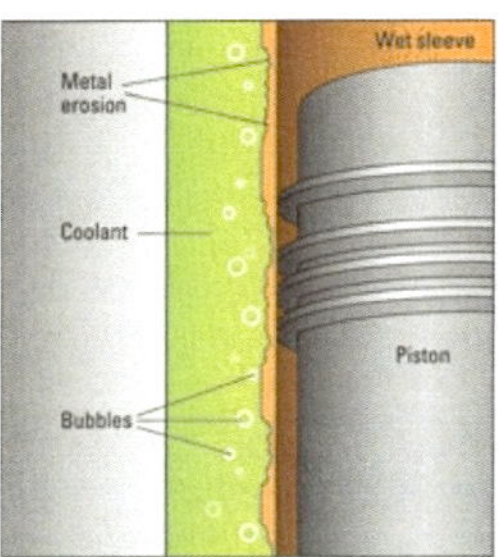

Good Coolant saves Wet-Sleeve Engines

As the diesel works, combustion forces set up a natural vibration in the sleeves, causing the sleeve walls to first pull rapidly away from the coolant, creating a low-pressure area in which the surrounding coolant boils and forms tiny air bubbles

Good Coolant saves Wet-Sleeve Engines

Then, as the sleeve springs back, it slams into the bubbles with a force estimated at up to 60,000 psi, causing the bubbles to collapse, or implode, violently

Coolant Filter for Diesel Engines

Diesel Coolant Filter

There are two types of filters: Filter with basic filtration element and filter with chemical release to extend coolant protection life

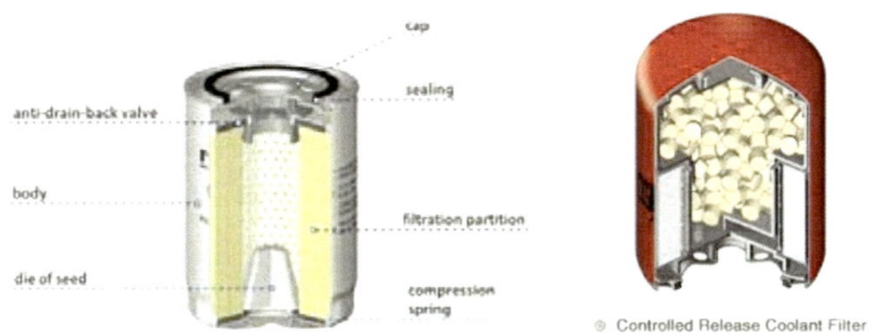

Diesel Coolant Additive

When a simple filtration element is used add DCA's (Diesel Coolant Additive) is recommended

Coolant Filter for Diesel Engines

While it is still a good idea to use coolant test strips to keep an eye on the level of DCA in the cooling system, one can simply replace the filter with a fresh filter with the proper amount of DCA already installed when it is time to replenish the DCA level

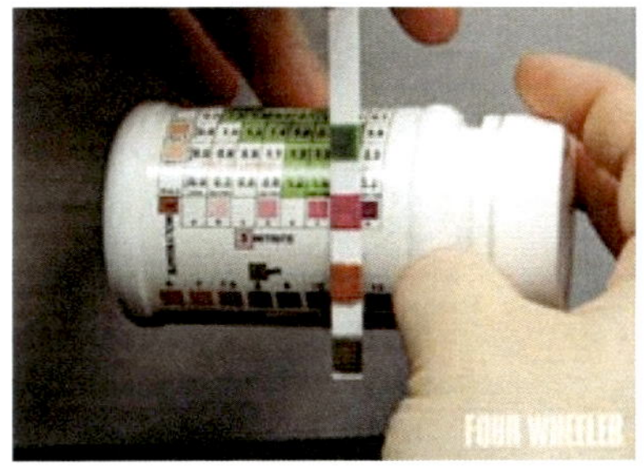

- Is defined as a 50% antifreeze - 50% water mixture, pre-charged with 0,4 units per liter of DCA4.

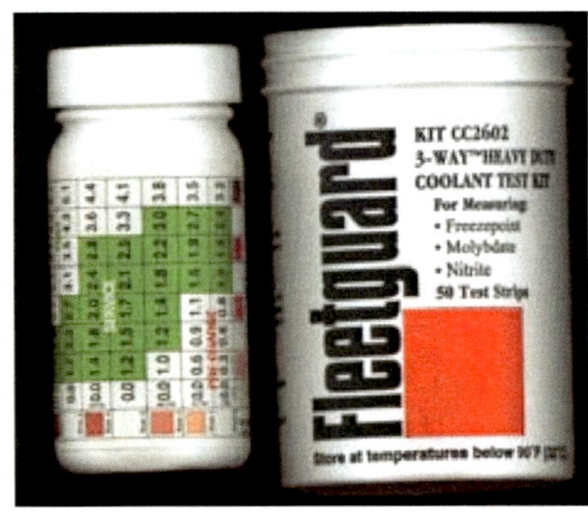

Heavy Duty Coolant (HDC)

Any coolant added to the engine must be **"Heavy Duty Coolant"** to maintain the correct balance of antifreeze, water and DCA4. Never add coolant which is not pre-charged with DCA4

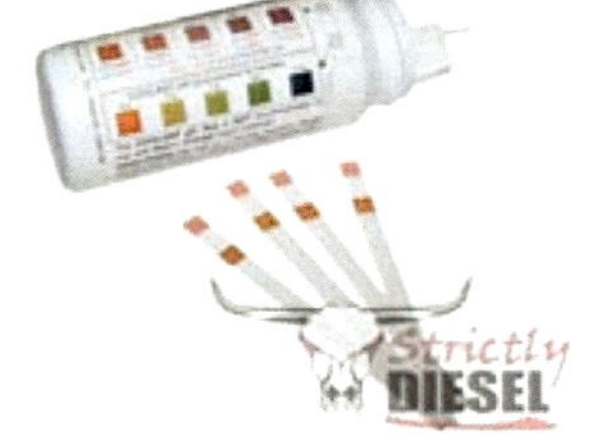

Additive Level Testing Point

Maintain additive level of Fully Formulated Coolant at optimum level (1.2 to 3.0 units per gallon for imperial measurement, 0.3 to 0.8 units per Liter for metric measurement) for maximum protection

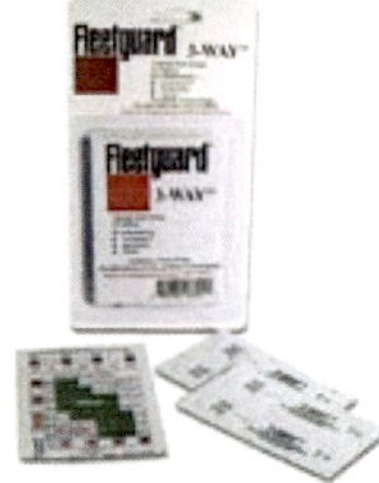

What are SCAs?

- **SCA's** – Supplemental Coolant Additives
- Used to prevent corrosion, formation of mineral deposits, cavitation erosion of the cylinder liners and foaming of the coolant.
- SCA's are not required with Heavy Duty Coolant (HDC).

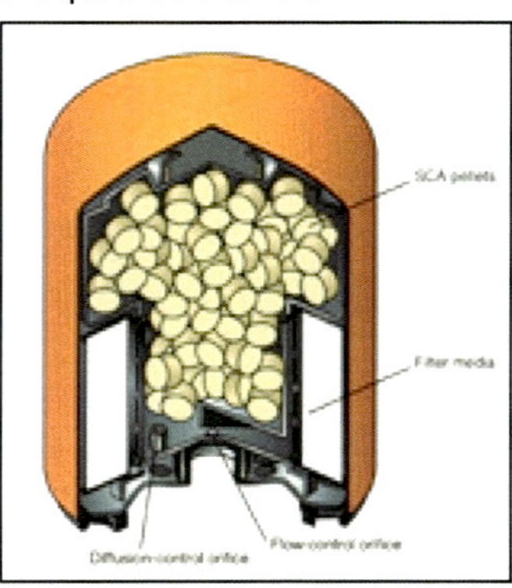

Coolant Filter with Additives

Many coolant filters come with a certain amount of **Supplemental Coolant Additive** (SCA) added in, negating the need to pour liquid additive in.

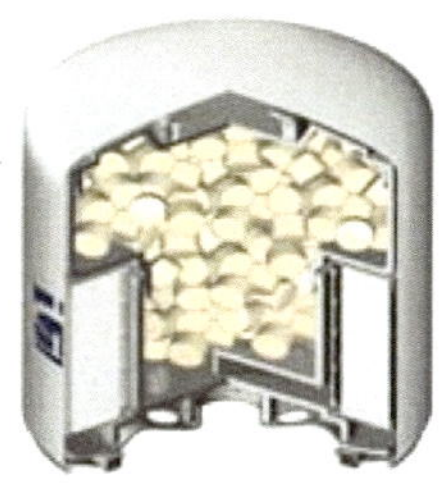

Supplemental Coolant Additive

Supplemental Coolant Additive (SCA) provides superior engine protection in heavy duty diesel applications where glycol based coolants are used, as well as marine and railroad diesel applications where water is used as the coolant

How often should system maintenance be performed?

- This is dependent on the type of SCA you have chosen to use. Refer to engine and additive manufacturer recommendations

Why doesn't a coolant filter come factory installed on some engines?

- Due to various engine designs, some engine and equipment manufacturers do not require coolant filtration. Coolant filtration can be added to these systems to prolong water life and/or aid with coolant maintenance

I've never had cooling system problems. Why do I need coolant additives and filters?

- It is very rare that a gasoline or diesel engine has "never" experienced a failure of a cooling system component, or a related part that couldn't have been prevented with the proper use of SCA's and a coolant filter

How often do I need to monitor the system? How do I control monitoring when vehicles are traveling nationwide?

- SCA level monitoring can be done very easily by using coolant testing. Testing should be done at the maintenance interval for the type of SCA being used to determine if more additives are actually needed to accurately track SCA depletion rates

Can liquid SCA's and filters with SCA's be used together?

- This depends on the total capacity of the cooling system. Most system capacities are of the size that either the liquid SCA or a filter with solid SCA is utilized. In larger capacity systems, however, both products are used for proper maintenance

What is the correct water and antifreeze mixture to be used in coolant systems?

- The ideal mixture is 50% water and 50% antifreeze. The coolant mixture should never contain less than 40% antifreeze or more than 60% antifreeze. The water used must meet engine manufacturer's guidelines for use in their coolant systems

Coolant seems to disappear from my system. Where does it go?

- Coolant can seem to "disappear" from the system due to the lack of a coolant recovery system, evaporation, hose and clamp leakage or seepage, water pumps and/or thermostats not functioning properly, improperly sealed, cracked or broken head gaskets, cracked cylinder heads or engine blocks, and leaking or seeping radiators, heater cores or oil coolers

Why does my coolant foam?

- Foam in coolant is usually the sign of trapped air in the system, a leak on the suction side of the water pump, an improperly functioning water pump, low or no coolant in the coolant recovery tank, the lack of a coolant recovery system, the coolant system lack of appropriate SCA's or the combining of incompatible chemicals in the coolant system

What happens if the coolant system is overcharged with additives?

- Over charging or over concentrating a coolant system with additives will result in the formation of solids. These solids will form deposits that drop out and clog passage ways in the system preventing proper heat transfer

How often should I change my antifreeze?

Antifreeze should be changed based on original equipment engine manufacturer's recommendations or with the use of full laboratory coolant analysis

Can I use a liquid SCA in either a gasoline or diesel engine with no coolant filter?

Yes. However we do recommend the use of an additive free filter on all coolant systems to remove all solid and liquid contamination. Coolant system maintenance should always be done as a complete package to be most effective

Is it better to use a filter with coolant additive or a liquid SCA with an additive free filter?

Which coolant maintenance set-up to use is entirely determined by user preference. When properly installed, pre-charged and maintained, both filters with SCA's and liquid SCA's used with additive free filters will offer the coolant system identical levels of protection

Where is located the Coolant Filter ?

Normally it is located before the thermostat housing or in the hose that enter to the coolant pump

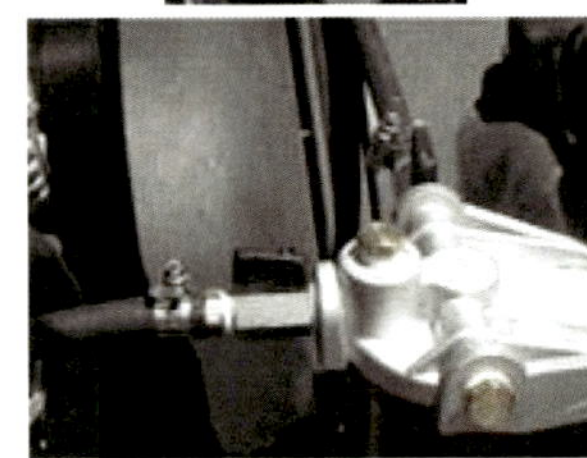

Coolant By-Pass valves

To remove the coolant filter I recommend install by pass valves in both sides of the coolant filter

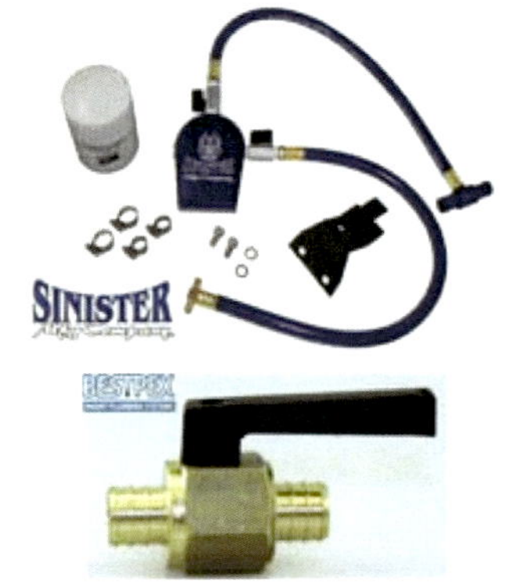

Coolant Filters

- Coolant filters are usually of the spin-on cartridge type connected in parallel to coolant flow. Coolant filters are used on most types of diesel engines
- Coolant filters mechanically filter the coolant through a fine media, removing impurities such as sand and rust particles suspended in the cooling system
- In some applications, a corrosion inhibitor (SCA) is placed in the coolant filter. This inhibitor dissolves into the coolant, forming a protective film on all metal surfaces in the cooling system.

OVERCOOLING

- The most common problem is a temperature above the thermostat range--<u>overheating.</u>
- The opposite may also happen and will create long term problems
- An overcooling problem therefore, has a single simple explanation. The thermostat is not functioning properly. Check to make sure that it is of the right type and properly installed. Do not assume that just because it fits it will function right. There is a lot more to thermostat design than most people realize. Do not experiment with unapproved thermostats.

OVERHEATING

- Overheating problems can be categorized into three basic problems that either alone or in combination with one another, will create overheating. They are:

    – lack of raw water flow

    – lack of coolant flow

    – heat exchanger defects

Lack of Raw Water Flow

- Lack of raw water flow will show up as an excessive increase of the raw water temperature as the raw water passes through the cooling system.

- Normal temp increase varies between different engine models but is usually in the range of 40 - 60 degrees F. In other words, if incoming raw water temperature is 70 degrees F, the outgoing water passing through the exhaust elbows will be in the range of 110 - 130 degrees F

Lack of Raw Water Flow

- This will create surface temperatures on the elbow that will be warm but not excessively hot. So the easiest way to identify a raw water problem is to check whether or not the engine overheating is combined with excessive temperatures on the outlet side of the raw water system.

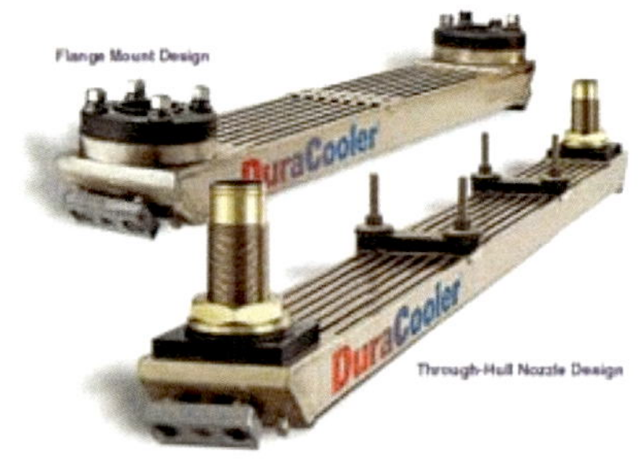

Keel Cooling Systems

Keel Cooler System

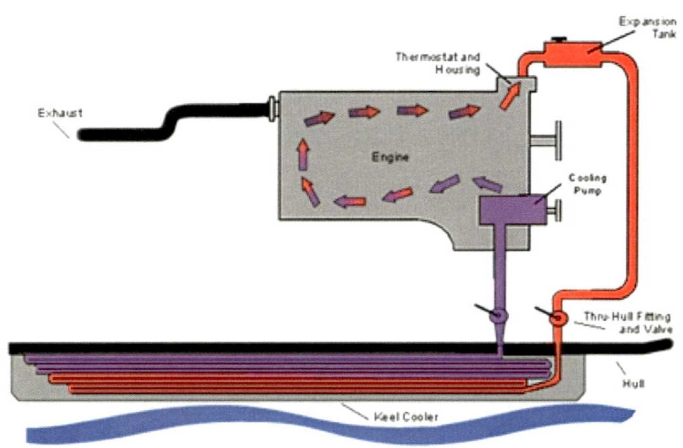

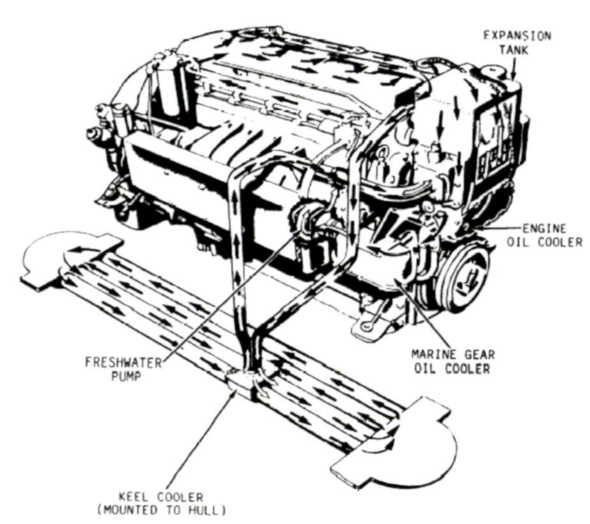

Keel Cooler System

A keel cooler is a closed circuit cooling unit mounted on a vessel's hull beneath the waterline

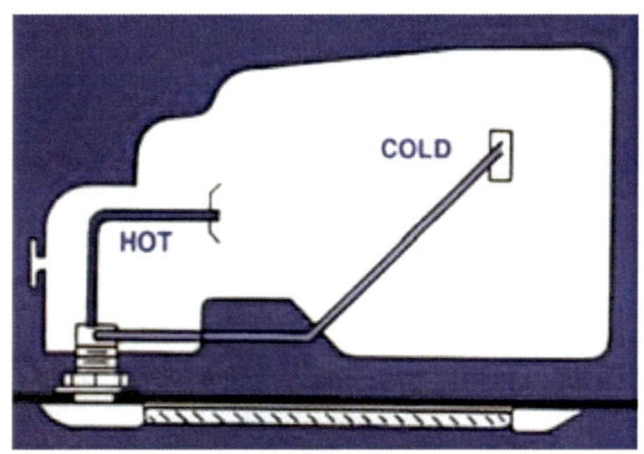

Keel Cooler System

This closed circuit cooling system eliminates the need for an inboard heat exchanger, raw water pumps and strainers, as well as the maintenance associated with them

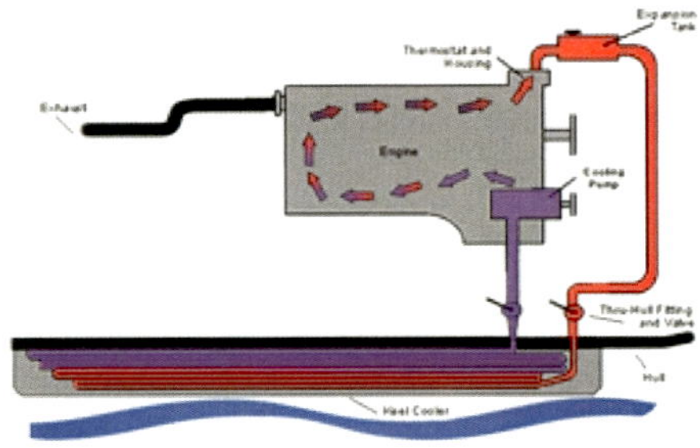

Keel Cooler System

Unlike inboard heat exchangers, Keel Coolers require no seawater inside the hull -- preventing the headaches of clogged sea strainers and raw water pump failures.

Keel Cooler Systems

This is done by eliminating the use of a heat exchanger. Instead of pumping raw water into the vessel's heat exchanger where it cools the coolant, the coolant is pumped through pipes or aluminum extrusions on the outside of the hull where the surrounding water (lake or ocean water) cools the coolant before it is pumped back into the engine

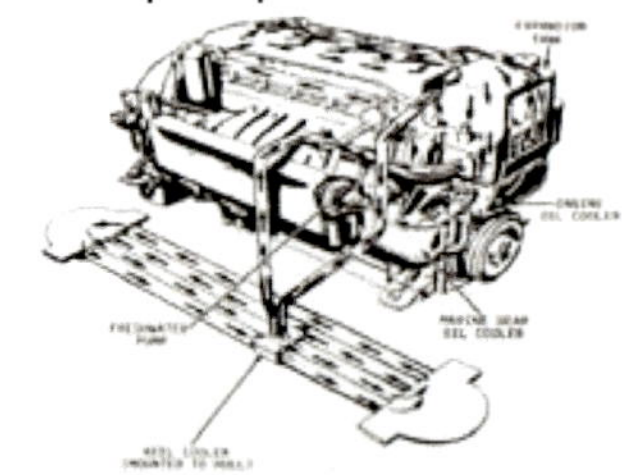

Keel Cooler Installation

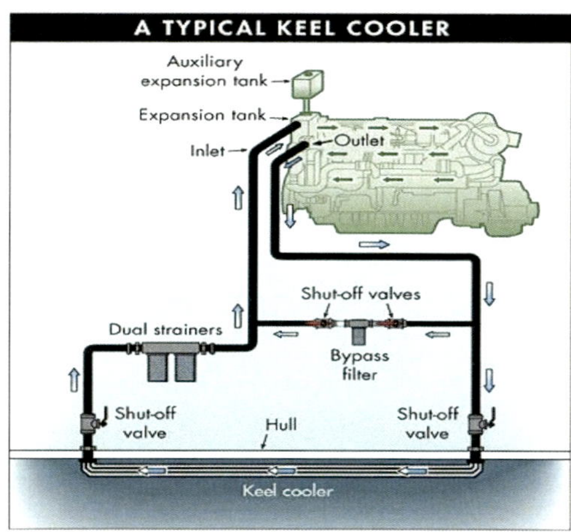

Single thru-hull

The single thru-hull fitting, located at only one end of the cooler, contains both the inlet and the outlet water connections

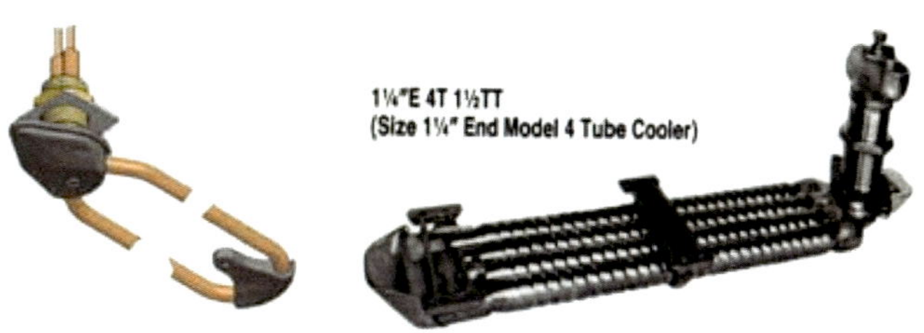

Double Stem Models

- The most economical Keel Cooler, "Double Stem" Models have two thru-hull fittings, one at each end of the cooler
- The thru-hull fittings have standard male pipe threads

Keel Cooler System

- Rather than pump corrosive raw water through your equipment for cooling, install a keel cooler - you will extend your components life and never again have to clean a plugged strainer!.
- Ideal for shallow coastal waters where seaweed, algae, mud, sand and silt are a constant problem.
- These coolers can be surface mounted, recessed or side mounted and have very little drag when properly installed

Keel Cooler System

- You can eliminate your troublesome water pump completely if you use a dry exhaust system , and actually have no raw water in your vessel at all!
- In most cases the engine's fresh water circulating pump is more than adequate for circulating water through the keel cooler by itself.

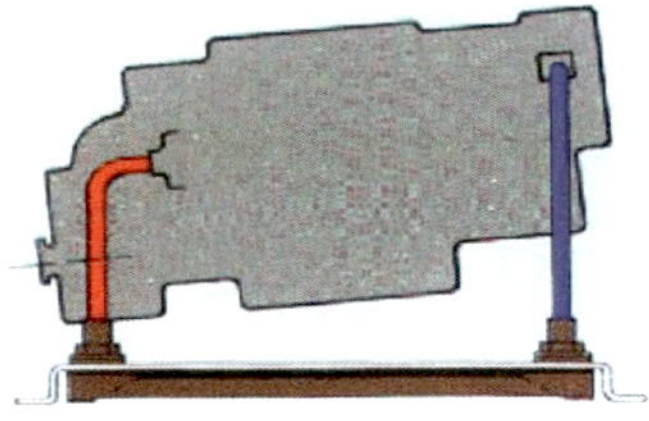

Keel Cooler System

Surface mounting on an existing hull with a wedge shaped fairing block as shown will not only protect the cooler from striking an object, but reduce the drag to an almost non existent level

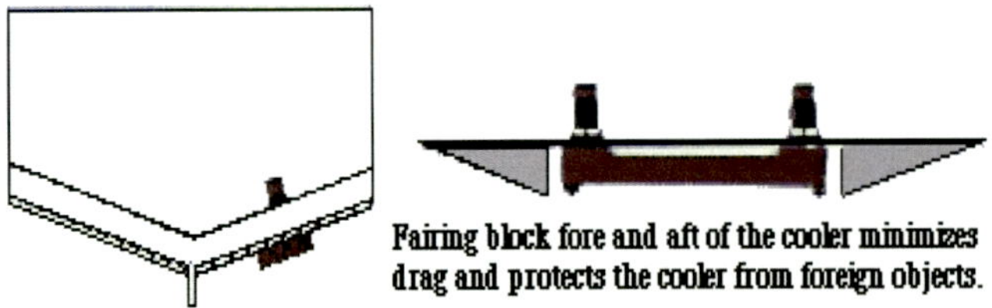

Fairing block fore and aft of the cooler minimizes drag and protects the cooler from foreign objects.

Keel Cooler System

Recessed mounted coolers offer the best protection, especially when hauling your vessel, and in new construction this is the preferred installation

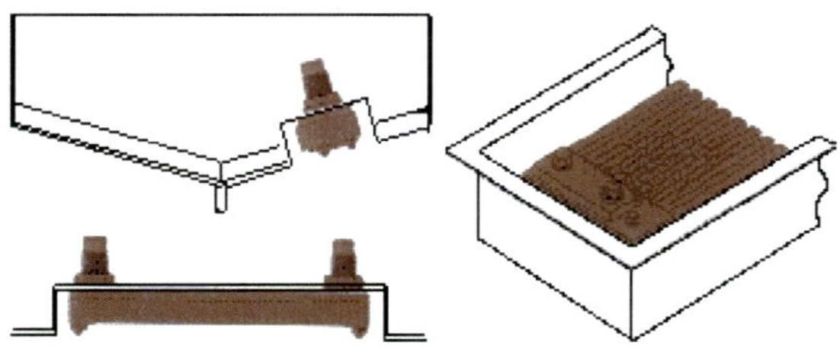

Keel Cooler System

For a shallow draft vessel, you may mount the cooler on either the side or stern of the vessel in a recessed or surface mounted installation. For HVAC cooling that is used dockside only, we actually mounted a cooler on the transom

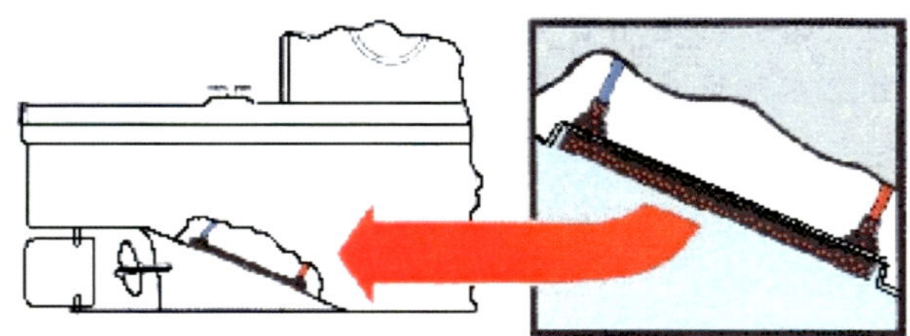

Keel Cooler System

Side mount shown in a recessed compartment

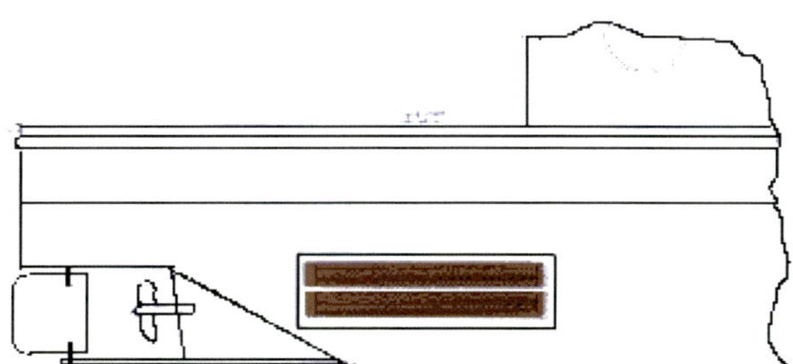

Keel Cooler System

Many different types of connections can be custom designed to exactly suit your requirements - sketch our what you need and our design team will prepare a plan for your consideration

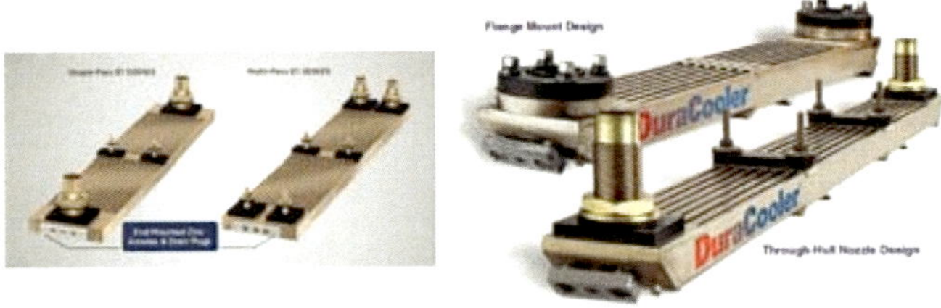

Keel Cooler - Materials

A standard series GRIDCOOLER is made of either heavy gauge 90/10 copper-nickel or 5000 series marine grade aluminum rectangular tubing

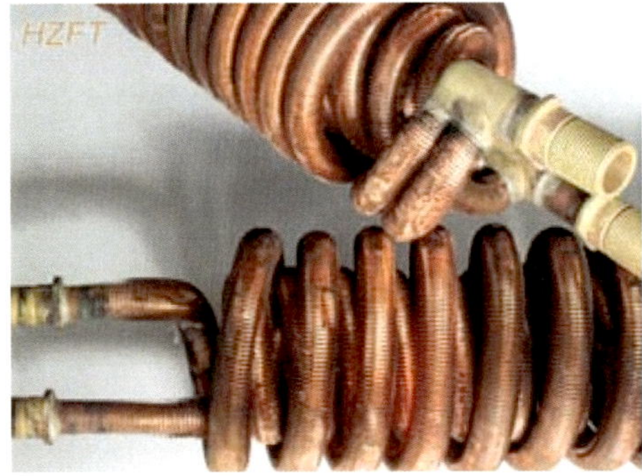

Keel Cooler - Materials

- These alloys were chosen because of their resistance to salt water corrosion and their heat transfer abilities.
- The copper/nickel units are silver brazed, and the aluminum units are welded. These construction materials and methods provide a level of durability far above any kit cooler

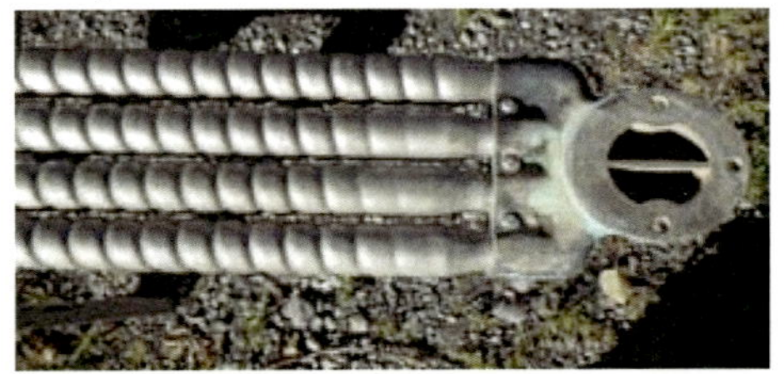

Fact & Fiction about Keel Cooler

- FICTION: A Keel Cooler can't be used in high speed applications

- FACT: Keel Coolers are currently being used on high speed crew and patrol craft . Keel Coolers creates a negligible amount of drag in high speed applications, and their sturdy design will hold up under the most demanding conditions

Fact & Fiction about Keel Cooler

- FICTION: A Keel COOLER must be mounted in a recess

- FACT: Any Keel COOLER may be mounted in a recess or just externally mounted on the side, bottom or rake of the hull, depending on the vessel design

Fact & Fiction about Keel Cooler

- FICTION: Keel COOLERS require too many through hull fittings

- FACT: Any Keel COOLER can use our L-Series option which requires only two through hull fittings, one for the inlet nozzle and one for the outlet nozzle

Fact & Fiction about Keel Cooler

- : Grooved round tubing is more efficient than smooth rectangular tubing
- : This statement might be true, depending on the cross-sectional area and perimeter of the two tubes used

Salt Water Temperature

The temperature differential between the hot coolant and the water that surrounds the keel cooler must be sufficient to carry away the engine's heat even when the dredge is floating in a puddle of hot water or the engine will overheat

Chapter 12
Fuel Systems

Fuel Systems & ABYC

- **Gasoline fuel systems** are covered by CFR (code of federal regulations) Title 33, Part 183, Sub Part J,and by ABYC Standards H-2, H-24, and H-25.
- **Diesel fuel systems** are covered by ABYC Standards H –32 and H-33 .
- **CFR's** (code of federal regulations) are actual LAW and to not comply is to violate Federal Law.
- The Federal Regulations do not address Diesel fuel systems

CFR's & ABYC Recommendations

- ABYC will not contradict Federal Law ,However:
  - Federal Laws were written a while ago and are not rewritten or reviewed often
  - ABYC reviews and adjusts regulations to keep up with changes in the industry.
  - ABYC address's things that CFR doesn't
  - Sometimes ABYC a little more demanding in requirements .
- Fuel System includes fuel fill, tank, tank vent, distribution and return system, pumps, valves, strainers and injection systems .

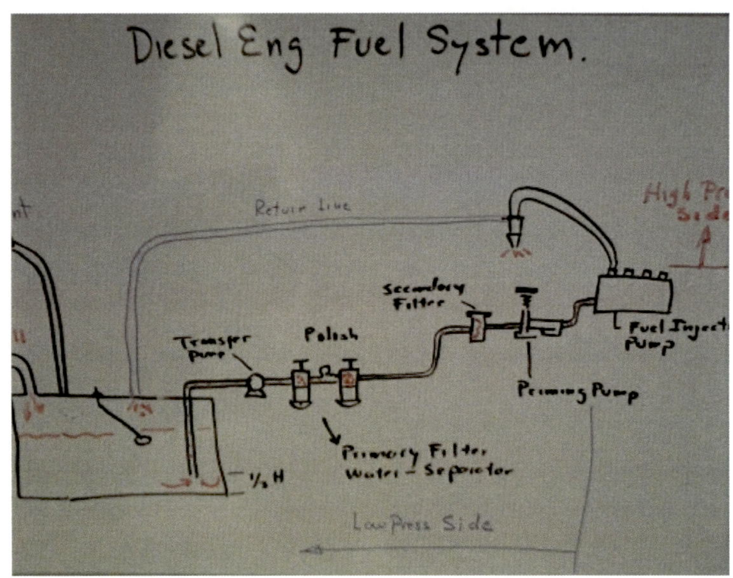

Double Fuel Tank

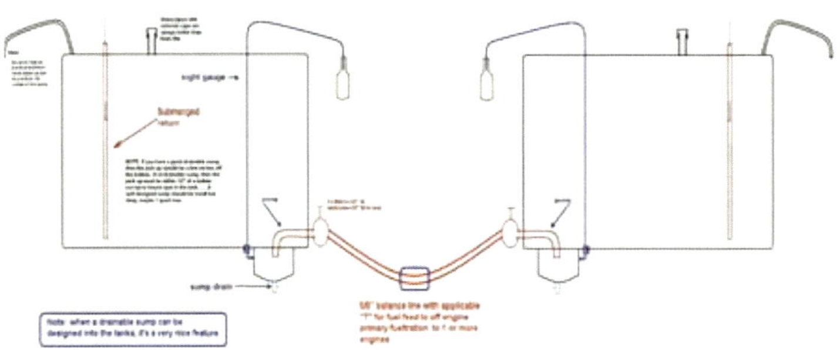

Bonding System

Equipment located in machinery spaces in contact with fuel lines, or that may be in contact with bilge water shall be designed so that the current carrying parts of the device are insulated from all exposed electrically conductive parts

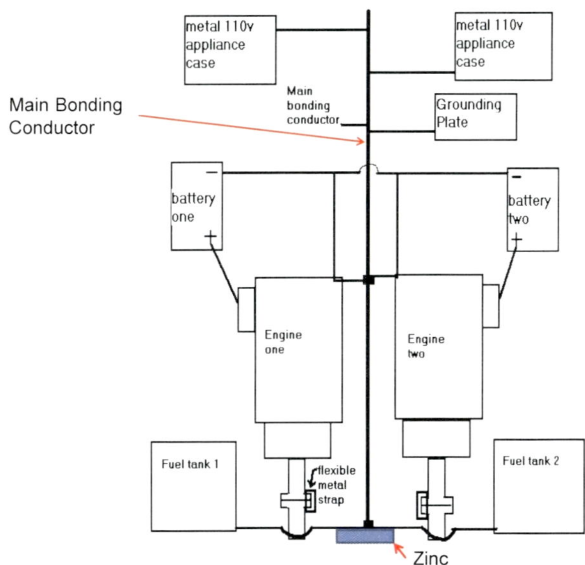

Common-Bonding Conductor

- The large common-bonding conductor should be bare or green.
- If wire , it may consist of bare , stranded, tinned copper or insulated stranded wire of minimum size 8 AWG
- If solid, it may be uninsulated copper , bronze strip, or copper pipe at least .030 in. thick and 0.5 in. wide.
- The copper , bronze strip or pipe may be drilled and tapped, provided it is thick enough to provide at least three full threads for terminal screws.

Common-Bonding Conductor

- Equipment to be connected to the boat's bonding system includes.
- Engines and Transmissions.
- Propellers and Shafts.
- Metal Cabinets and control boxes.
- Electronics Cabinets
- Metal fuel and water tanks, and electrical fuel pumps and valves
- Metal battery boxes
- Metal conduit or armoring
- Trim Tabs flaps

Electrical System Conductor

- As defined by ABYC Standard E-9.
- DC Grounding Conductors are normally noncurrent-carrying conductors. They are used to connect metallic noncurrent-carrying parts of DC devices to the engine negative terminal or its bus
- DC Grounded conductors are current-carrying conductors connected to the side (usually negative) of the source
- Ground is established by a conducting connection with the earth, including any conductive part of the wetted surface of a hull

Gasoline Fuel Systems

All gasoline fuel systems must be pressure tested by the Boat Manufacturer.

- Test to 3 psi or 1.5 X pressure at lowest point in fuel system.
- Test all parts of fuel system

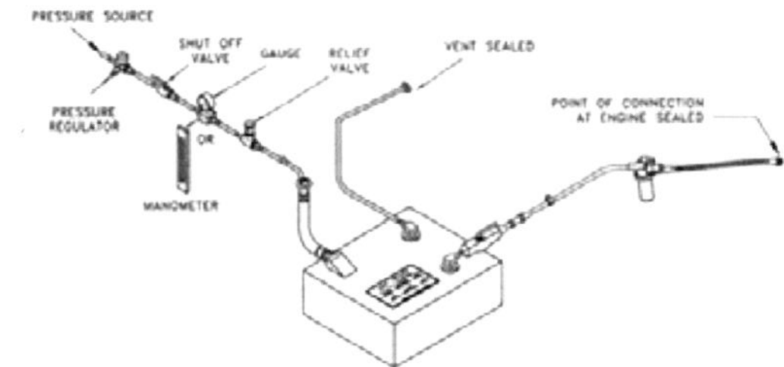

A Typical Fuel System

- There must not be any drain plugs anywhere in the fuel system.
- If a filter or drain plug has a service fitting, it must either be tapered pipe plug or screw type with lock device other than split lock washer .
- All gasoline Fuel systems must be pressure-tested boat the Boat Manufacturer

Fuel Hose

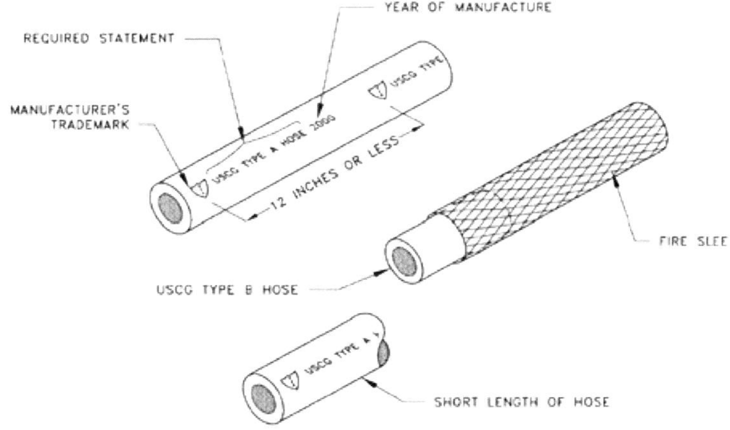

Marine Fuel Hose

- All Fuel hose must have markings on it stating:
  - USCG TYPE A1
  - USCG TYPE A2
  - USCG TYPE B1
  - USCG TYPE B2.

Fuel Hose

- It must also state:
  - Year of manufacture
  - Manufacturers Name or Logo
- • It must be in 1/8" block letters

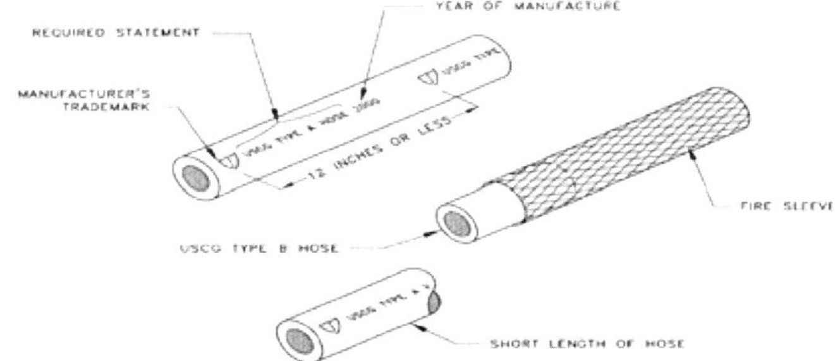

Fuel Hose Type

- **A1**- Can be used anywhere in Fuel System.
- **B1**- Usually ok for vents and fills if the hose will leak less than 5 oz in 2.5 minutes if severed at lowest point.
    - "1" denotes low permeability, required for fuel supplies or where it always has fuel in it.
    - "A" denotes: 2 ½ minutes fire test required.
    - "2" denotes a higher level of permeability.
    - "B" denotes no fire test has been made.

Hose Clamps

- Clamps not to overlap each other.
- Clamps not mounted beyond bead.
- ¼" of hose beyond clamp
- Hose clamps not to rely on spring tension
- Clamps to be reusable:

FIGURE 28 Hose Clamp Types

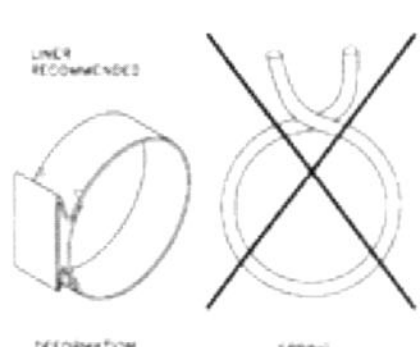

Fuel fill system

- Fuel fill must be located in a way that an overflow of up to 5 gals/ min for 5 seconds will not enter the boat.

FIGURE 30 Fuel Fill Locations

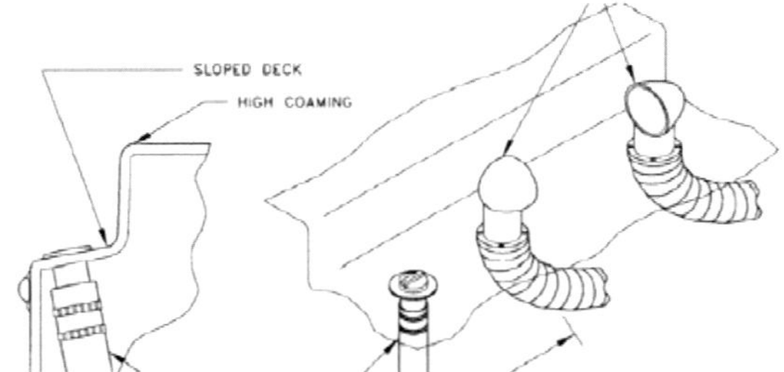

Fuel Fill System

- Must be at least 15" from any vent and may have to be further some times.
- Fuel fill hose should be A1 type

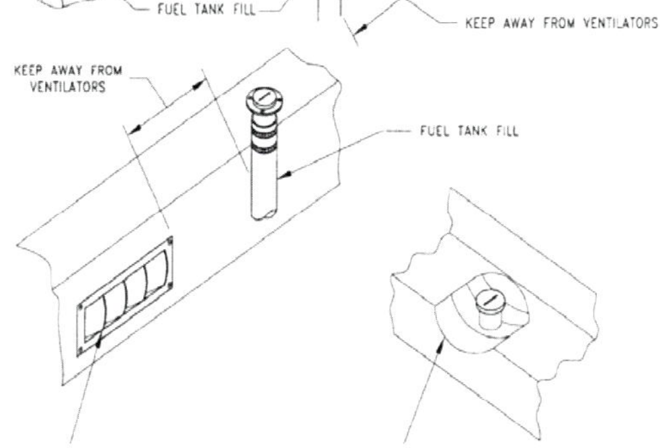

Fuel Fill System

- Should be double hose-clamped with ½"wide clamps

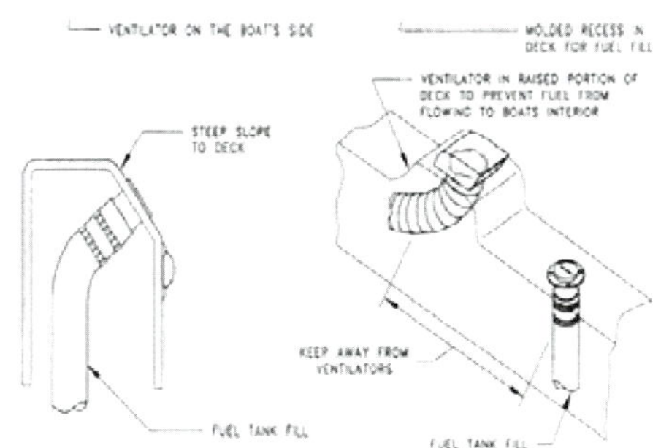

Tank Vent

- Must prevent tank from reaching 80% of tanks pressure rating
- Each tank vent must have a flame arrester that can be cleaned
- Tank Vent must not allow fuel to enter boat during an overflow.

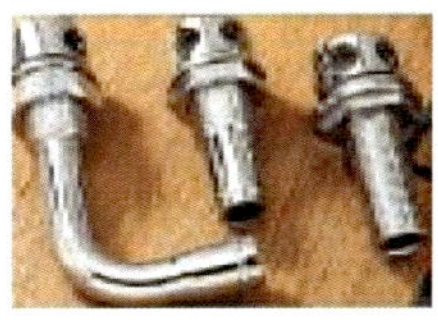

Tank Vent

- Must not allow fumes to enter boat. Keep at least 15" away from other vents. .
- There should be no traps in hose that would impair its ability to vent tank .

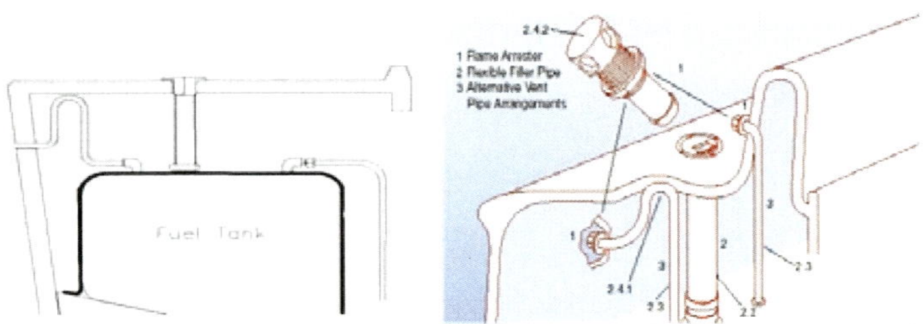

Tank Vent

- Generally, a 9/16" I.D. Hose with fittings that are at least 7/16 I.D. are minimum .
- Type A1 or A2 hose is ok

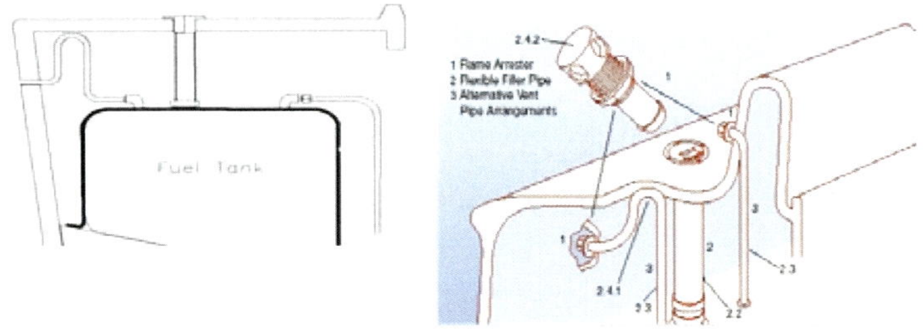

Fuel Tanks

- Both **Steel** and **Aluminum** have been widely used for Diesel applications
- Wrong installations affect alloys such as <u>Mild steel</u> specially if the surface of the tank held standing water on their surface from deck leaks and water migration into the boat

Aluminum Tanks

- Aluminum is vulnerable to Poultice corrosion and anything damp on its surface in an oxygen starved environment will induce this type of corrosion and it will eventually perforate the tank causing a fuel leak

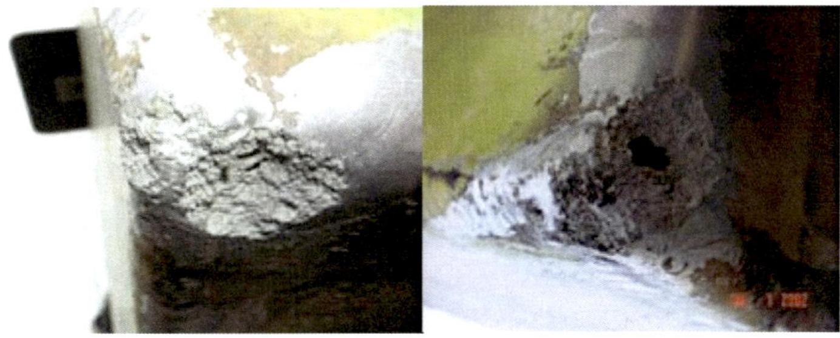

Fuel Tank

- Must be pressure tested by tank manufacturer.
    - Tested to 3psi when entire fuel system less than 6.4' top to bottom.
- Must be pressure tested for shock, surge and pressure impulse (when appropriate)
    - If only tested to 25G Vertical Acceleration (Sec 183.584) tank must state: "To be used aft of boats half length" .

Fuel Tanks

- Boat Builder must pressure test tank along with entire fuel system .
- Steel Tanks must never be encapsulated in foam or fiberglass .

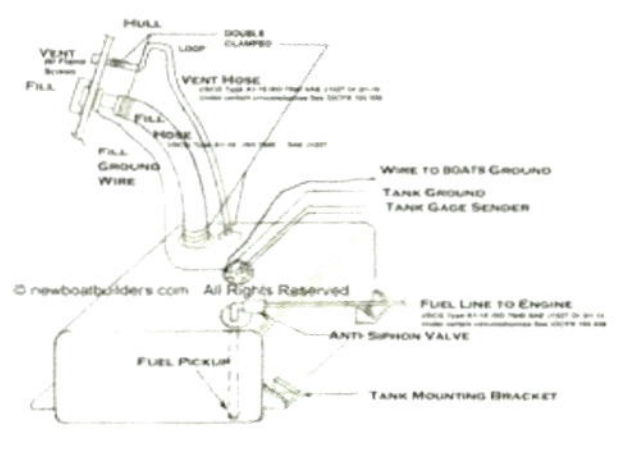

Fuel Tanks

- Gasoline tanks must never be integral.
- They must never share a bulkhead or be part of any structural member.
- They must never support a deck

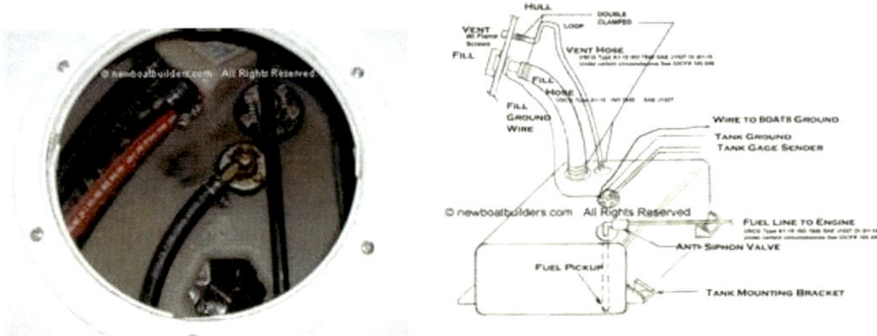

Fuel Tanks

- Label and all fittings must be accessible for inspection and for service.
- Every Fuel tank must have a label that states the following.

FIGURE 8   Fuel Tank Label

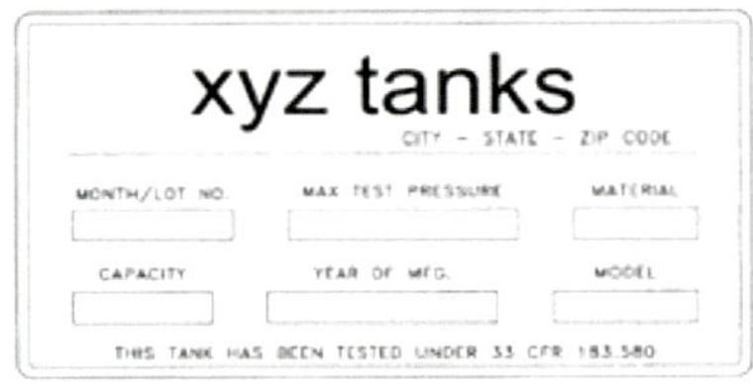

Fuel Tanks

- Water must drain from all surfaces of a metal tank
- Any chock, strap or support that's not part of tank must be insulated from it. Non-absorbent material. (Poly urethane sealant would work for this) .

Metallic Tank

- Metal must be resistant to diesel and other typical liquids, such as lube oil, grease, bilge solvents, salt water, etc
- If tank is galvanized steel, only outside of tank galvanized.

Metallic Tank

- If aluminum, use: 5052, 5083, 5086, 6061, or 6063
- If Stain Steel, use 300 series

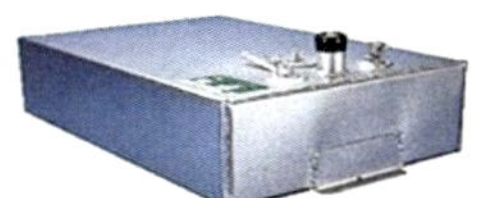 

Terneplate Steel

- Fuel Tanks shall not be constructed of terneplate steel
- Terneplate is an Steel sheet with a coating of terne metal, an alloy of lead and tin.
- Over time the coating will form a thick gooey paste that will definitely clog up Diesel fuel systems

Metallic Tank

- Never use self-wicking material or carpet or foam rubber
- Tanks mounted above flat surfaces must be separated by ¼" air space when tank is full. surface to self-drain
- Must not be reached by normal accumulation of bilge water

Metallic Tank

- Metals and alloys matched to minimize galvanic action
  - No copper based fittings into aluminum tank
- Steel pipes should be zinc plated & yellow dichromate dipped after machining

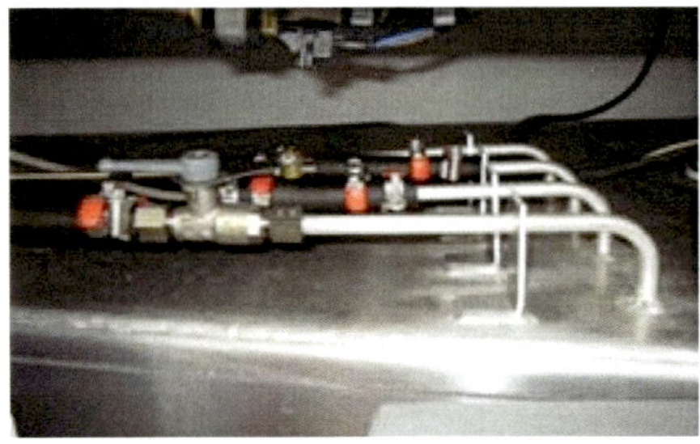

Metallic Tank

- Fasteners in aluminum tanks should be 300 series stain-steel
- Metal parts of tank should self drain water
- from surfaces

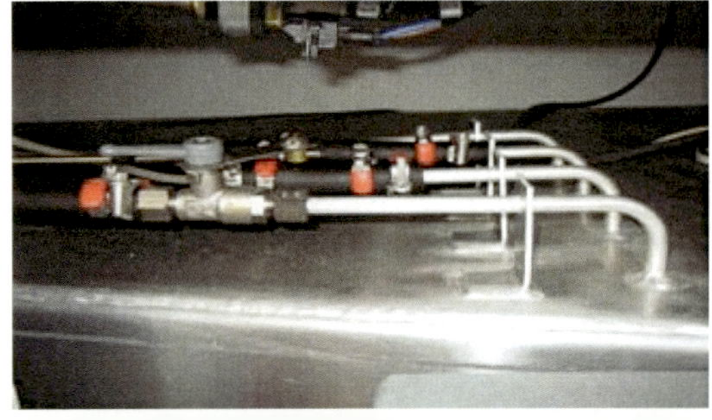

Fuel Tanks

- Tank must sit on non-absorbent material.
- Should not sit right on glass (1/4 inch separation is a good practice)
- No copper based fittings in aluminum tanks

Baffles

- Baffles must not trap fluid on bottom or vapor at top .
- If baffled openings in baffles max of 30% of tanks cross section

Plastic Tank

Foam material is not recommended like a sole support for bond !!!!

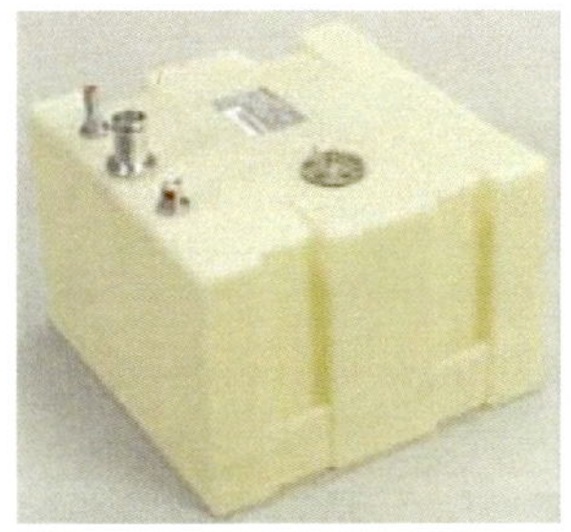

Plastic Tank & Foam

All Foam must never change volume by more than 5% or dissolve after being immersed for 24 hrs in fuel, oil, or 5% solution of sodium phosphate and water

Fuel Tank & Foam

- Foam must not absorb mare than .12 lb of water/sq ft of cut surface
- If in an engine room, or low in bilge, must resist oil, fuel and not absorb water

Foam & Ferrous Alloys

- Foam can't be sole support for non-Ferris metal tank.
- If Non-Ferris metal tank foamed in place,top must drain. Can be no gaps anywhere for water to be trapped between foam and tank.

Foam & Ferrous Alloys

- In metal tank foam must never fail at tank
- If foam not bonded to tank maintain at least ¼"space around tank

Professional Installations

- Believe it or not, this is a fuel tank for a Generator on a fishing boat!
- Note that the "tank" is "secured"

Professional Installations

- Here's the boat it's on:

Fuel Lines Support

- Hose secured and routed to prevent chafe
- Clips, straps, etc ., corrosion resistant

Fuel Lines Support

Valves must either have own support or not place strain on hose

Fuel Tank anti-siphon

Fuel distribution systems shall be provided with anti-siphon protection by keeping all parts of fuel distribution and return lines above the level of the tank top from the tank to the carburetor/ Fuel Injection inlet or its equivalent

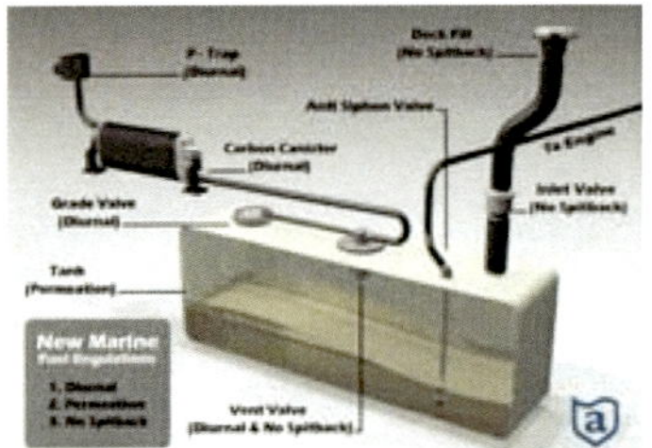

Fuel Tank anti-siphon

Or by installing an electrically operated valve at the tank withdrawal fitting, or along the fuel distribution line, connected to be energized open only when the engine ignition switch is on and the engine is running

Anti-siphon Devices

If any fuel feed is not above the top surface of the fuel tank all the way to the prime pum , it will require some sort of shut off or anti-siphon device

Manual Fuel Stop

- Must pass 2 ½ minutes fire test.
- Must be located right at tank connection point, above top surface
- Must be usable in some way from outside compartment

12V/24 V Shut off valve

This type of shut off valve is normally open. Will be closed only when the emergency stop button is applied

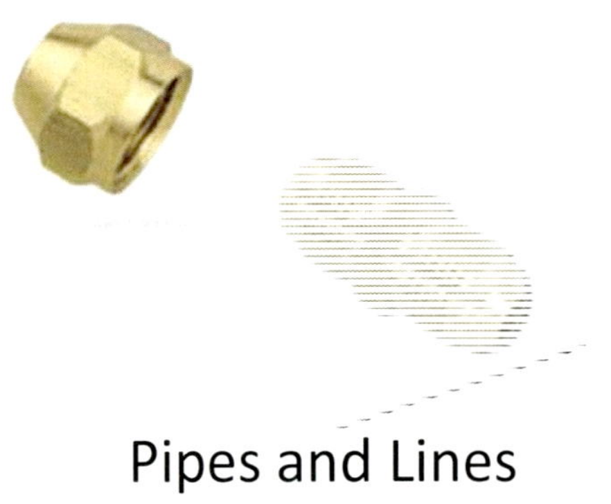

Pipes and Lines

Flare Fittings

Flare Nut & Fitting

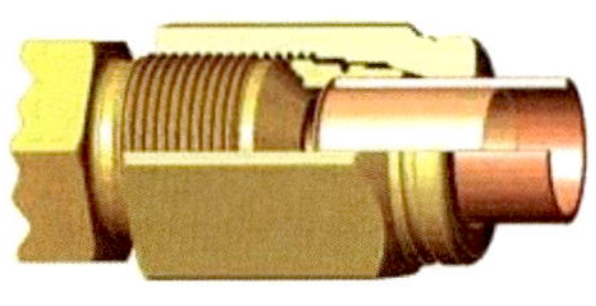

Flare Nuts

Flare Fittings

Flare configurations

 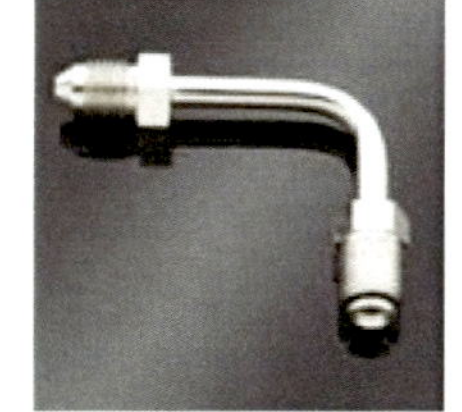 

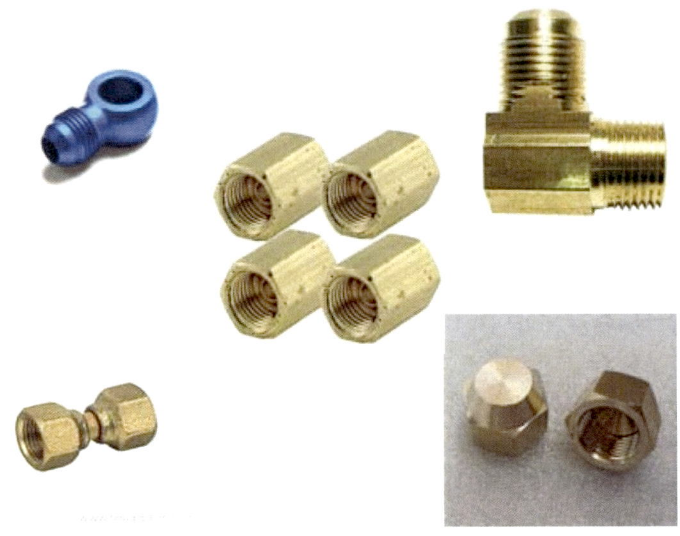

Soldering Pipes

Fuel Valves

- Must have stops at full open and full closed positions . Multi-purpose valves must indicate positions

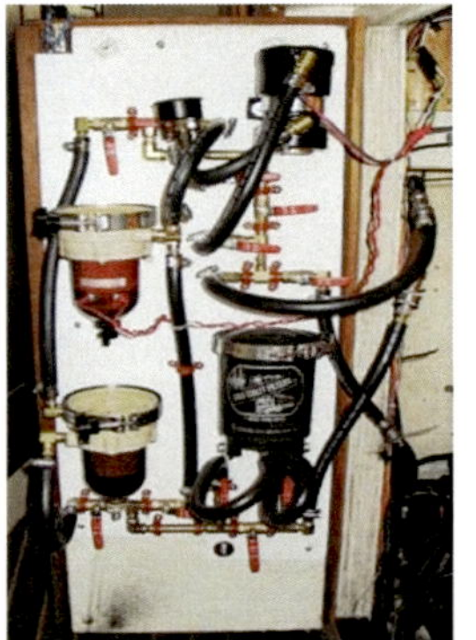

Solenoid valves

Solenoid Valves

- Solenoid valves are the most frequently used control elements in fluidics. Their tasks are to shut off, release, dose, distribute or mix fluids. They are found in many application areas.

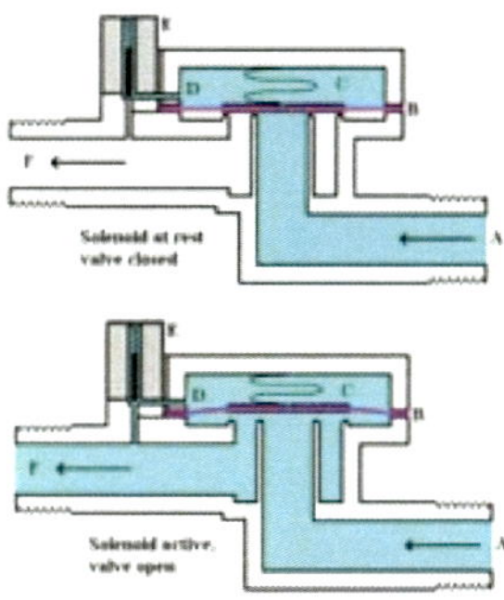

2 Way Solenoid Valve

Two-way valves are shut-off valves with one inlet port and one outlet port. In the de-energized condition, the core spring, assisted by the fluid pressure, holds the valve seal on the valve seat to shut off the flow

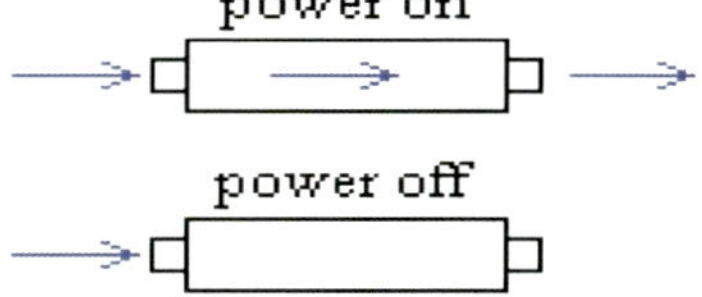

Flow Meters

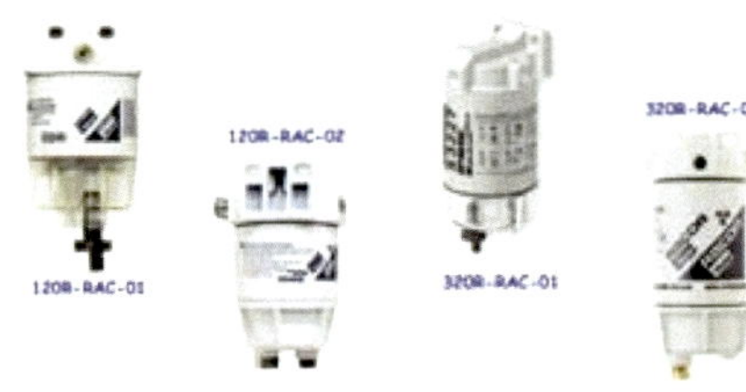

FUEL FILTERS

Properties

Marine Fuel Filters

In a typical installation there are two filters. The primary filter or Water separator is located between the fuel tank and the priming pump and the secondary filter is located before the fuel injection pump (Bolted on the block of the engine)

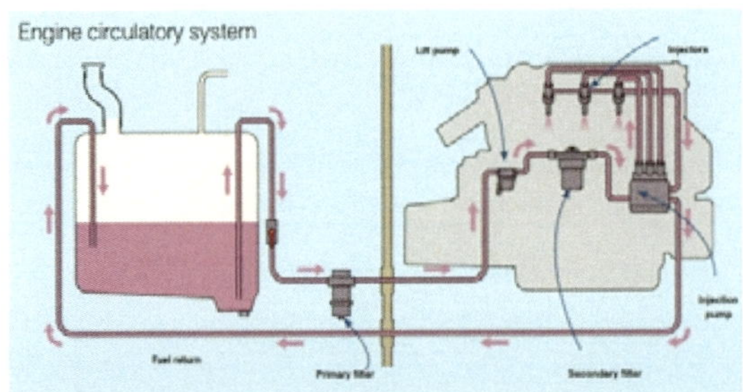

Fuel Filters

- Normally I recommend install a transfer pump (12/24V DC) between the fuel tank and the primary filters to fill the system quickly and to bleed the system easily
- This pump should be activated from the ignition position at the ignition switch
- Use the formula recommended in the next chapter (Pag), to calculate the pump capacity

Water Contamination

- The presence of water in diesel fuel systems may cause the following problems
  - Water causes rust and corrosion of iron components
  - Biodiesel and bio-oil fuels contribute to the more aggressive growth of microbes
  - The microbes form a sludge that can plug filters and hinder injector performance
  - Moisture on the fuel can cause premature wear and injector failure

Fuel Filter Specification

The unit used to denominate the filter are the Microns (one millionth of a meter). A micron rating for a fluid filter is a generalized way of indicating the ability of the filter to remove contaminants by the size of the particles

- **Red** = 30 micron, primary filtration.
- **Blue** = 10 micron, secondary filtration.
- **Brown** = 2 micron, final filtration.

Fuel Filter Specification

- The average cross-section of a human hair is 40-90 microns. The human eye cannot see anything smaller than 40 microns in size.
- The lower the micron rating, the greater the efficiency and hence the amount of dirt that is captured

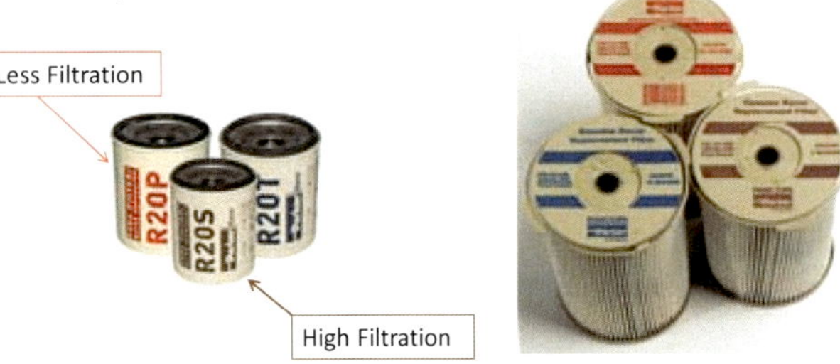

Fuel Filter Fittings

- All threaded fittings, tapered pipe thread NPT fine.
- Use appropriate thread sealant for fuel systems . Never use Teflon tape

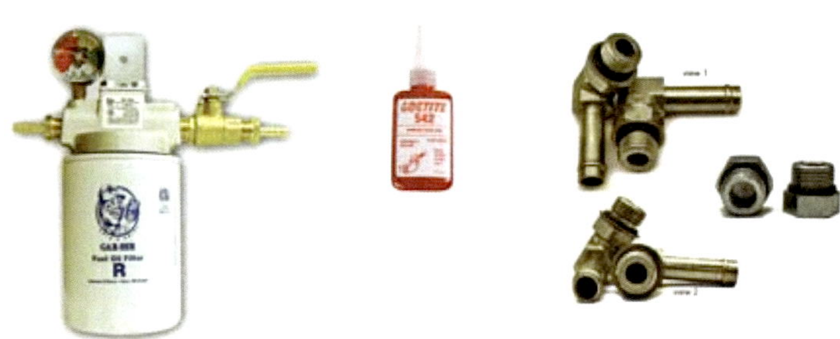

Water Separator

- The water separator is a vital item in the fuel system of gasoline and diesel engines.
- The life of the engine and their performance depends of the fuel quality

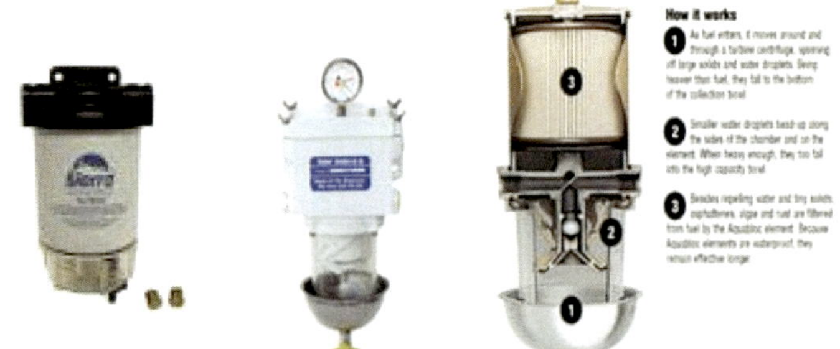

This gauge permanently replaces the T-handle on the top of the turbine series filters

The gauge measures the restriction caused by the filter and indicates when the filter element should be replaced. Includes needle follower

At the first indication of decreased performance, note the dial reading or apply the 'red line' decal provided with most kits. This will assist in knowing when to change the filter at the next interval

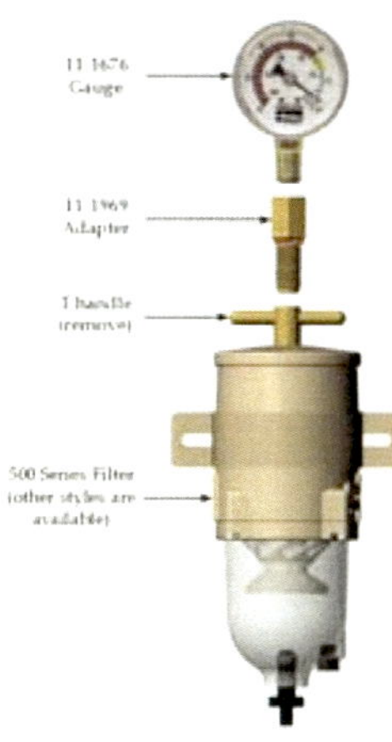

Fuel Restriction Indicator

Vacuum restriction indicators monitor element condition as the filter slowly becomes clogged with contaminants. As the element gets dirty, restriction increases and less fuel is delivered to your engine

Reading The Vacuum Gauge

The gauge's main function is to monitor the condition of the primary fuel filter element. As the element becomes impacted with dirt and debris, the level of effort required for the engine's fuel lift pump to pull fuel through the filter increases, as does the vacuum imparted on the filter

Primary Filter Service

- (1).- Isolate the primary filters by closing the valves in both sides of the filters.
- (2) .- Drain the filters by means of the drain plug

Primary Filter Service

- (3).- Remove carefully the plastic bowl in order to unscrew the float and the check valve
- (4) .- Take special care with the aluminum ball and the lips seal. If the ball drops on the bilge shall be replaced

Primary Filter Service

- (5).- After a detailed clean using mineral spirit , replace the whole kit of seals using vaseline to avoid future leaks
- (6) .- Take care with the orientation of the ball seal. It shall be with the lips in permanent contact with the ball

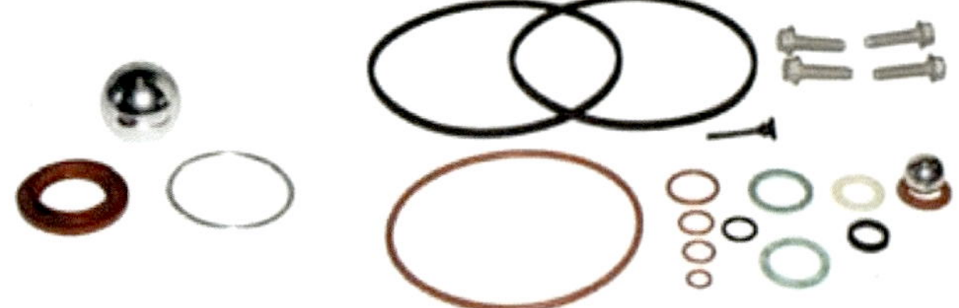

Primary Filter Service

- (7).- Once the procedure has been done, and without the upper cap; open the valves slowly until the fuel spill from the top of the filter
- (8) .- The fuel transfer pump can be used to fill the filters or manually you can add extra fuel

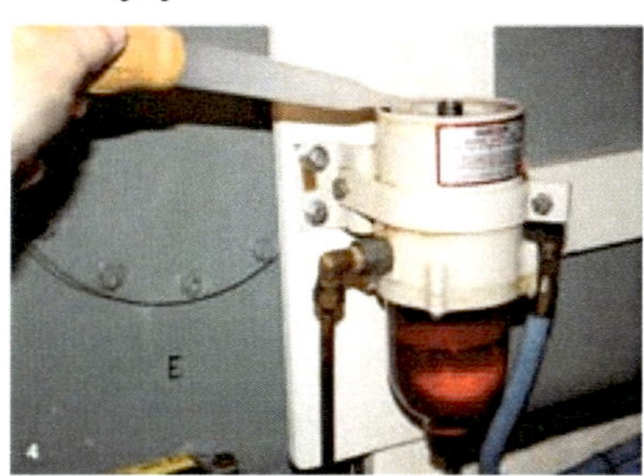

Primary Filter Service

- (9).- Close the top covers , start the engine and check through the plastic cup the presence of air bubbles
- (10) .- If there is air in the cup, then release the top cover gently until a controlled fuel spillage. In this way the air will come out of the system (With the engine running)

Primary Filter Service

- (9).- Once the procedure has been done, and without the upper cap; open the valves slowly until the fuel spill from the top of the filter
- (10) .- The fuel transfer pump can be used to fill the filters or manually you can add extra fuel

Secondary Fuel Filter

- The secondary filter is a cellulose encapsulated element that can be bolted in an engine block bracket
- The most common faults are due to air trapped in between the filter and the bracket, during the installation of this filter. To avoid that issue I recommend fill 1/3 of the capsule with engine oil and the rest with diesel

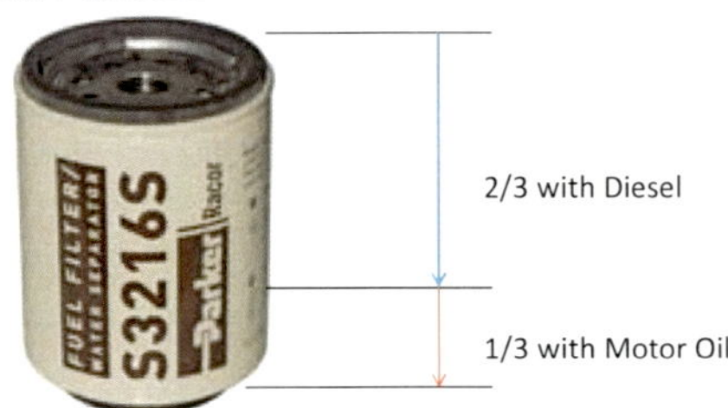

Secondary Filter Installation

- Under pressure conditions, the engine oil have a higher expansion coefficient than the diesel fuel; in this way the air bubbles will be eliminated. The engine oil will emulsified with the fuel slowly acting like a lubricant on the fuel system
- In normal conditions using this technique the bleeding procedure is less complicated and faster

Fuel Pumps

Low Pressure Side

Transfer and Priming Pumps

Fuel Transfer Pumps

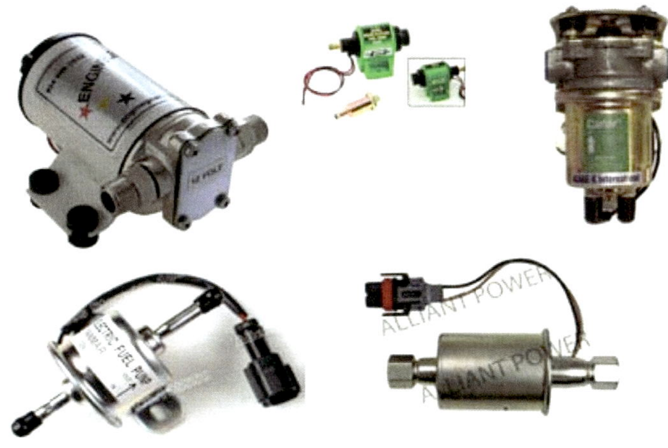

Electric Pump

Submersible pumps

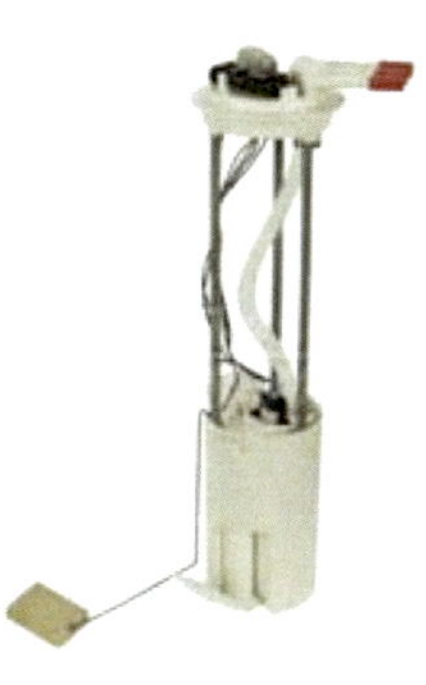

CHARACTERISTICS

Electric Submersible Pump

- 12/24V DIESEL / FUEL PRIMING PUMP 12/24 Volt submersible
- 4.5 Gallons Per Minute (17 LPM) capacity flow
- They are basically transfer pumps located inside of the fuel tank at the low pressure side

Feed Fuel Engines

- Gravity

- Mechanical Pump

- Electrical Pump (12 V & 24 V).

Mechanical Fuel Pump

Mechanical pump

- Most mechanical fuel pumps are diaphragm pumps, which are a type of positive displacement pump

- Diaphragm pumps contain a pump chamber whose volume is increased or decreased by the flexing of a flexible diaphragm, similar to the action of a piston pump

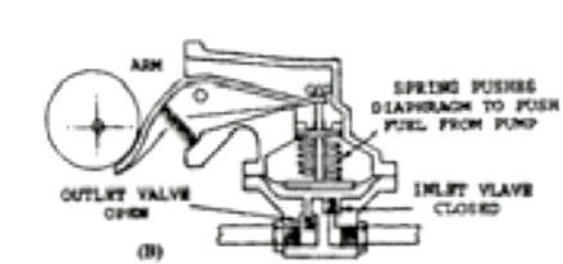

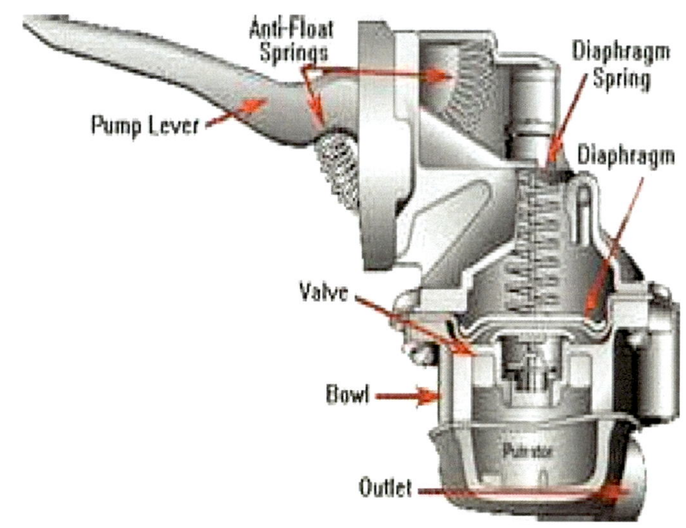

Mechanical pump

- A check valve is located at both the inlet and outlet ports of the pump chamber to force the fuel to flow in one direction only

- In the most common configuration, these pumps are typically bolted onto the engine block or head, and the engine's camshaft has an extra eccentric lobe that operates a lever on the pump

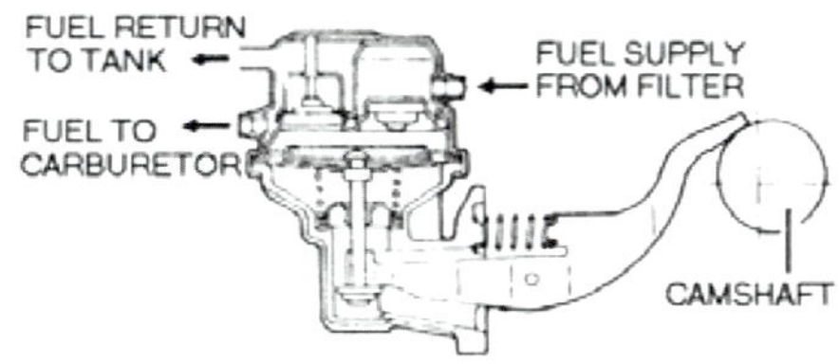

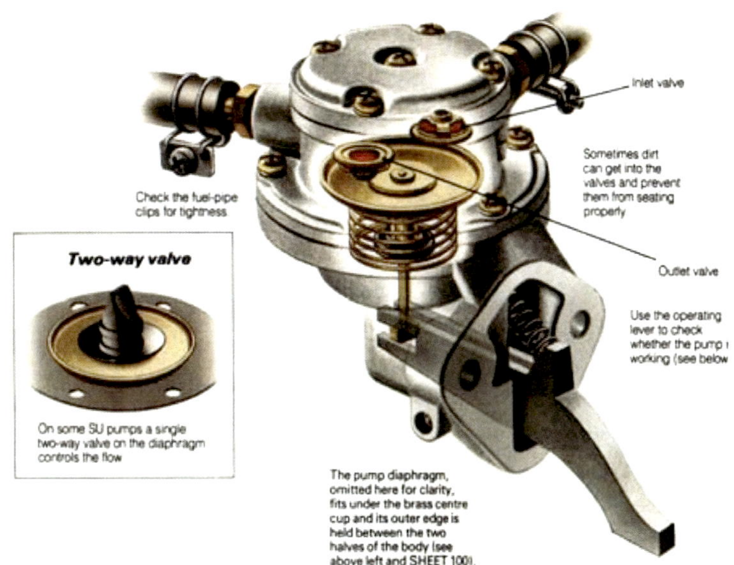

Mechanical pump

- Either directly or via a pushrod, by pulling the diaphragm to bottom dead center. In doing so, the volume inside the pump chamber increased, causing fuel to be drawn into the pump from the tank
- The return motion of the diaphragm to top dead center is accomplished by a diaphragm spring, during which the fuel in the pump chamber is squeezed through the outlet port and into the fuel injection pump

Fuel Pump & Priming Pump

- At this point, any remaining fuel inside the pump chamber is trapped, unable to exit through the inlet port or outlet port
- The diaphragm will continue to allow pressure to the diaphragm, and during the subsequent rotation, the eccentric will pull the diaphragm back to bottom dead center, where it will remain until the inlet valve to the carburetor reopens

Diesel Priming Pump

It is moved by the camshaft of the fuel injection pump

FUEL PUMP CALCULATION

Naturally Aspirated Engines
Fuel Injected Engines

How much fuel your fuel pump needs?

- To make this horsepower, your engine will consume a certain amount of fuel

- That amount is referred to as the "Brake Specific Fuel Consumption", or BSFC. The BSFC is generally estimated to be between 0.30 and 0.37 for most naturally-aspirated / Carbureted (non-turbo/super-charged) engines, and between .45 and .55 for turbo/super-charged engines

How much fuel your fuel pump needs?

- Multiply horsepower by .38 (naturally-aspirated motors) or .47 (force-induction motors) to come up with a fairly accurate guide to how many liters per hour of fuel you will need to feed the engine

- For example, if you are building a really hot little 4 cylinder turbocharged engine and plan to make about 450HP, you would need a fuel pump that can produce about about 212 liters per hour (450 * 0.47)

How much fuel your fuel pump needs?

- It is critical that the fuel pump in your fuel-injected engine is able to produce at least as much or more volume over time than the engine requires
- If the fuel pump is unable to meet the fuel requirements then the fuel mixture will become lean and the engine will go into pre-detonation and will eventually destroy itself

DIESEL FUEL TERMS

Terminology and definitions

Diesel Fuel

- Diesel fuel, which is less volatile than gasoline and is made up of heavier petroleum fractions, ignites by compression in the cylinder rather than by a spark.
- In the same way that gasolines are labeled with an octane rating, the ignition performance rating of diesel fuel is called the <u>cetane number</u>.

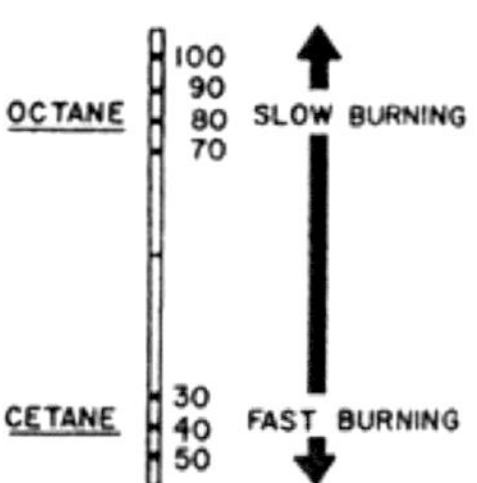

Cetane Number

- **Cetane number** or CN is an indicator of the combustion speed of diesel fuel. It is an inverse of the similar octane rating for gasoline
- In a particular diesel engine, higher cetane fuels will have shorter ignition delay periods than lower Cetane fuels
- Cetane numbers are only used for the relatively light distillate diesel oils. For heavy (residual) fuel oil two other scales are used CCAI and CII.

Diesel Fuel

Cetane is often used as a short-hand for cetane number, a measure of the combustibility of diesel fuel. Cetane ignites very easily under compression

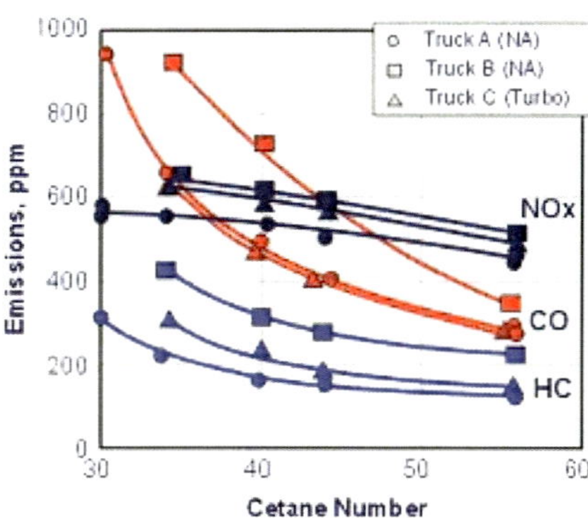

Cetane Number

- A measure of the starting and warm-up characteristics of a fuel
- In cold weather or in service with prolonged low loads, a higher cetane number is desirable
- Legislation dictates the Cetane index should be 40 or above.
- A Cetane rating of at least 40 is recommended at temperatures above 32 degrees
- A Cetane rating of at least 45 is recommended at temperatures below 32 degrees

Viscosity

- Affects injector lubrication and atomization

- The injector system works most effectively when the fuel has the proper "body" or viscosity

- Fuels that meet the requirements of 1-D or 2-D diesel fuels are satisfactory with Cummins and CAT fuel systems

Water & Sediment

- Affect the life of fuel filters and injectors. The amount of water and solid debris in the fuel is generally classified as water and sediment
- More water vapor condenses in partially filled tanks due to tank breathing caused by temperature changes

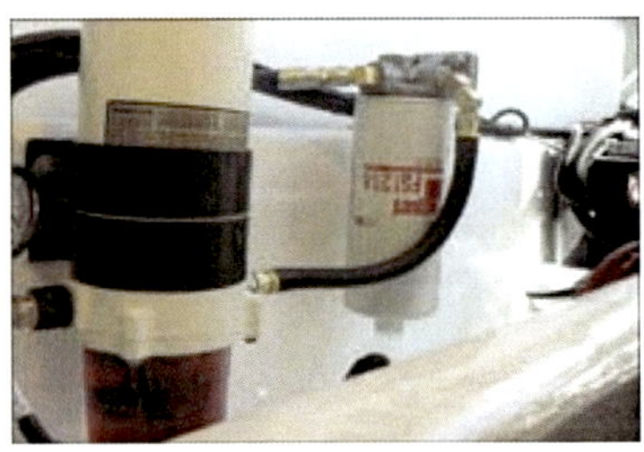

Water & Sediment

- Filter elements, fuel screens in the fill pump, and fuel inlet connections on injectors must be cleaned or replaced when they become dirty
- These screens and filters, in performing intended function, will become clogged when using a poor or dirty fuel and will need to be changed more often. Water and sediments should not exceed 0.1 volume percent

Carbon Residue

- Measures residue in fuel - can influence combustion
- The tendency of a diesel fuel to form carbon deposits in an engine can be estimated by various tests to determine the carbon residue after 90% of the fuel has been evaporated

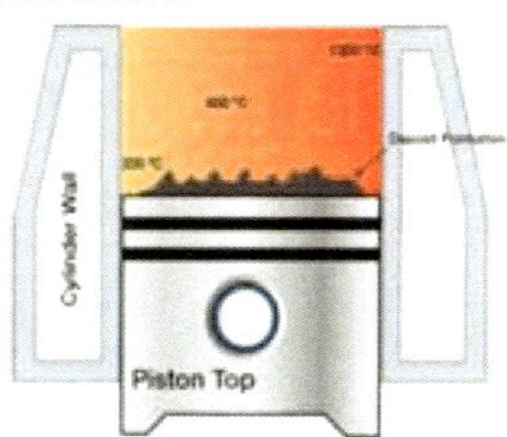

Important Terms

- **Boiling Point** - The temperature at which a liquid changes to a vapor or gas.
- **Viscosity** - The degree to which a fluid resists flow under a standard force. The more viscous a liquid, the more resistant it is to flow. Generally, viscosity decreases as the temperature of a liquid increases
- **Volatility** - A measure of how quickly a substance forms a vapor at ordinary temperatures and pressure
- **Sulfur** compounds can be corrosive to metals in fuel systems and are controlled by the total sulfur content limits found in the fuel specification

Diesel Fuel Components

Petroleum diesel, also called **Petrodiesel**,or fossil diesel is produced from the fractional distillation of crude oil between 200 °C (392 °F) and 350 °C (662 °F) at atmospheric pressure, resulting in a mixture of carbon chains that typically contain between 8 and 21 carbon atoms per molecule

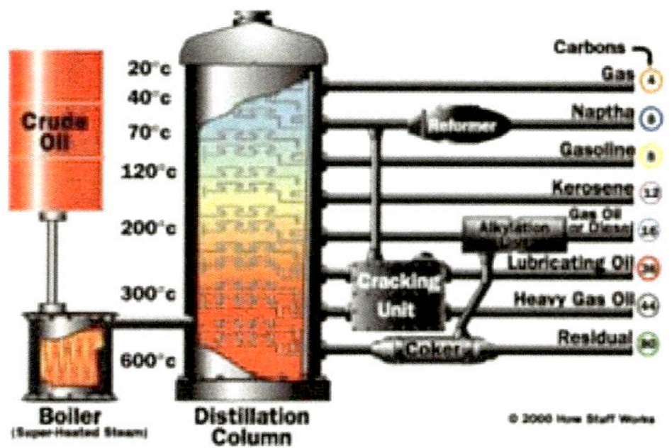

Diesel Fuel Components

The density of petroleum diesel is about 0.832 kg/l (6.943 lb/US gal), about 12% more than ethanol-free petrol (gasoline), which has a density of about 0.745 kg/l (6.217 lb/US gal)

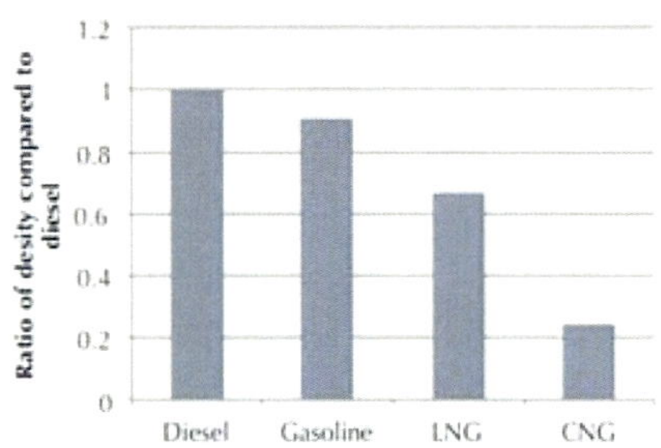

Chemical Composition

- Petroleum-derived diesel is composed of about **75%** saturated hydrocarbons and **25%** aromatic hydrocarbons
- The average chemical formula for common diesel fuel is $C_{12}H_{23}$.

Chemical Composition

The one composition requirement common to all petroleum fuels is that they consist entirely of hydrocarbon molecules (**hydrogen and carbon**) except for small amounts of impurities and/or additives

Diesel Fuel Components

Diesel is generally simpler to refine from petroleum than gasoline, and contains hydrocarbons having a boiling point in the range of 180-360°C (360-680°F)

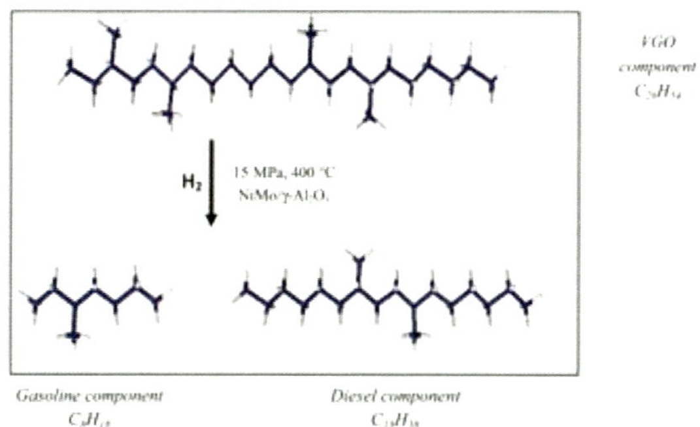

Diesel Fuel Components

Because of recent changes in fuel quality regulations, additional refining is required to remove sulfur, which contributes to a sometimes higher cost. In many parts of the United States

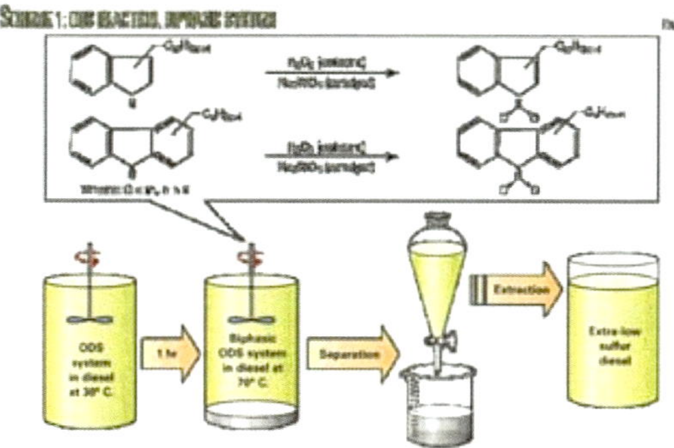

Diesel Fuel Components

Reasons for higher-priced diesel include the shutdown of some refineries in the Gulf of Mexico, diversion of mass refining capacity to gasoline production, and a recent transfer to ultra-low sulfur diesel (ULSD)

Sulfur Content

- Affects wear, deposits, and particulate emissions
- Diesel fuels contain varying amounts of various sulfur compounds which increase oil acidity

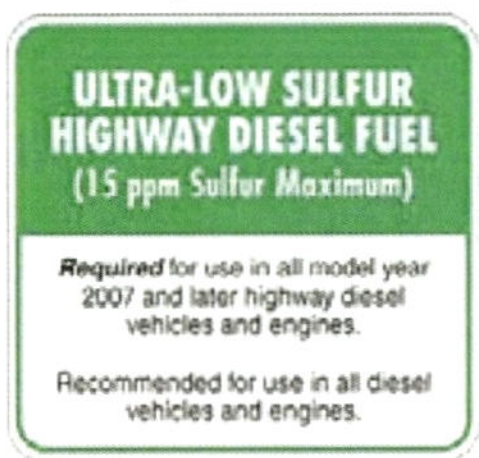

Sulfur Content

- Legislation has reduced the sulfur content of highway fuel to 0.05% by weight .
- Off road fuel has an average of 0.29% sulfur by weight
- **WHAT'S THE DIFFERENCE BETWEEN HIGHWAY AND OFF-ROAD FUEL?**
- Differ only in the tax applied at the time of sale. One is red and the other is yellow

Off Road Vs Highway Fuel

- **WHAT'S THE DIFFERENCE BETWEEN HIGHWAY AND OFF-ROAD FUEL?**
- Off road diesel is colored usually red tint and has no highway taxes in the price other than that there is no difference .
- It should be noted that if you are considering using off-road diesel in your pick-up or other on-road vehicles, DON'T .
- It is highly illegal and if you get caught there will be harsh fines and penalties .

Off Road Vs Highway Fuel

- One caveat is that is certain states like New York. The state requires you to purchase on road fuel and pay the tax up front
- Even if you use the vehicle exclusively off road. You are supposed to file for a rebate the end of the year, but since most people don't bother the state makes a fortune
- To further expound on the taxes The Federal tax on diesel is $.24 a gallon, The state tax varies by state but NY state charges 22.65 cents per gallon highway tax and 8 cents per gallon sales tax

Questions & Answers

- *Q:* **How much does a gallon of diesel fuel weigh?**
- It varies somewhat, but averages around 7.1 lb per US gallon.
- *Q:* **What is premium diesel fuel?**
- There is no definition of premium fuel, so the term really means nothing and is over used by suppliers. For a description of the characteristics proposed for "premium diesel"

Fuel Additives

- **Anti-oxidants** are primarily used to prevent gum formation in gasolines and aviation fuels.
- **Detergent additives** prevent the buildup of gum deposits in engines and extend fuel injector life. They also help keep fuel filters clean. Detergent additives are primarily found in diesel fuels and automotive gasolines.
- **Icing inhibitors** are used primarily in aviation fuels to prevent the formation of ice crystals from entrapped water in the fuel at freezing temperatures encountered during high altitude flight

BIODIESEL /BIOFUEL

Biodiesel is a cleaner burning, renewable fuel for diesel engines made from oilseed crops (like canola or soybean) or from used cooking oil and other fats

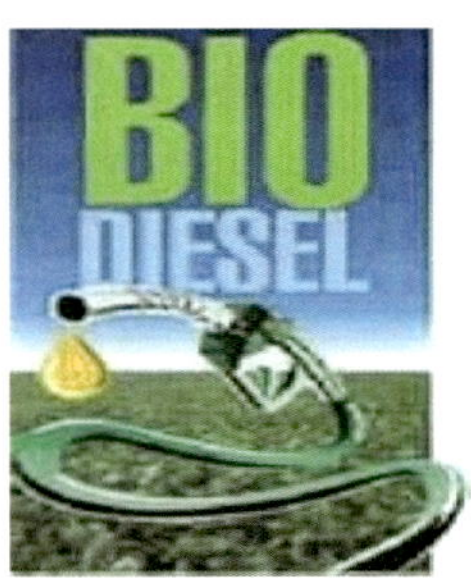

BIODIESEL /BIOFUEL

- Biodiesel has many benefits. It's simple to use, biodegradable and nontoxic.
- Has physical and chemical properties similar to petroleum diesel and can be used in most diesel applications with little or no modification to the engine or fueling system

BENEFITS OF BIODIESEL

- Biodiesel delivers the highest health and air quality benefits of any fuel.
- Biodiesel can be blended with petroleum diesel at any ratio.
- Biodiesel is a cleaner fuel than petroleum diesel .
- Biodiesel burns significantly cleaner than regular petroleum diesel and reduces polycyclic aromatic hydrocarbons and other toxic carcinogenic compounds found in diesel exhaust

What is biodiesel?

- Biodiesel is made in a chemical process called transesterification, where organically derived oils (vegetable oils, animal fats and recycled restaurant greases) are combined with alcohol (usually methanol) and chemically altered to form fatty esters such as methyl ester
- The process results in two products -- methyl esters (the chemical name for **biodiesel**) and glycerine.

What about the environment?

Biodiesel is simple to use, biodegradable, nontoxic, and essentially free of sulfur and aromatics

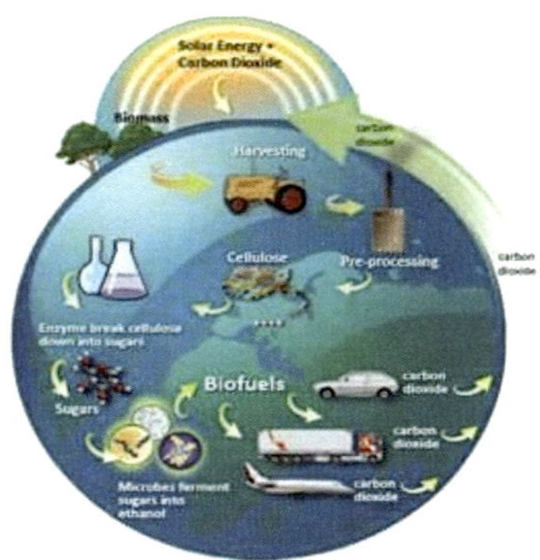

Blended Biodiesel

- **Biodiesel contains no petroleum**, but it can be blended at any level with petroleum diesel to create a biodiesel blend
- B10 =10% Biodiesel +90% Petro-Diesel
- B20 =20% Biodiesel +80% Petro-Diesel

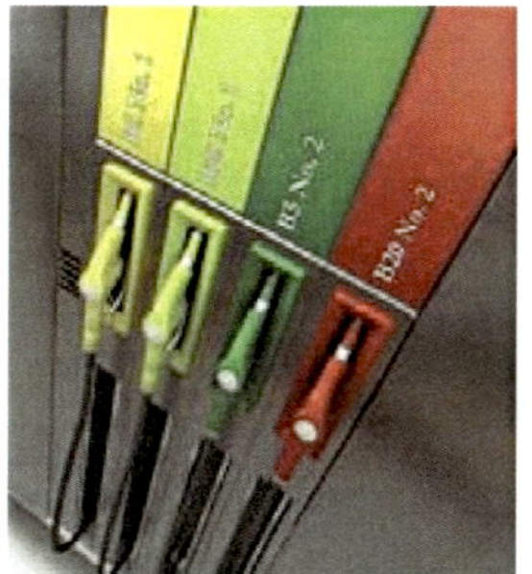

About the Emission Reduction

- When using Biodiesel, you substantially reduce the amount of harmful emissions released into the air. Emission reductions achieved by using Biodiesel:
- Carbon dioxide -78% (lifecycle)
- Carbon monoxide -48%
- Hydrocarbons -67%
- Particulates -47%
- Air toxics -60 to 90%
- Sulfates -99%
- Mutagens -89%

Engine Performance

- Biodiesel has superior lubricity to petroleum diesel. Increased lubricity enhances engine performance and can prolong engine life and decrease fleet operating costs .
- Biodiesel has a higher cetane rating (47-70) than petroleum diesel (42-44). **Biodiesel's high cetane rating results in a more complete combustion of the fuel .**
- This increased cetane also aids in self-ignition of the fuel for easier starting, smoother running engine performance and quieter operation .

How is biodiesel made?

Biodiesel is made through a chemical process called transesterification whereby the glycerin is separated from the fat or vegetable oil

How is biodiesel made?

The process leaves behind two products -- methyl esters (the chemical name for biodiesel) and glycerin (a valuable byproduct usually sold to be used in soaps and other products)

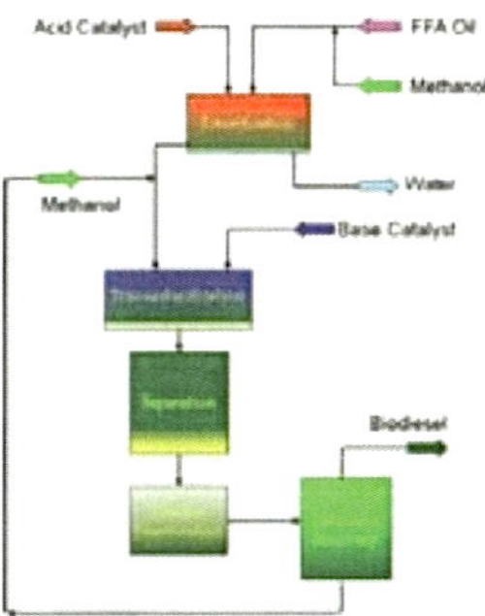

Is Biodiesel the same thing as raw vegetable oil?

- No! Fuel-grade biodiesel must be produced to strict industry specifications (ASTM D6751) in order to insure proper performance.
- Biodiesel that meets ASTM D6751 and is legally registered with the Environmental Protection Agency is a legal motor fuel for sale and distribution.
- **Raw vegetable oil cannot meet biodiesel fuel specifications**, it is not registered with the EPA, and it is not a legal motor fuel.

Why should I use biodiesel?

245

Make your own biodiesel

- Anybody can make biodiesel. It's easy, you can make it in your kitchen -- and it's better fuel than the petro-diesel the oil companies sell you
- Your diesel motor will run better and last longer on your home-made fuel, and it's much cleaner -- better for the environment and better for health

Make your own biodiesel

- If you make it from used cooking oil it's not only cheap but you'll be recycling a troublesome waste product that too often ends up in sewers and landfills instead of being recycled.

Three choices

- There are at least three ways to run a diesel engine on vegetable oil.
  - Mix it with petroleum diesel fuel, or with a solvent.
  - Use the oil just as it is -- usually called SVO fuel (straight vegetable oil) or PPO fuel (pure plant oil)
  - Convert it to biodiesel
- The first two methods sound easiest, but, as so often in life, it's not quite that simple.

Vegetable Oil

Vegetable oil is much more viscous (thicker) than either petro-diesel or biodiesel

Mixing it

- Vegetable oil is much more viscous (thicker) than either petro-diesel or biodiesel
- The purpose of mixing or blending straight vegetable oil (SVO) with other fuels and solvents is to lower the viscosity to make it thinner, so that it flows more freely through the fuel system into the combustion chamber
- If you're mixing SVO with petro-diesel you're still using fossil-fuel -- cleaner than most, but still not clean enough, many would say

Mixing it

- Still, for every gallon of SVO you use, that's one gallon of fossil-fuel saved, and that much less climate-changing carbon dioxide in the atmosphere
- People use various mixes, ranging from 10% SVO and 90% petro-diesel to 90% SVO and 10% petro-diesel
- Some people just use it that way, start up and go, without pre-heating it (which makes veg-oil much thinner)

Mixing it

- You might get away with it in summer time with something like an older '80s Mercedes 5-cylinder IDI diesel, which is a very tough and tolerant motor -- it won't like it but you probably won't wreck it. Otherwise, it's not wise.
- To do it properly and safely you'll need what amounts to a proper SVO system with at least fuel pre-heating. In which case there's no need for mixes, you can just use 100% SVO.

Blends with solvents

- Blends of SVO with various solvents, such as mineral turpentine (white spirit), or with various "secret" ingredients, such as naphthalene (mothballs) and xylol (paint-stripper), or with unleaded gasoline, are experimental at best .
- little or nothing is known about the effects of these additives on the combustion characteristics of the fuel or their long-term effects on the engine. Not recommended -- use such blends at your own risk.

Oil Viscosity

- Higher viscosity is not the only problem with using vegetable oil as fuel. Veg-oil has different chemical properties and combustion characteristics from the petro-diesel fuel that diesel engines and their fuel systems are designed to use
- Diesel engines, especially the more modern, cleaner-burning diesels, are high-tech machines with precise fuel requirements. They're tough, but they'll only take so much abuse

Mixing it

- There's no guarantee of it, but using a blend of up to 20% veg-oil of good quality with 80% petro-diesel is said to be safe enough for older diesels, especially in summer
- Otherwise using vegetable oil as fuel requires a professional SVO solution -- or convert the veg-oil biodiesel
- Mixes and blends are generally a poor compromise. But mixes can have an advantage in cold weather: as with biodiesel, some winterised petro-diesel mixed with straight vegetable oil lowers the temperature at which the SVO starts to gel

Straight vegetable oil

- Straight vegetable oil fuel (SVO) systems can be a clean, effective and economical option.
- Unlike biodiesel, which runs in any diesel without modification, you have to modify the engine to use SVO.
- The best way is to fit a professional single-tank SVO system with replacement injectors and glow-plugs optimized for vegetable oil, as well as fuel heating
- With the Elsbett single-tank SVO system you can use petro-diesel, biodiesel or SVO, in any combination

Recommendations

- There are also two-tank SVO systems which pre-heat the oil to make it thinner.
- You have to start the engine on ordinary petro-diesel (or biodiesel) in one tank and then switch to SVO in the other tank when the veg-oil is hot enough (i.e. thin enough), and switch back again to the petro-diesel tank before you stop the engine, or you'll coke up the injectors

Biodiesel or SVO?

- Biodiesel has some clear advantages over SVO
  - It works in any diesel engine, without any conversion or modifications to the engine or the fuel system -- just put it in and go.
  - It also has better cold-weather properties than SVO (but not as good as petro-diesel
  - Unlike SVO, it's backed by many long-term tests in many countries, including millions of miles on the road

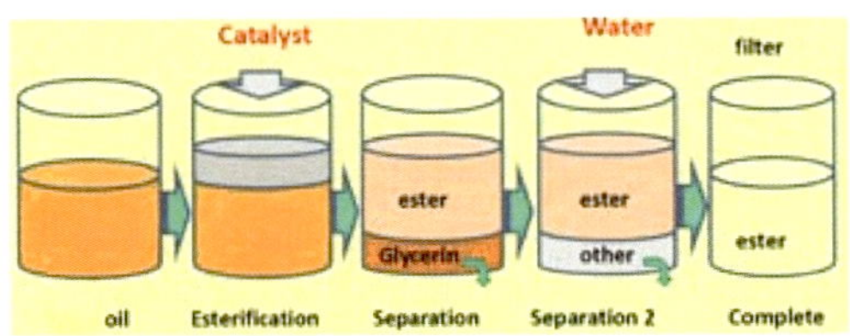

How to Make your own Biodiesel

Biodiesel is a clean, safe, ready-to-use, alternative fuel, whereas it's fair to say that many SVO systems are still experimental and need further development.

Biodiesel or SVO?

- Biodiesel can be more expensive, depending on how much you make, what you make it from and whether you're comparing it with new oil or used oil. And unlike SVO, it has to be processed first

- But the large and rapidly growing worldwide band of biodiesel homebrewers don't mind that -- they make a supply every week or once a month and they soon get used to it. Many have been doing it for years

Biodiesel or SVO?

- Anyway you have to process SVO too, especially WVO (waste vegetable oil, also called UCO, used cooking oil), which many people with SVO systems use because it's cheap or free for the taking.
- With WVO, food particles and impurities and water must be removed, and it probably should be deacidified too.
- Biodieselers say, "If I'm going to have to do all that I might as well make biodiesel instead." But SVO types scoff at that -- it's much less processing than making biodiesel, they say.

Where do I start?

- Start with the **process**, **NOT** with the processor. The processor comes later.
- Start with **fresh unused oil**, **NOT** with waste vegetable oil (WVO), that also comes later.
- Start by making a small, 1-litre test batch of biodiesel using fresh new oil. You can use a spare blender, or, better, make a simple **Test-batch mini-processor**.
- Once you've mastered small test batches with new oil that pass the quality control checks provided, you'll learn how to make test batches with used oil (WVO) that also pass the quality checks

The Process

- Biodiesel is made from vegetable and animal oils and fats, or **triglycerides**. Biodiesel cannot be made from any other kinds of oil (such as used engine oil).
- Chemically, triglycerides consist of three long-chain fatty acid molecules joined by a glycerine molecule.

The Process

- The biodiesel process uses a catalyst (lye) to break off the glycerine molecule and combine each of the three fatty-acid chains with a molecule of methanol, creating mono-alkyl esters, or Fatty Acid Methyl Esters (FAME) – biodiesel
- The glycerine sinks to the bottom and is removed

What is (LYE)

Lye, also known as Sodium Hydroxide, NaOH, and caustic soda , is the third ingredient used to make biodiesel

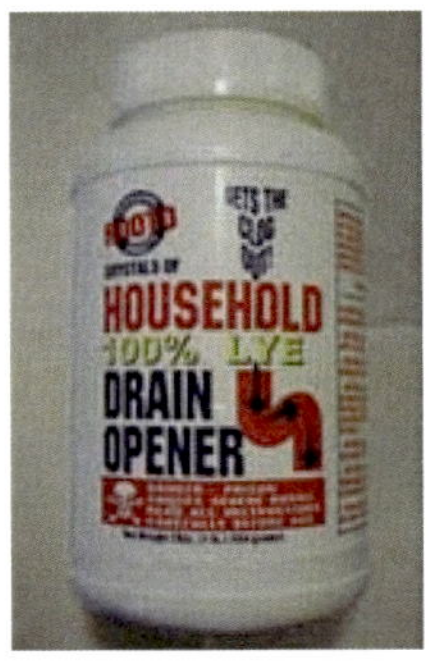

What is (LYE)

Look for it at plumbing supply houses or from chemical suppliers on the internet

LYE - [K(OH) or Na(OH)]

Accurately measuring the appropriate amount of lye is critical to a successful biodiesel reaction. Having a measurement that is off as few as a couple grams can make the difference between success and failure

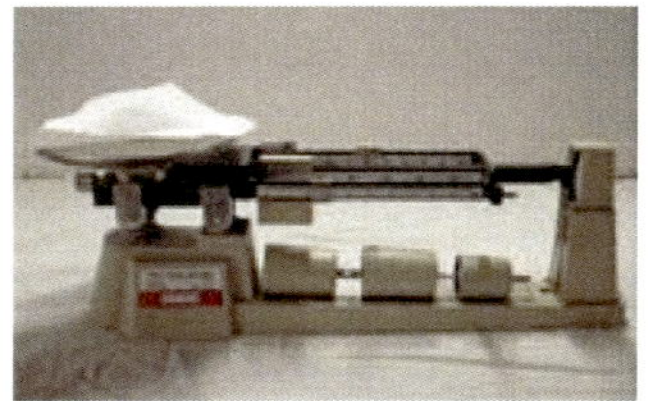

Methanol

You can buy methanol from a local race shop. Methanol is extremely volatile and flammable. Be sure to wear heavy duty synthetic rubber gloves and use an approved respirator when working with methanol

The Procedure

- Vegetable oil (500 ml) + Temp= Oil @ 110 F.
- After the oil is heated, pour it into the mixing bucket. The bucket must be completely dry and free of any residue. The residue of any substance left behind can upset the delicate reaction and ruin the batch of biodiesel.

The Procedure

- Methanol (100 ml) + Na(OH)(3.5 gms) = Sodium Methoxide (Shake vigorously).
- Adding Sodium Methoxide to the Oil in the Mixing Bucket
- Any undissolved lye crystals can upset the reaction

The Procedure

Finally, all of the sodium methoxide has been added to the oil and it is a rich chestnut color

The First Minute of the Mixing Process

- As you can see the first minute of the reaction looks like a muddy, cloudy-looking mixture
- What's happening is that the mixture is thickening slightly just before the main chemical reaction starts to take place, as the glycerin begins to separate out from the vegetable oil

Final Step

- Set the bucket aside, place a lid on it and let it stand overnight. It will take at least 12 hours for the glycerin to settle out.
- The glycerine sinks to the bottom and is removed

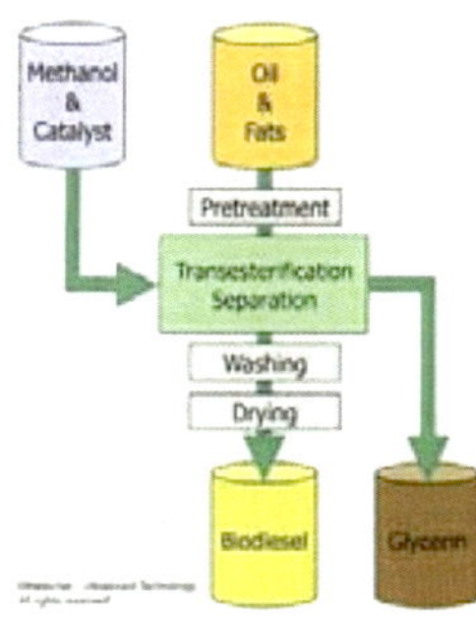

Costs and prices

- Most people in the US use about 500 gallons of fuel a year (about 10 gallons a week), costing about US$1,300 a year at the fuel pump (Nov 2013 prices).
- Biodiesel homebrewers using waste vegetable oil as feedstock make biodiesel for 50 cents to US$1 per US gallon, so their 500 gallons a year costs them $250-500, while a good processing system can be set up for around $100 and up.

Costs and prices

- An SVO system costs from about $500 to $1,200 or more. So, if the vegetable oil is free, with an SVO system you'll probably be saving on fossil-fuel prices within a year, not a long time in the life of a diesel engine..
- But you'll probably still be saving less than the biodieselers.
- Will the engine last as long with SVO? Yes, if you use a good system

Recommendations

- If you plan to use or store Biodiesel, first clean the fuel system, including fuel tanks, where sediments or deposits may occur .

- Once you've started using Biodiesel, make sure to monitor both your vehicle and dispensing filters and change them as needed until the sediment build-up is eliminated .

Recommendations

- Oil changes. Biodiesel may make its way past the piston rings and into the oil pan. This is due to the slightly higher viscosity and density of biodiesel compared to petroleum diesel .
- High levels of biodiesel present in the engine oil may polymerize over time and cause some engine oil sludge.
- High levels of biodiesel present in the engine oil may polymerize over time and cause some engine oil sludge..

Recommendations

- Do not use raw vegetable oil in a Modern diesel engine (With electronic Fuel Injection).
- Fats and oils (triglycerides) are much more viscous than biodiesel, and low-level vegetable oil blends can cause long-term engine deposits, ring sticking, lube oil gelling, and other maintenance problems that can reduce engine life.

Chapter 13
Injection System

Types of Fuel Injection Systems

- **High Pressure Injection Systems**
  - Individual Jerk Pumps
  - Multi-Plunger Inline Pumps
  - Distributor Pump Injection

Fuel Metering and Timing is performed inside the Inj. Pump

Deliver Pressures Over 17000 PSI

- **Low Pressure Injection Systems**
  - Pressure/Time Inj. System
  - Unit Injectors

A Rocker arm activated plunger then raises this low supply fuel between 2000-3000 PSI

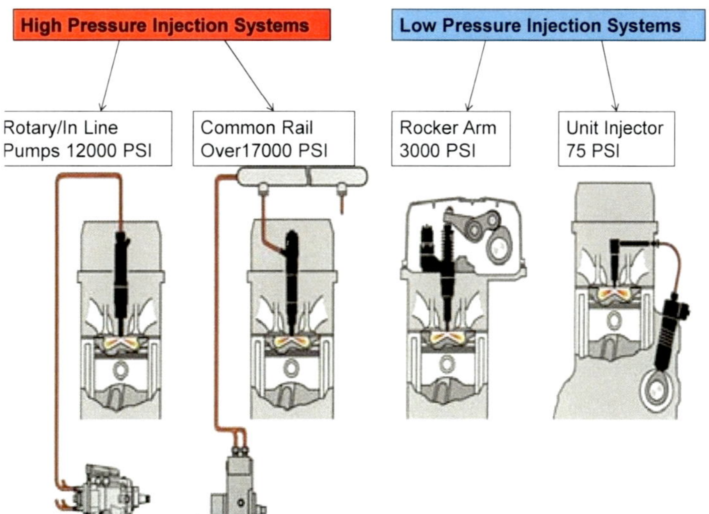

Fuel Injection Systems

- All Diesel fuel Injection Systems must perform the following functions
  - Accurately meter fuel to the injectors
  - Time the injection at the proper point in the combustion cycle
  - Control the rate of injection
  - Properly atomize the fuel
  - Start and Stop injection quickly
  - Generate the proper injection pressure

Diesel Fuel Injectors

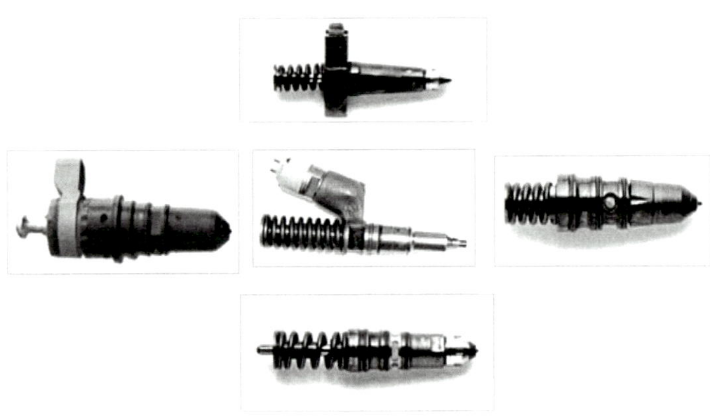

Injector Parts

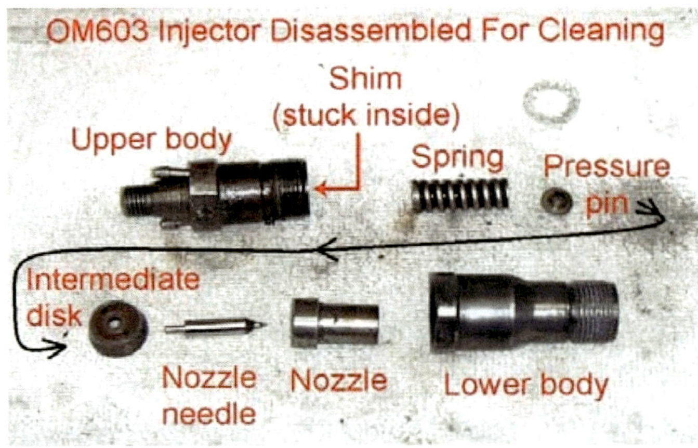

Mechanical Fuel Injectors

- The main purpose of the diesel injector is to deliver a fine mist of diesel regardless of the temperature and pressure inside the cylinder

Mechanical Fuel Injectors

They are designed to operate under pressures ranging between 1700 and 2200 pounds per square inch (psi) (For Mechanical Fuel Inj. Systems)

Electronic Fuel Injectors

They are designed to operate under pressures ranging between 18000 and 25000 pounds per square inch (psi) (For Common Rail Systems)

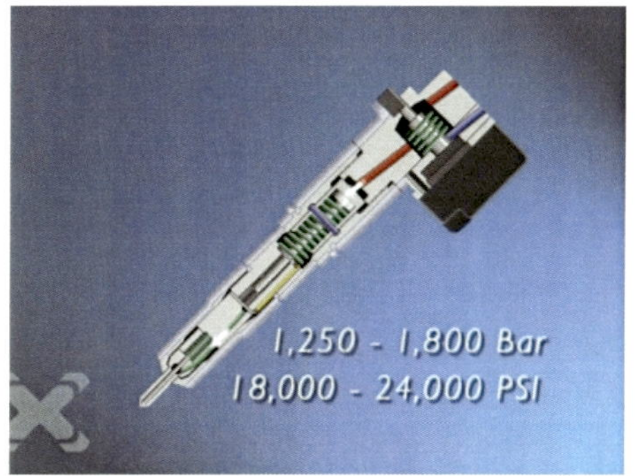

Diesel Injectors

- There is one fuel injector for every cylinder in your diesel engine, threaded into a special bore just like the regular spark plugs in a fuel gasoline engine

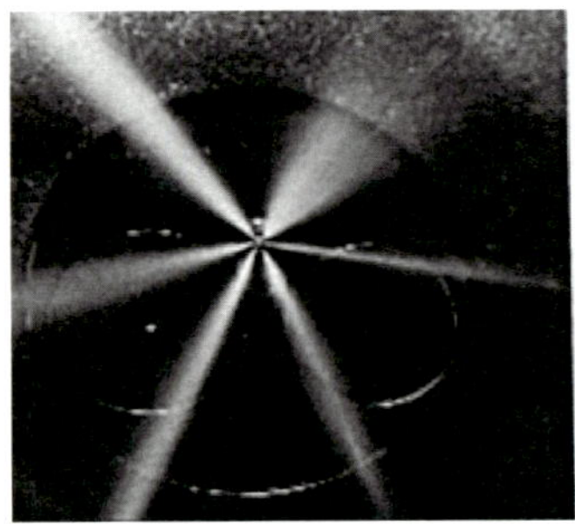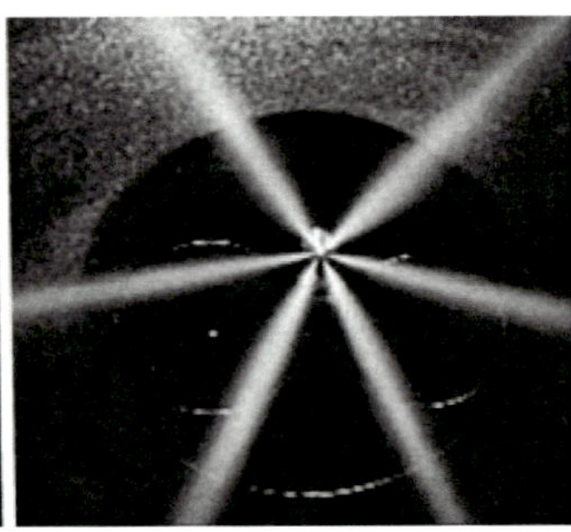

Diesel Injectors

- If you could make a transversal cut through a common diesel injector, you will find a thin spring right in the middle of the injector's body
- Under this spring, you will see a thin needle extending to the bottom of the body and blocking the nozzle at the bottom.

Diesel Injectors

- The fuel passes through a drilling in the nozzle body, to a chamber above where the needle-valve seats in the nozzle assembly

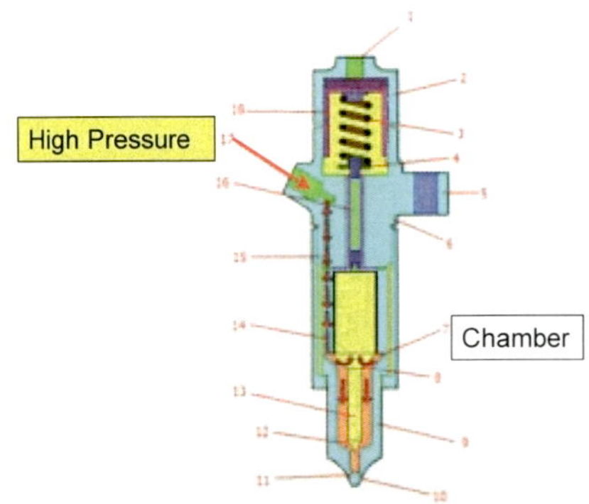

Diesel Injectors

As fuel pressure in the injector gallery rises, it acts on the tapered shoulder of the needle valve, increasing the pressure until it overcomes the force from the spring, and lifts the needle valve from its seat

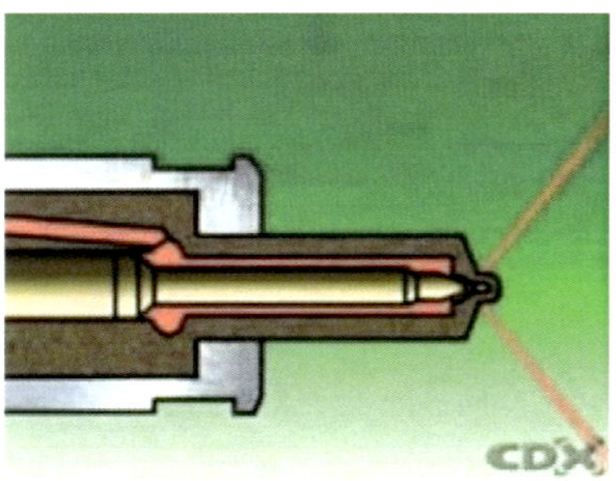

Nozzle – Fuel Inlet

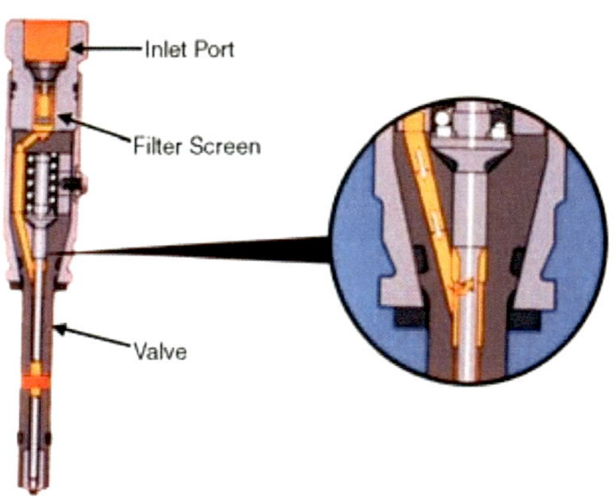

Nozzle – Valve Opening Pressure and Injection

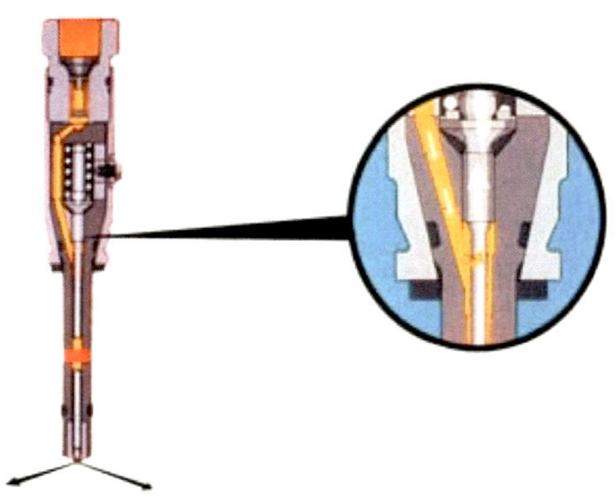

End of Injection

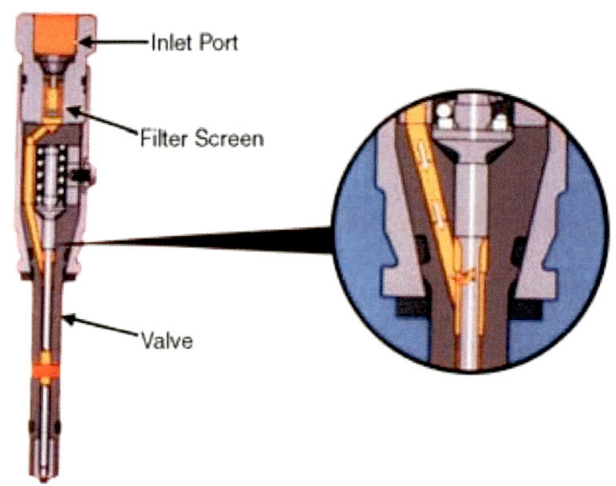

Pencil Type Injector

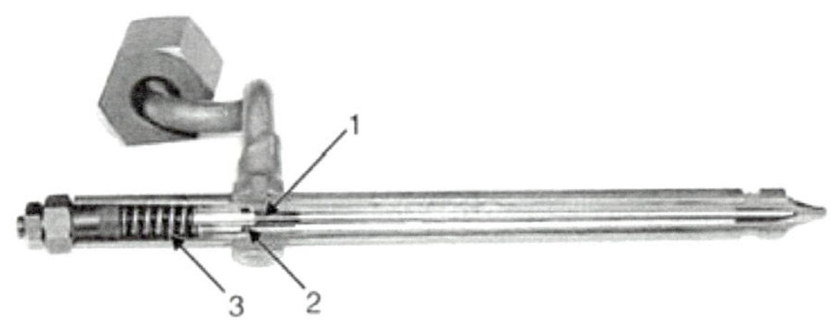

1. Nozzle Volume Area
2. Valve Check and Guide
3. Valve Opening Spring

Electronic Fuel Injector

Instead of an spring, a solenoid control the movement of the needle. The power for the solenoid comes from the ECM

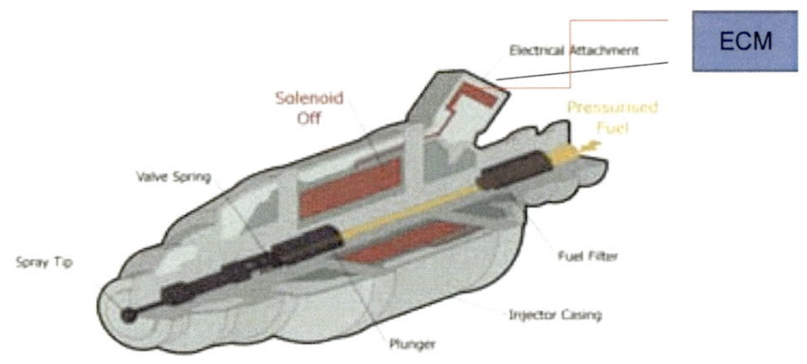

262

Fuel Injection Nozzles

787

Types of injector nozzle

- There are two main types of injector nozzle, **hole (open)** and **pintle (close)**

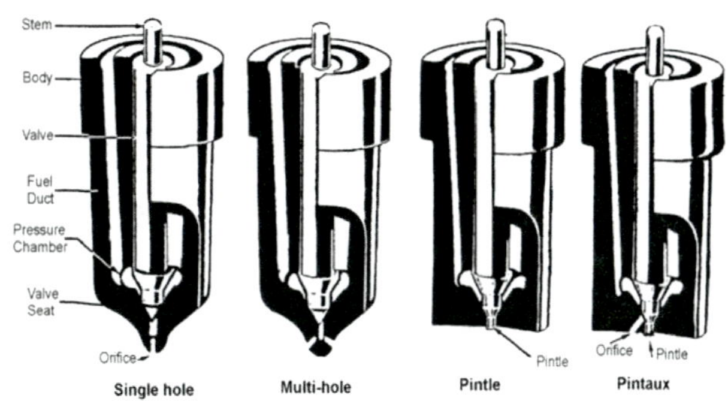

Hole-Type nozzle (High Press)

Hole-type nozzles give a hard spray, which is necessary to penetrate the highly compressed air. The fuel has a high velocity and good atomization which is desirable in open combustion chamber engines

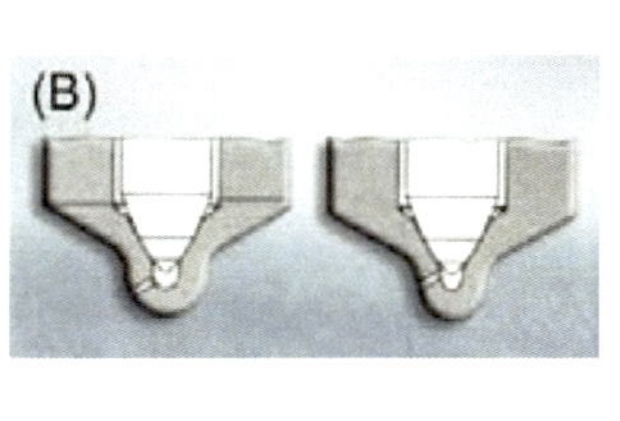

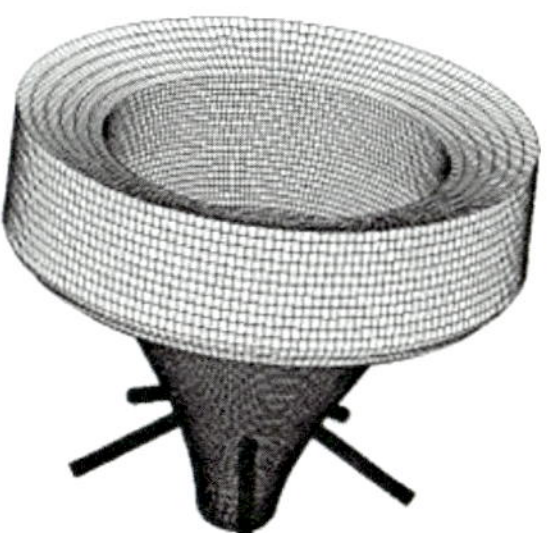

263

Hole-Type nozzle

- Commonly used in Direct Injection systems and open combustion chamber diesel engines
- A hole nozzle has one or more spray holes or orifices to provide good atomization into the combustion chamber

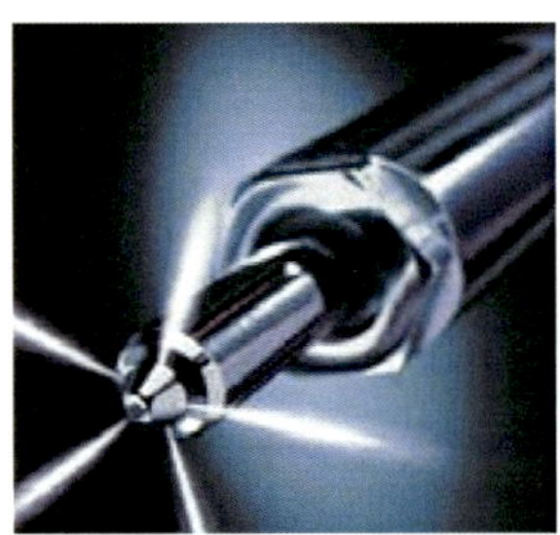

Pintle-Type nozzle (Low Press)

In **pintle**-type nozzles, a pin, or pintle, protrudes through a spray hole. The shape of the pintle determines the shape of the spray, and the atomization of the spray pattern. **Pintle nozzles open at lower pressures than hole-type nozzles**

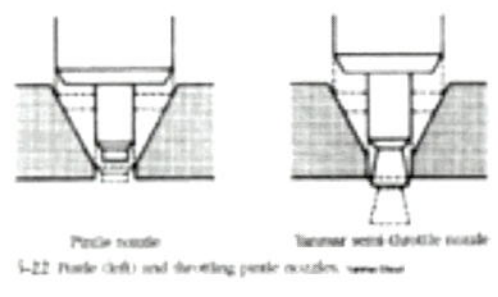

5-22 Pintle (left) and throttling pintle nozzles.

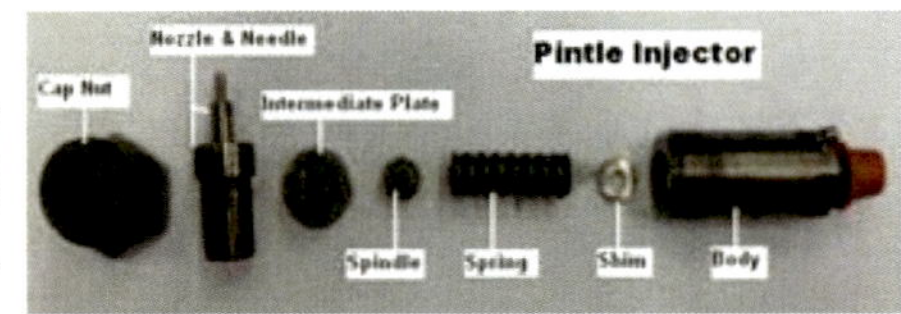

Pintle-Type nozzle

Are used in small bore , high speed engines

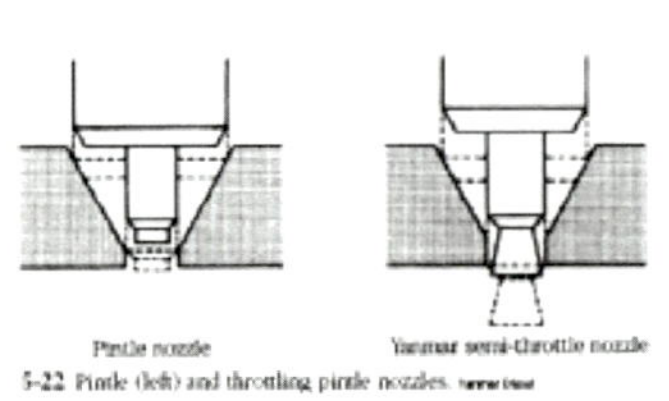

5-22 Pintle (left) and throttling pintle nozzles.

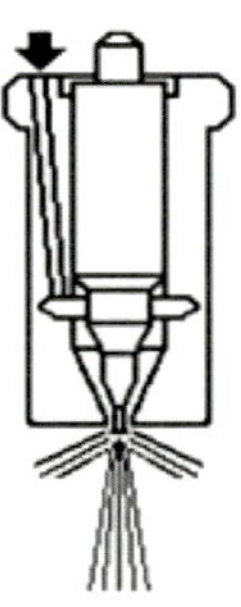

Pintle-Type nozzle

Are used in engines having pre-combustion, divided, air cell or energy-cell combustion chambers

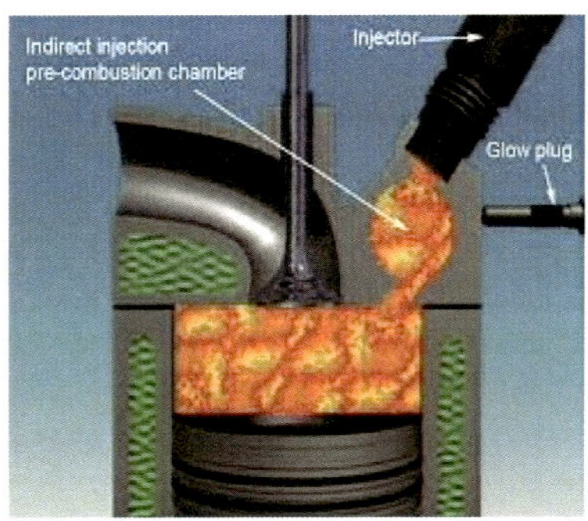

Spray Patterns

Spray cones are precisely calculated to produce maximum fuel efficiency

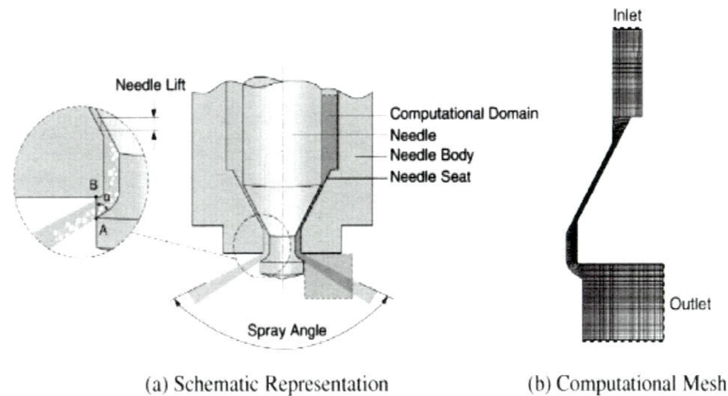

(a) Schematic Representation   (b) Computational Mesh

Servicing Fuel Injectors

Due to poor servicing the length of the spray cone would increase and the spray angle would decrease

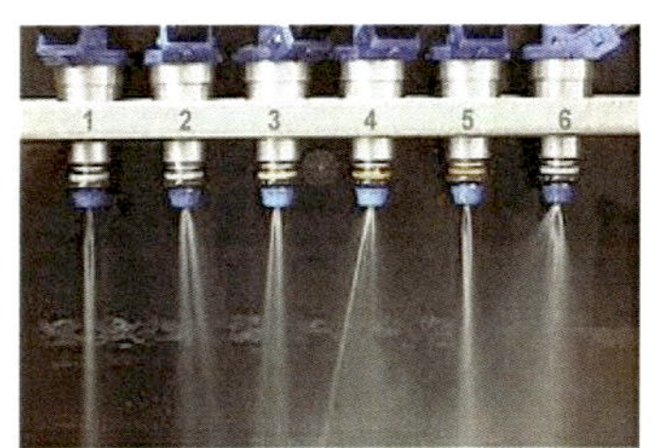

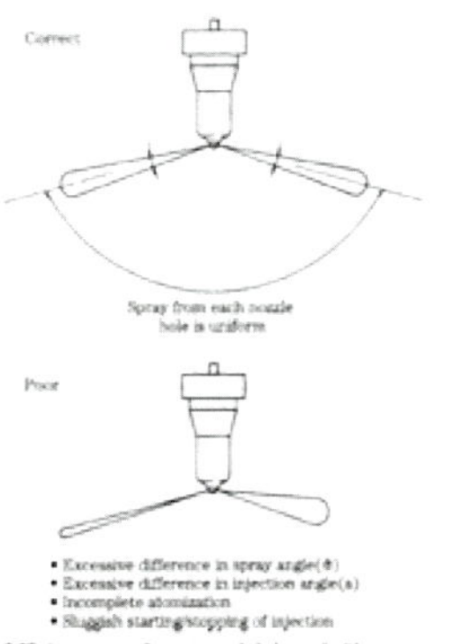

Servicing Fuel Injectors

The service includes a complete cleaning and inspection at manufacturer recommended intervals usually every 600-1000 Hrs

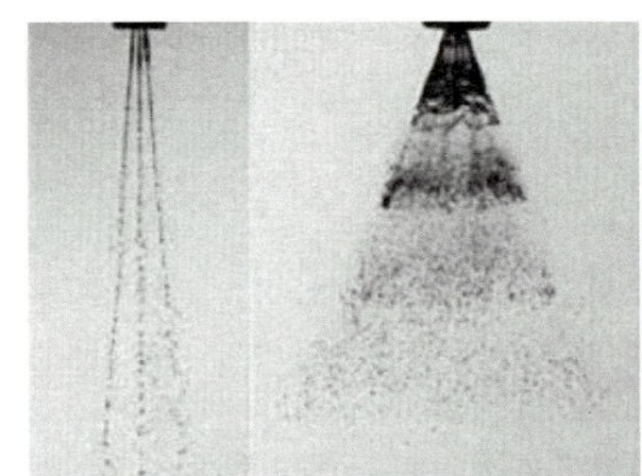

Diesel Injectors

The nozzle is exposed to high-combustion temperatures and combustion by-products

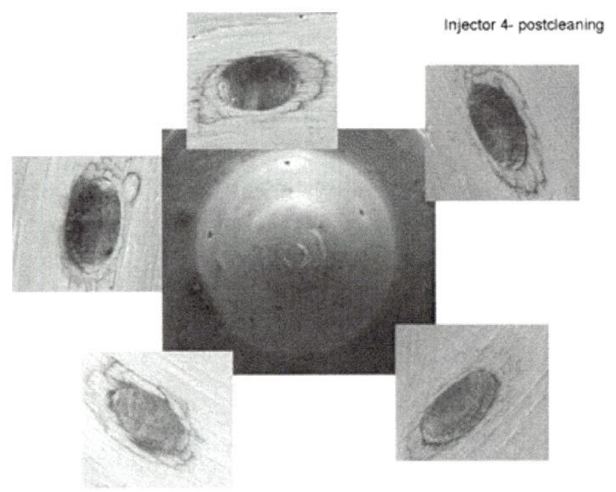

Tip Erosion by Dirty Fuel

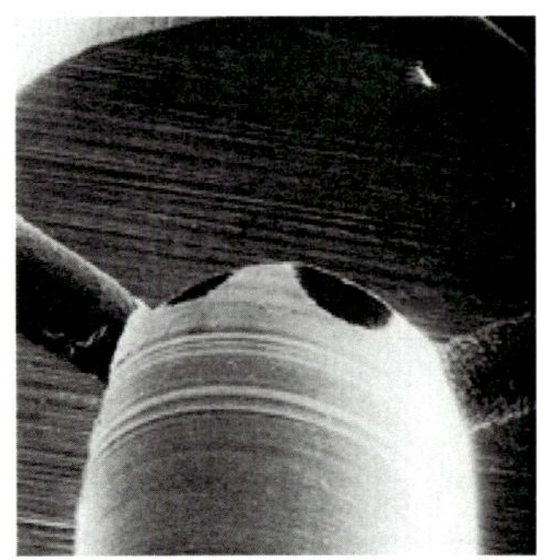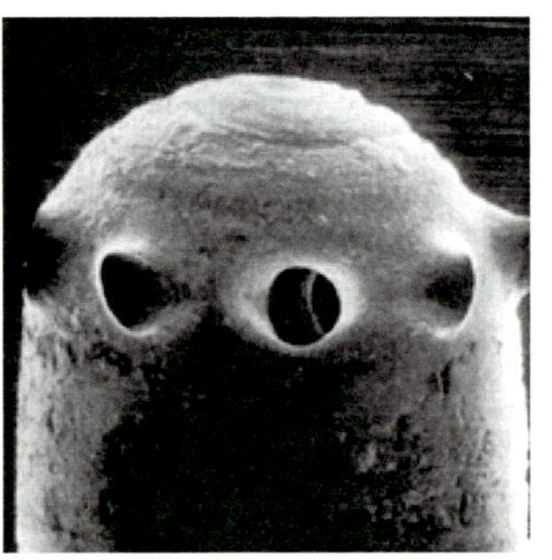

Diesel Injectors

If not serviced periodically, your diesel fuel injectors will cause your engine to miss--fail to feed fuel to the combustion chamber--once the small passages become clogged

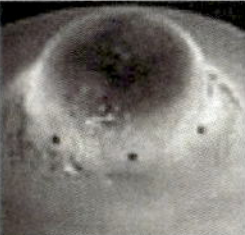

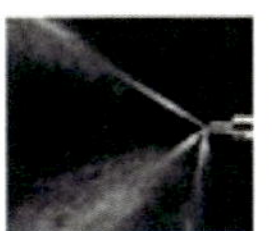

Diesel Injectors

Diesel fuel injectors are engineered to react to specific opening fuel pressures. It is a delicate balance between the electrical fuel pump on one end and the fuel injector at the other end to keep the fuel injection system working properly

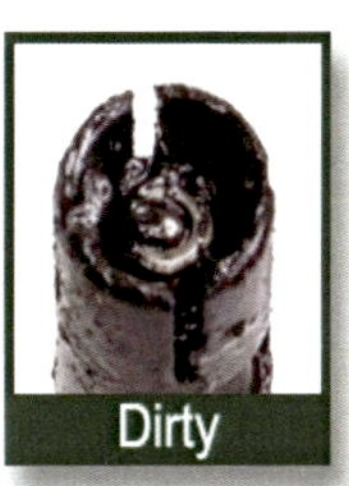

Inspect the Fuel

Injector Testing

Use gloves and wear safety glasses

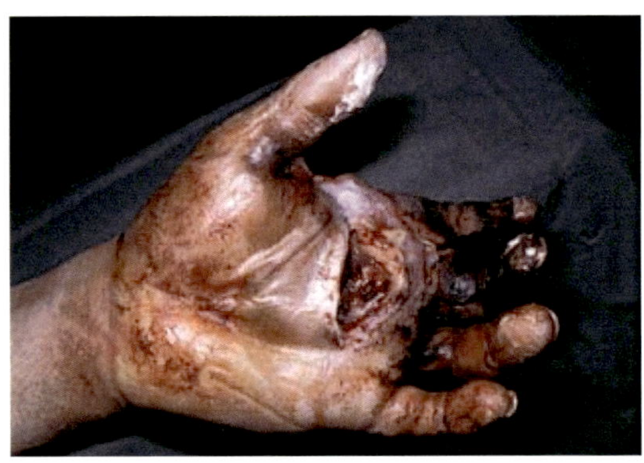

The Bottle Jack Testes

Many of the commercial injector testers you can buy seem to come with a 5000PSI (350 bar) gauge

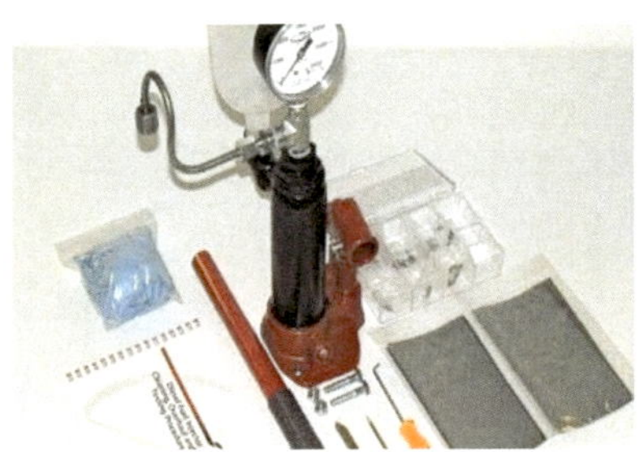

The Bottle Jack Tester

A two ton bottle jack can be converted into a good injector tester

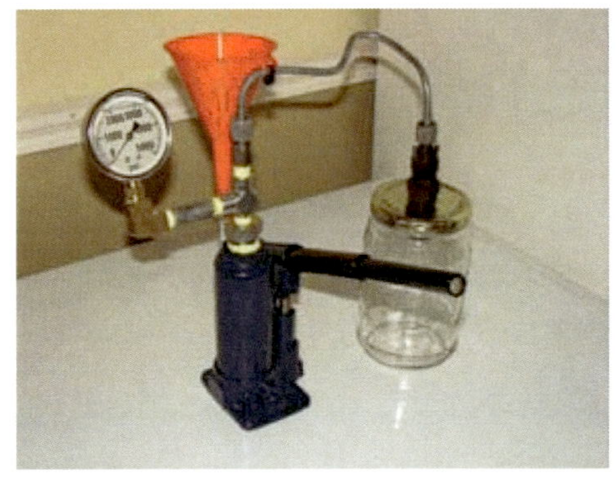

The Bottle Jack Tester

The first step is to disassemble the bottle
jack by loosening the giant nut on the top

The Bottle Jack Tester

The giant nut from the top of the jack contains
an o-ring that you should remove. Then you
must somehow attach a threaded fitting to the
top of the jack nut

The bottle Jack Tester

- Take the outer cylinder of the jack and attach
  some sort of supply line where you can add
  diesel fuel

The bottle Jack Tester

To make a fluid reservoir, You can use an old food can. Inside the can, the PEX fitting extends up about 3/8" and I attached a piece of cloth to the top of it

The bottle Jack Tester

- The valve on the tester was intended for disconnecting the gauge during rapid pumping so it doesn't get damaged by banging around with wild pressure changes

The bottle Jack Tester

- You should also prepare a jar into which the injectors can spray. I drilled out the lid of a tall glass food jar so that it would thread onto the injector

Installation

Valve Opening Pressure (VOP) Test

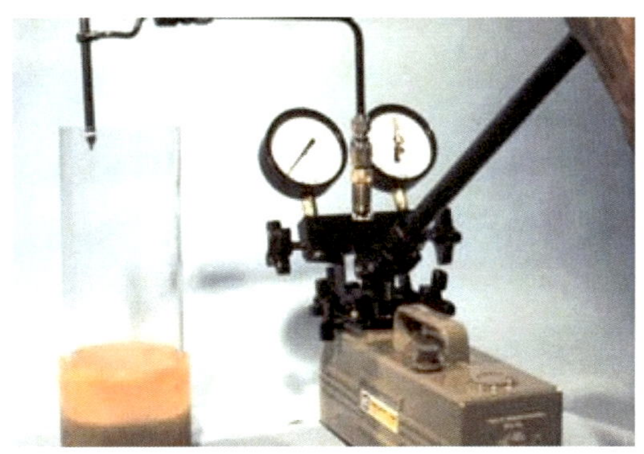

Flush the Nozzle

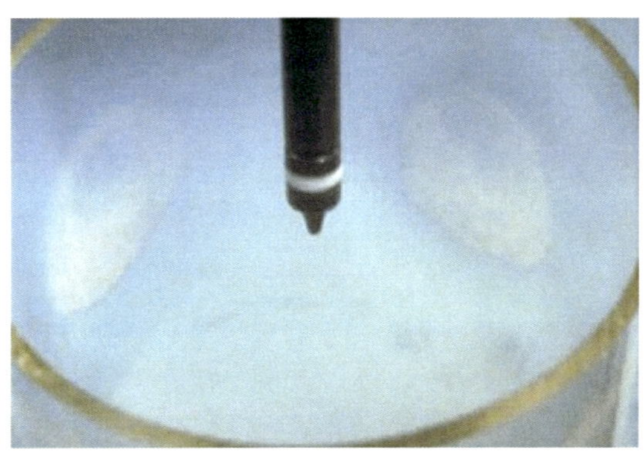

Tip Leakage Test

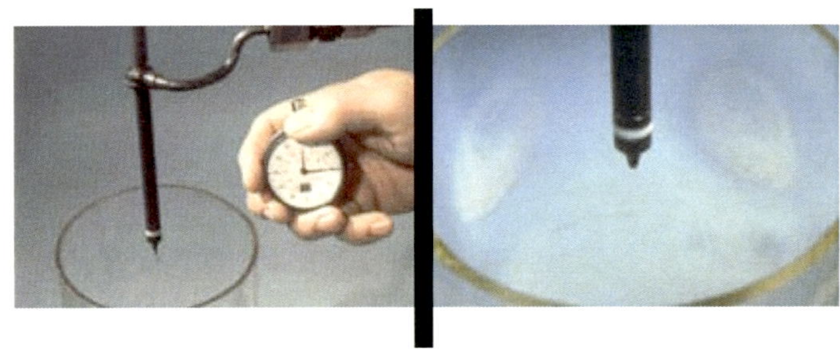

Microscope

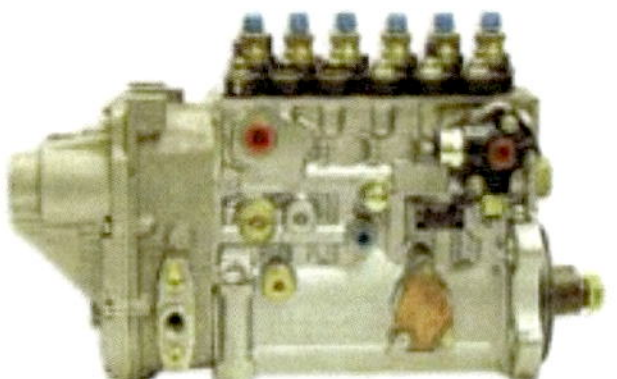

Diesel Fuel Injection Pump

Diesel Injection Pump

An **Injection Pump** is the device that pumps fuel into the cylinders of a diesel engine

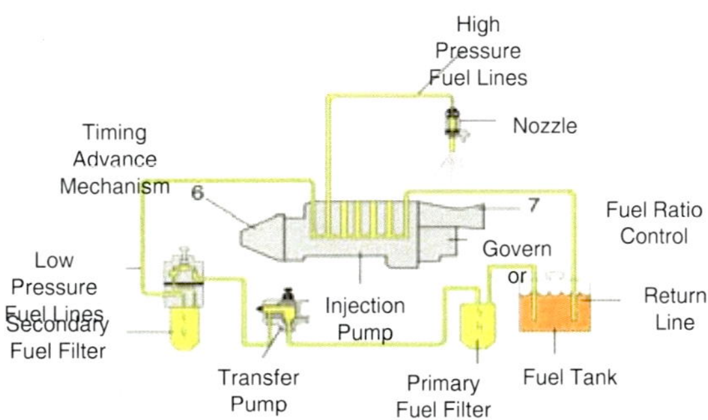

Diesel Injection Pump

An **Injection Pump** is the device that pumps fuel into the cylinders of a diesel engine.

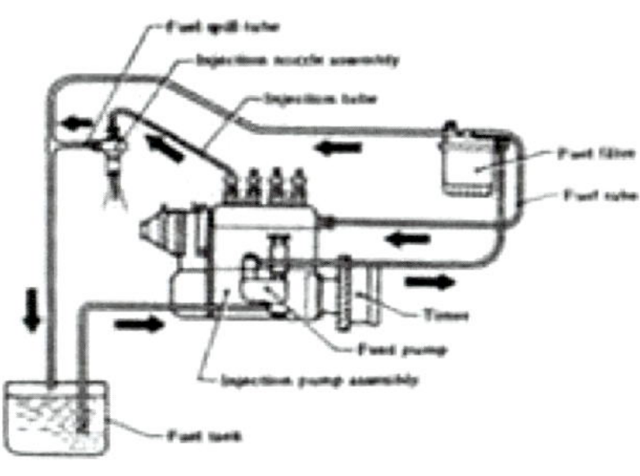

Fuel Injection Pumps

- The purpose of the fuel injection pump is to deliver an exact metered amount of fuel, under high pressure, at the right time to the injector
- There are many different features on an injection pump that provide certain characteristics on the engine some of which are
  - The aneroid which limits black smoke at low engine speeds on turbocharged engines
  - The governor which controls engine speed
  - The torque control which accurately matches fuel output with engine requirements.

Fuel Injection Pumps

- The injection pump is the heart of the diesel engine
- It is responsible for delivering the fuel to the engine and for controlling the speed of the engine (governing)
- The injection pump does the job of both the throttle and the ignition system needed in gasoline engines
- For many home mechanics the diesel injection pump is a bit of a mystery

Diesel Injection Pump

The injection pump does the job of both the throttle and the ignition system needed in gasoline engines

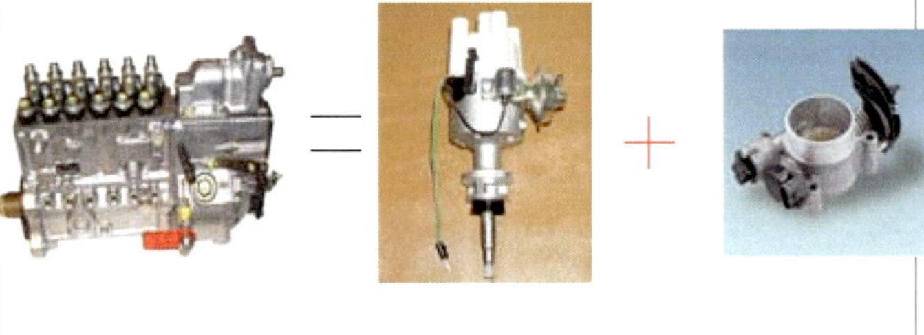

Injection Pump Functions

- **Timing** The timing is adjusted in response to engine RPM
- At higher RPM s, the fuel pressure from the vane transfer pump is higher. Pressure changes effects a spring loaded plunger, and the resulting movement will move the cam rollers to either advance or retard the timing
- There is also a cold start device which advances the idle timing manually

274

Injection Pump Functions

- **Governor** A mechanical governor limits the maximum speed of the engine to 2200 RPM in a regular inboard Boat
- **Stop** A magnet valve or solenoid opens and shuts off the fuel channel between the feed pump and the metering pump
- **Aneroid** An air inlet pressure sensor is used to determine maximum amount of fuel delivered on injection pumps for turbo engines

In Line Fuel Injection Pumps

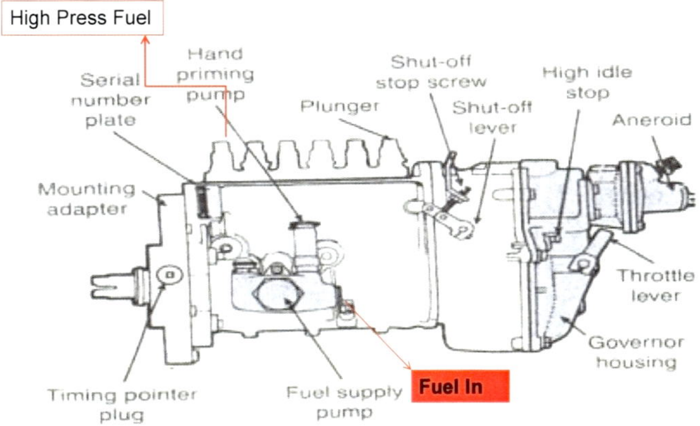

Figure 20-2. Components of a typical inline injection pump assembly. (Navistar International Transportation Corp.)

Distributor Fuel Injection Pump

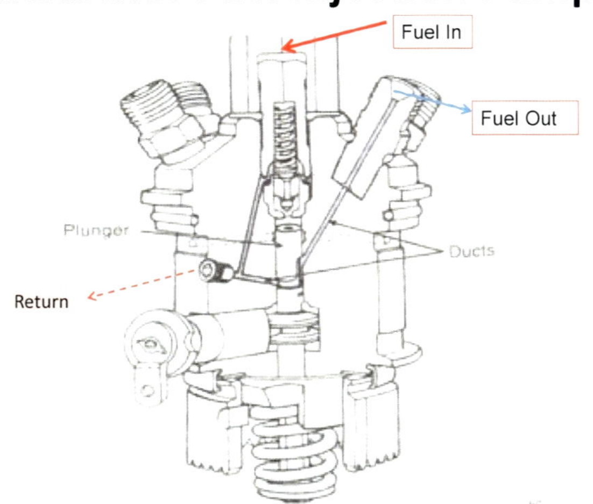

Mechanical Injection

Diesel engines make use of a camshaft, rotating at half crankshaft speed, lifted mechanical single plunger high-pressure fuel pump driven by the engine crankshaft

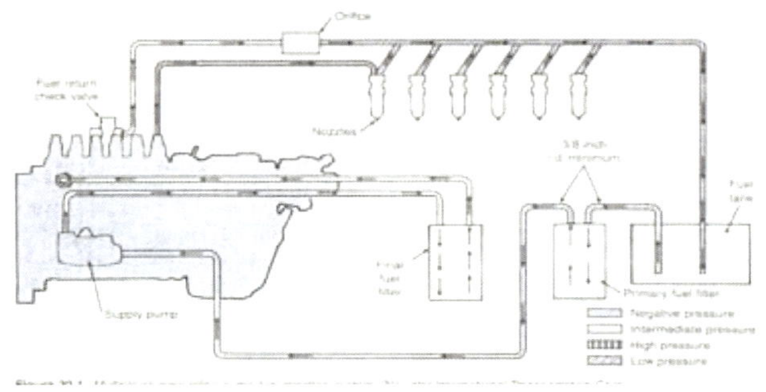

Diesel Injection Pump

Still, the pump has the ability to vary timing, although not to the sophistication of an electronically controlled system. Inline injection pumps look like mini inline engines

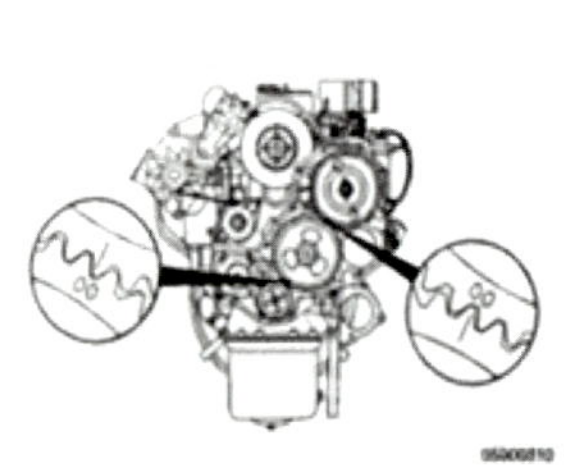

Fuel Injection Pump Camshaft

Both engine camshaft and Injection Pump camshaft have the same configuration according with the angle lobes . They work in parallel

276

Internal Construction

Earlier diesel pumps used an in-line layout with a series of cam-operated injection cylinders in a line, rather like a miniature inline engine

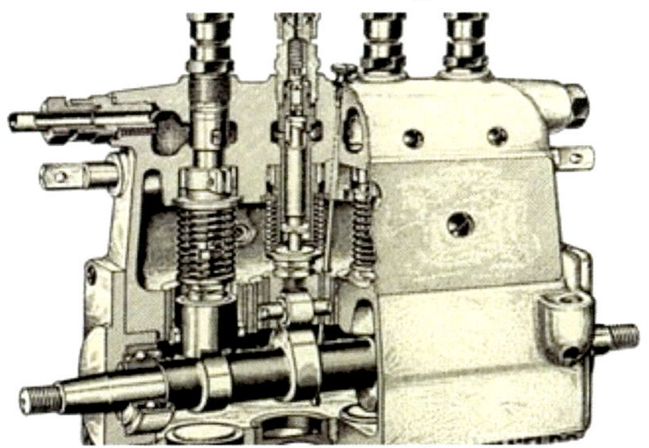

Internal Construction

The pistons have a constant stroke volume, and injection volume (ie, throttling) is controlled by rotating the cylinders against a cut-off port that aligns with a helical slot in the cylinder

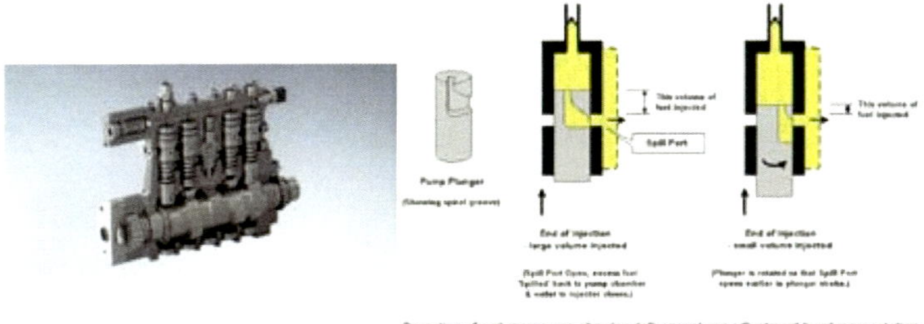

Internal Construction

When all the cylinders are rotated at once, they simultaneously vary their injection volume to produce more or less power from the engine

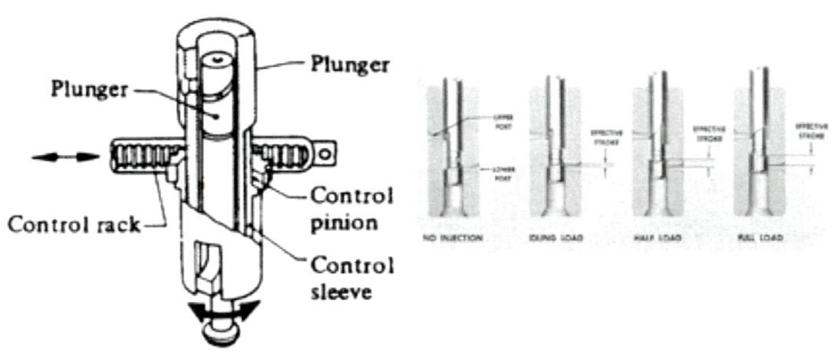

Mechanical Injection

For each engine cylinder, the corresponding plunger in the fuel pump measures out the correct amount of fuel and determines the timing of each injection

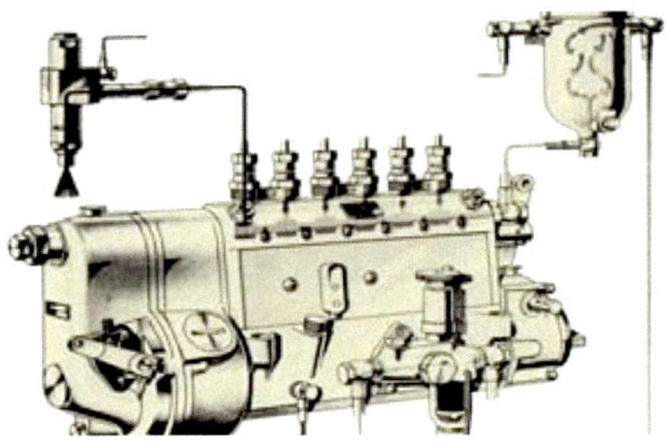

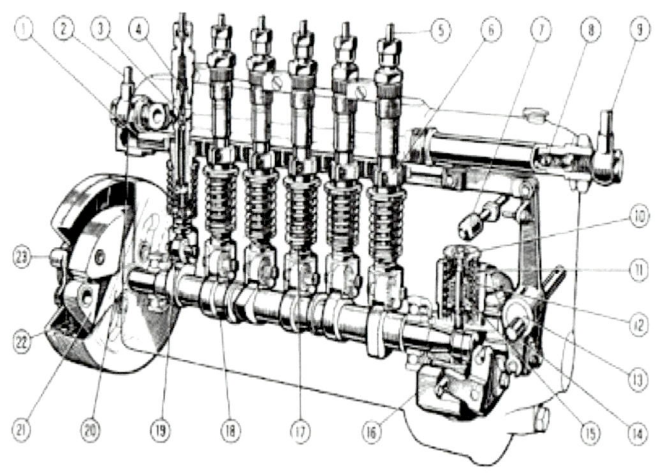

Diesel Injection Pump

- The basic principle is for a plunger to act on a column of fuel, to lift an injector needle off its seat

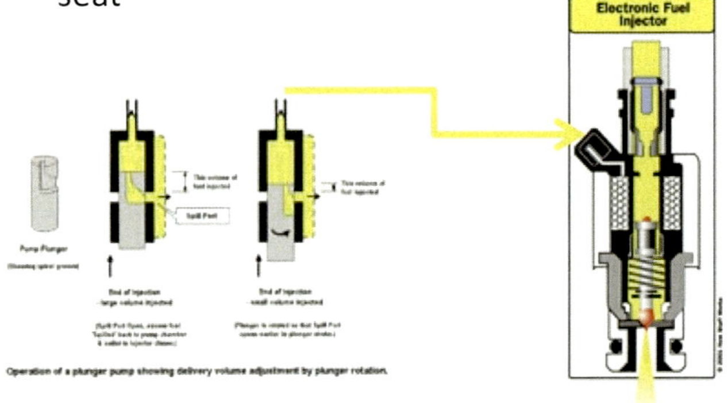

Fuel Metering and Delivery

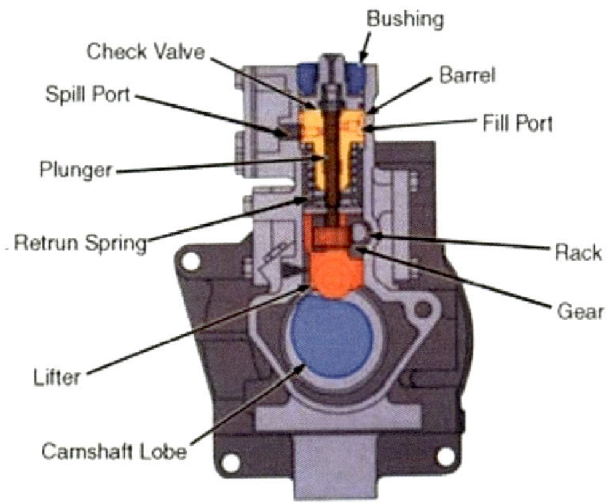

How the plunger works

- Inside the pump is a pumping element, and a delivery valve for each cylinder of the engine. The element has a barrel, and a plunger that fits inside it.

- The plunger is rotated by a control sleeve, a rack, and a pinion. Moving the rack rotates the pinion, the control sleeve, and then the plunger. The rack's movement is controlled by the governor

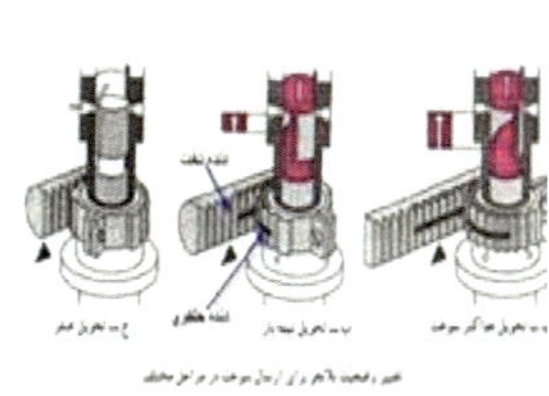 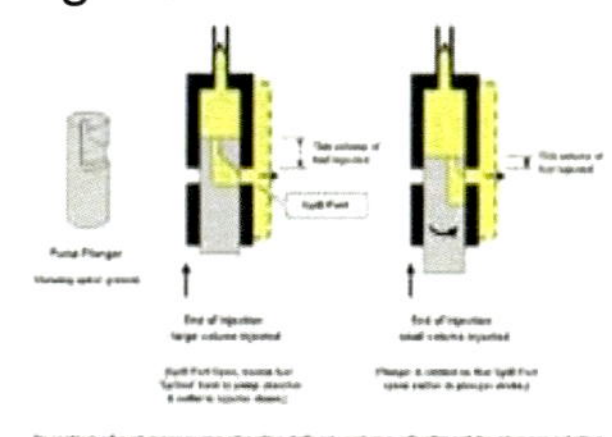

Mechanical Injection

Fuel volume for each single combustion is controlled by a slanted groove in the plunger which rotates only a few degrees releasing the pressure and is controlled by a mechanical governor, consisting of weights rotating at engine speed constrained by springs and a lever

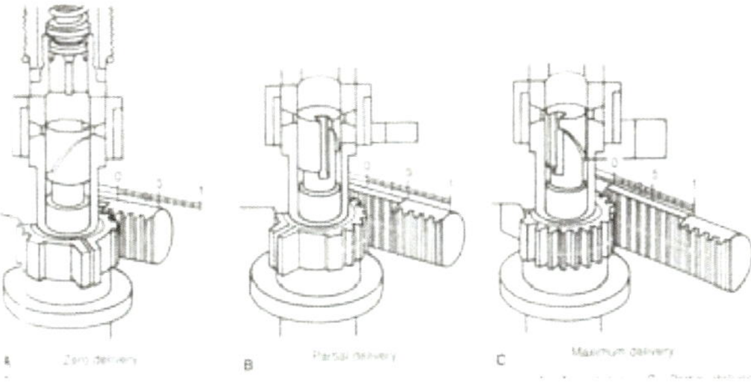

Pumping Elements

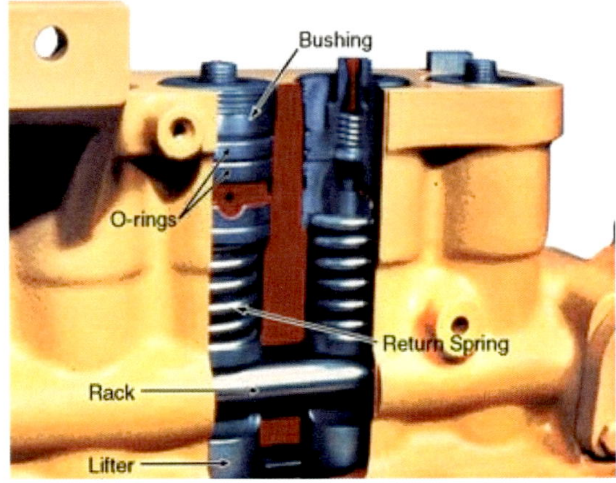

The Fuel Rack Controller

The movement of the rack is controlled by the governor. This movement controls the amount of fuel delivered by the plunger

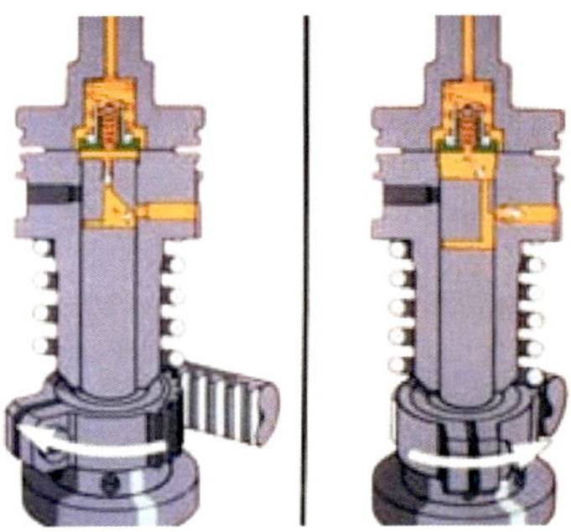

Fuel Injection Pump Operation

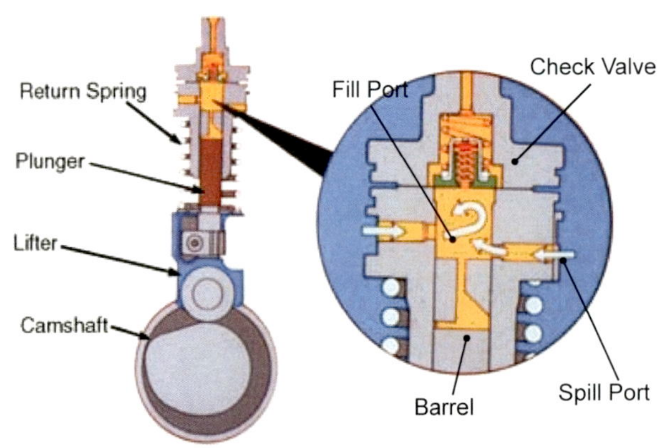

Fuel Injection Pump Operation

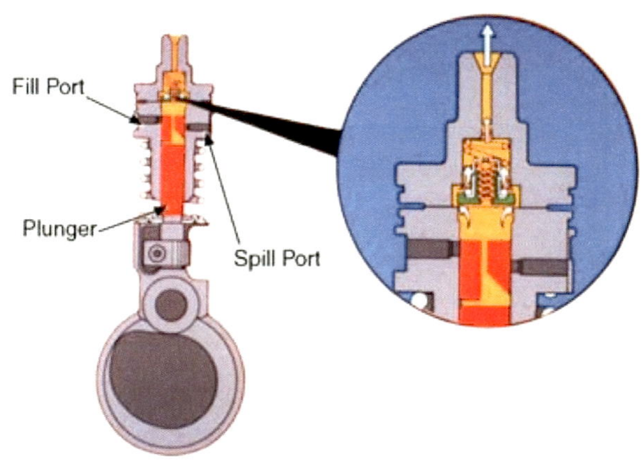

Fuel Injection Pump Operation

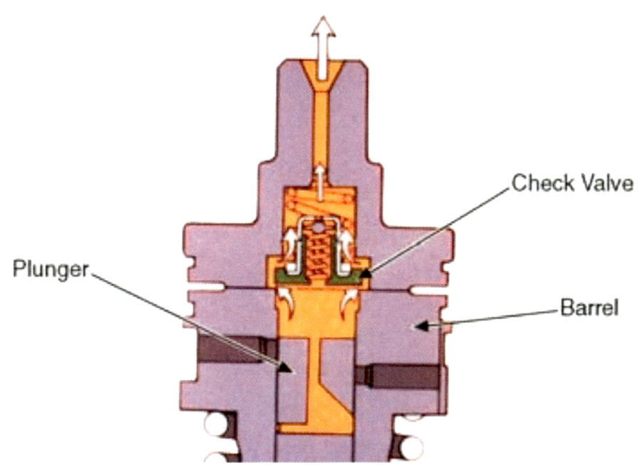

Fuel Injection Pump Operation

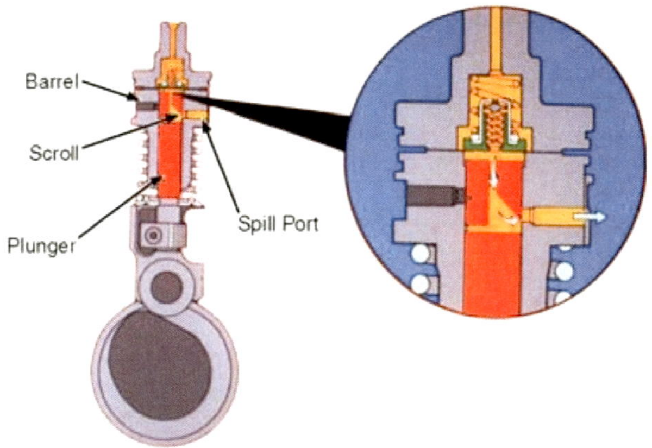

Return Line

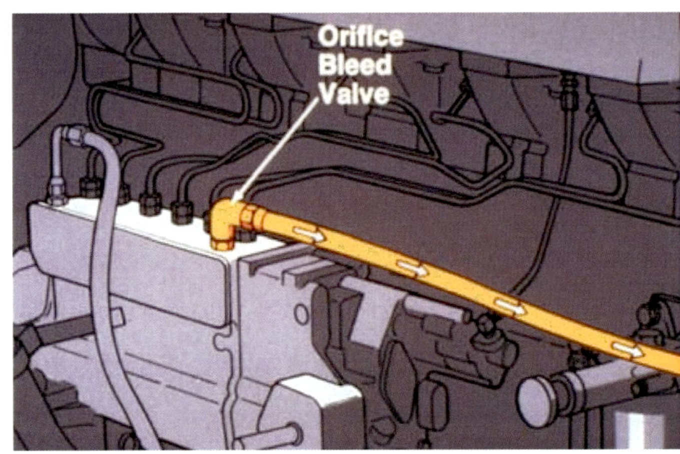

Fuel Injection Pump Operation

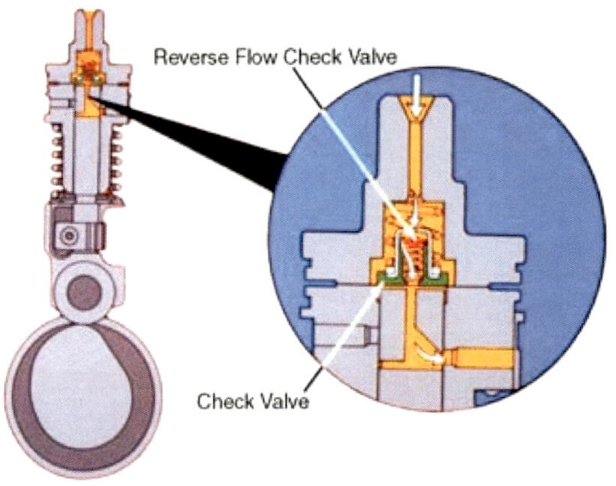

Fuel Injection Pump Camshaft

- The engine camshaft and the fuel injection pump camshaft work together in parallel.
- If the piston No 1 is in TDC then the plunger No 1 will be at TDC

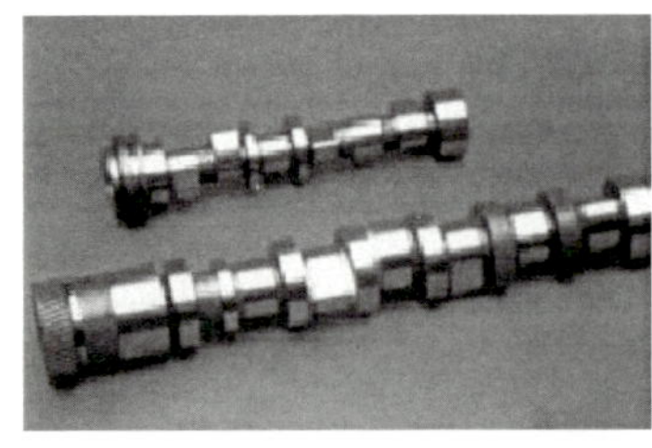

Hydraulic Timing Advance Unit

Fuel Pressures

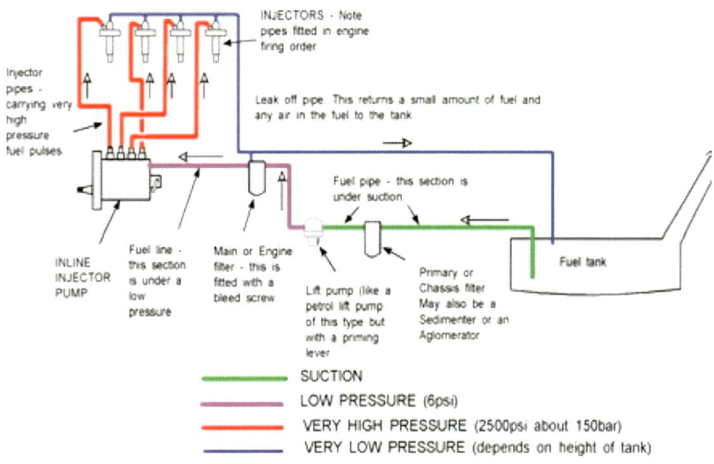

Mechanical Injection

Separate high-pressure fuel lines connect the fuel pump with each cylinder

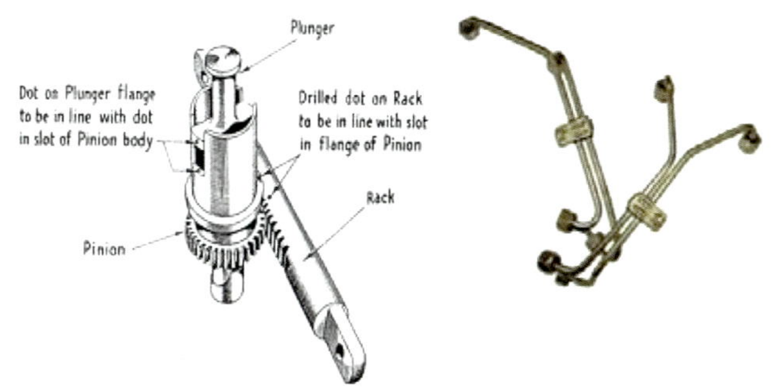

Spill & Dowel Port

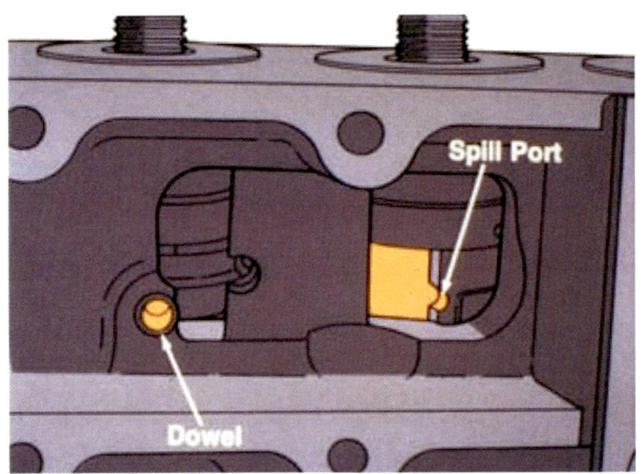

Check Valve Bleed Off

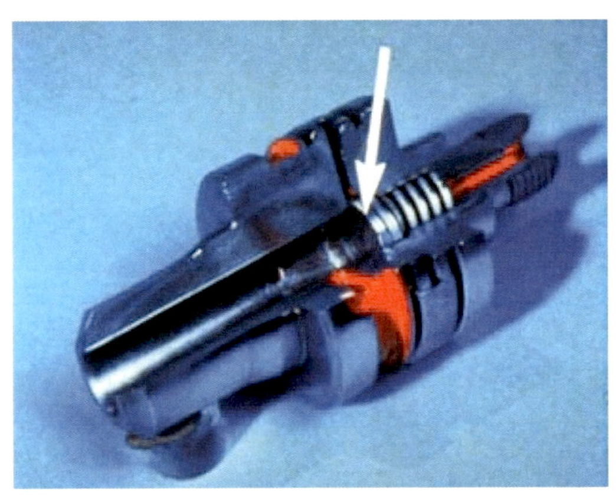

Guide Pin and Spacer Ring

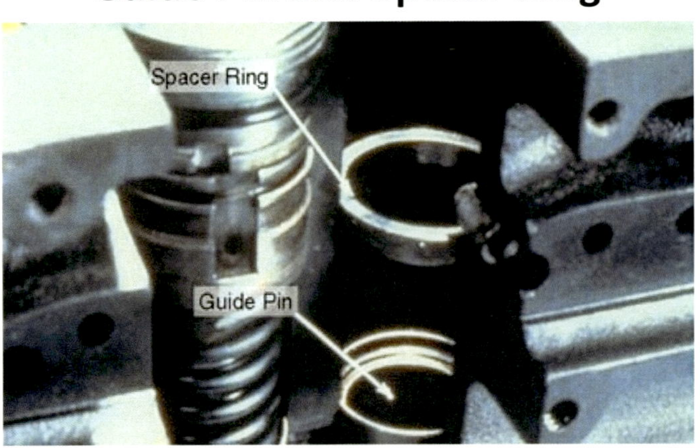

853

The Governor

A vital component of all diesel engines is a mechanical or electronic governor which regulates the idling speed and maximum speed of the engine by controlling the rate of fuel delivery

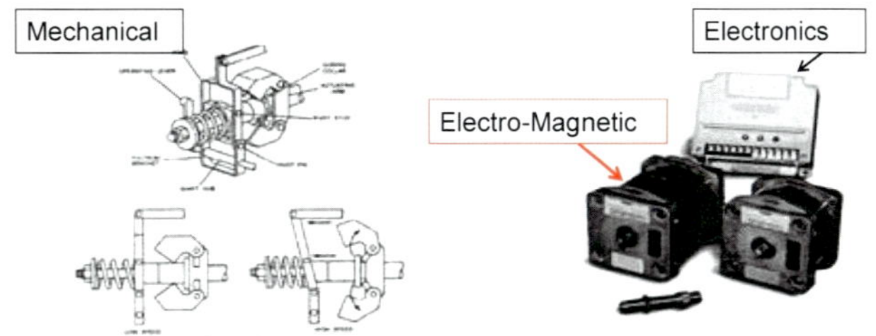

Mechanical Governor

Diesel engine mechanical governors consist of two basic mechanisms: the speed measuring mechanism and **the fuel changing mechanism**

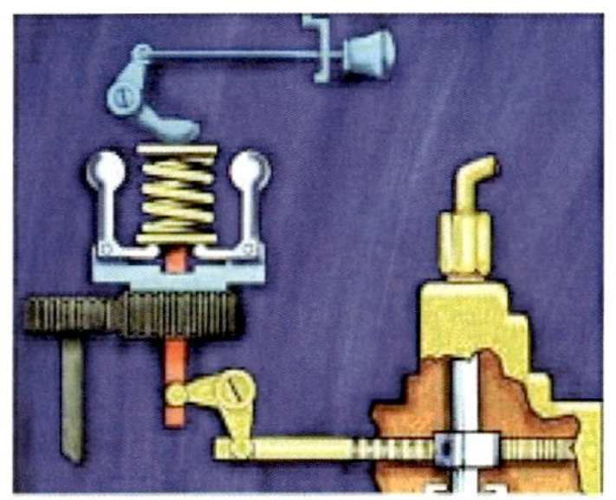

Mechanical Governor

- The speed measuring mechanism senses engine speed changes

- fuel changing mechanism increases or decreases the amount of fuel supplied the engine to correct these changes.

The Mechanical Governor

- In summary, the basic governor consists of the: drive gears, flyweights, spring, and control lever of the speed measuring mechanism, and the connecting linkage, rack and fuel injection pump of the fuel changing mechanism.

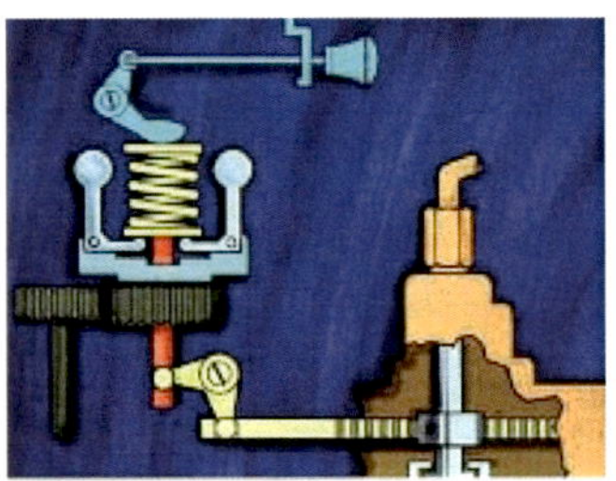

The Basic Governor

The rack which meshes with the injection pump plunger gear segments extends from the injection pump housing into the governor. The rack and fuel injection pumps are parts of the fuel injection pump housing assembly

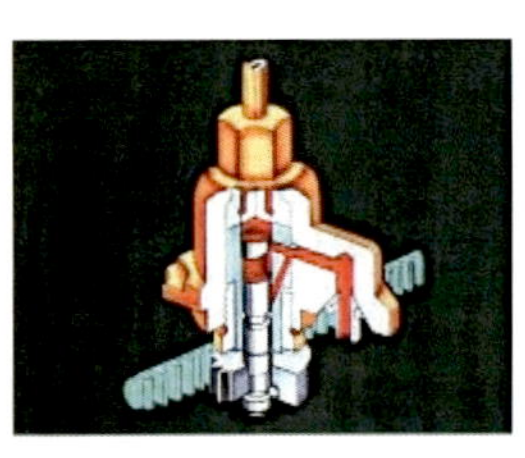

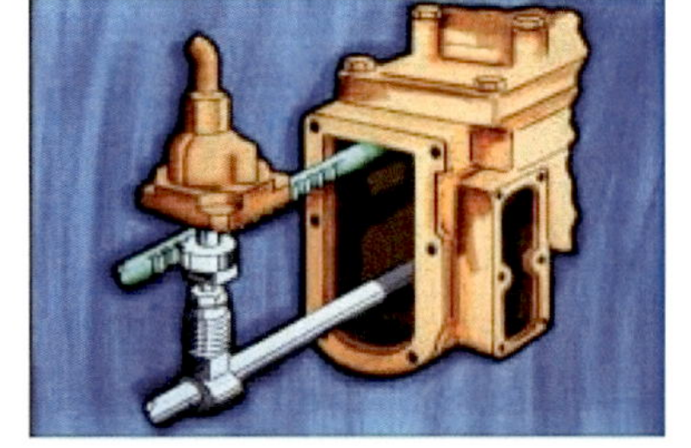

Fuel Changing Mechanism

In this cutaway governor and fuel injection pump housing, we see that the rack extends into the governor. Rack movement controls the amount of fuel injected in each cylinder

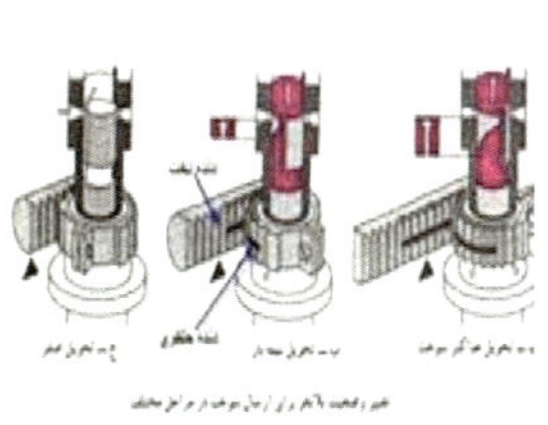 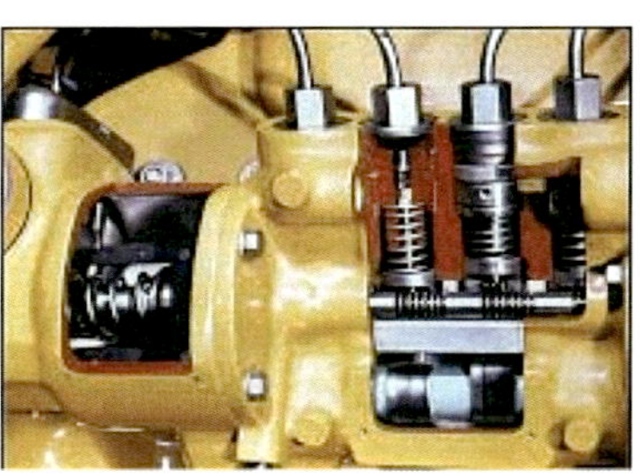

The Speed Controls

The flyweights, spring, spring seat and thrust bearing. The thrust bearing (not previously mentioned) is an anti-friction bearing between the flyweight ballarms which rotate and the spring seat which, of course, does not rotate

The Speed Control

- The governor is driven by the lower gear bolted to the fuel injection pump camshaft
- Looking closer, we can see (from right to left) the drive gear ,flyweights , spring, spring seats, control lever and the collar and bolt which connects to the rack

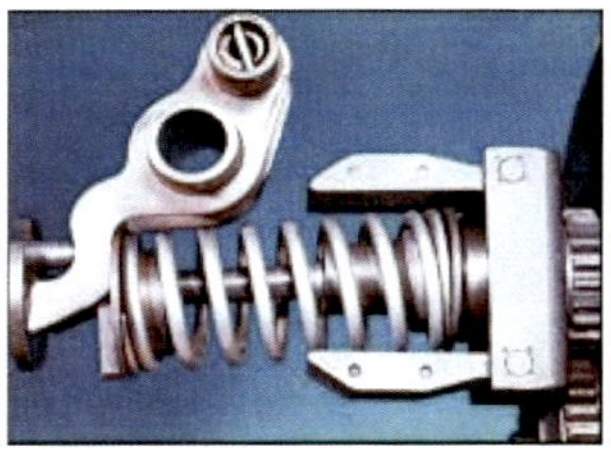

The Governor Cross section

This governor cross section illustrates: (1) lever, (2) spring seat, (3) spring, (4) spring seat and thrust bearing and (5) flyweight assembly. The arrows indicate drive gear rotation and rack movement

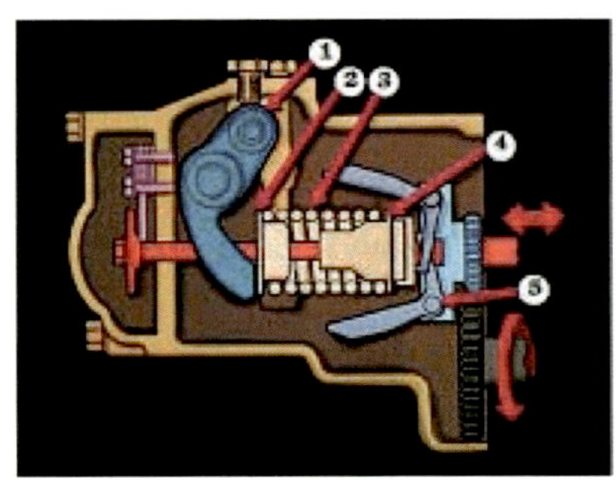

Fuel & Speed Mechanism

- Flyweight movement - outward in this example - due to engine speed changes, are transferred through the simple linkage to the rack and, therefore, to the fuel injection pump plunger

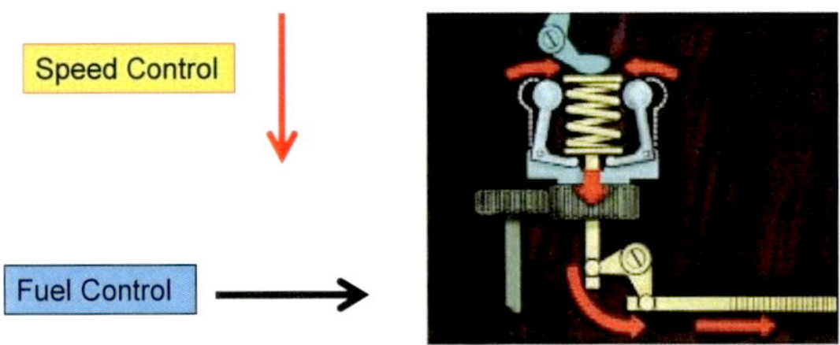

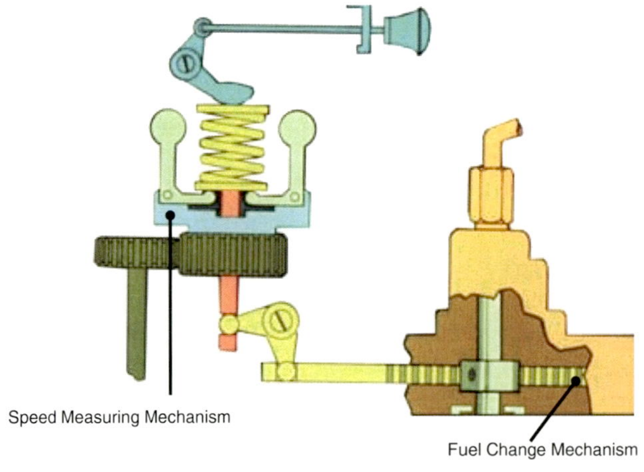

288

864

Fuel Changing Mechanism

When the engine load increases - as when a dozer digs in – the speed decreases. The flyweight force decreases, and the spring moves the linkage and rack to increase the fuel to the engine. The increase fuel position is held until the engine speed returns to the desired setting, and the flyweight force again balances the spring force

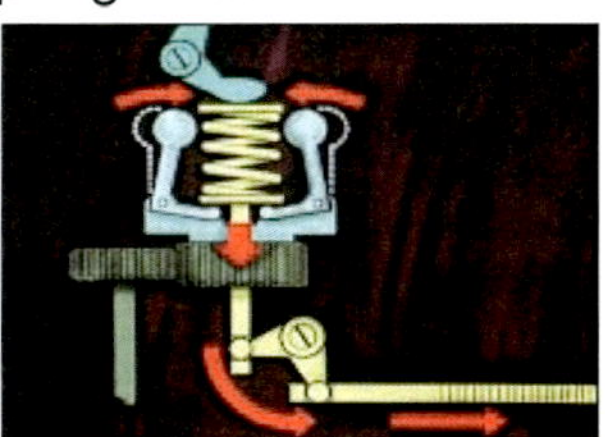

Fuel Changing Mechanism

When the engine load decreases, the speed increases. The flyweight force increases, overcoming the spring force, moving the rack to decrease fuel to the engine. The decrease fuel position is held until engine speed returns to the governor control setting, and the spring force again balances the flyweight force

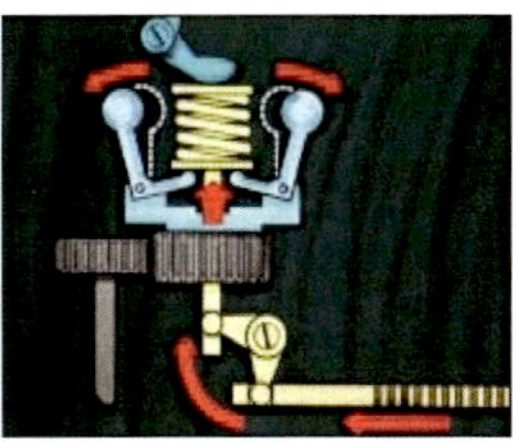

The speed measuring mechanism

- The speed measuring mechanism is simple, has few moving parts and measures engine speed accurately. The main parts are:
- 1) gear drive from the engine,
- 2) flyweights, and
- 3) spring.

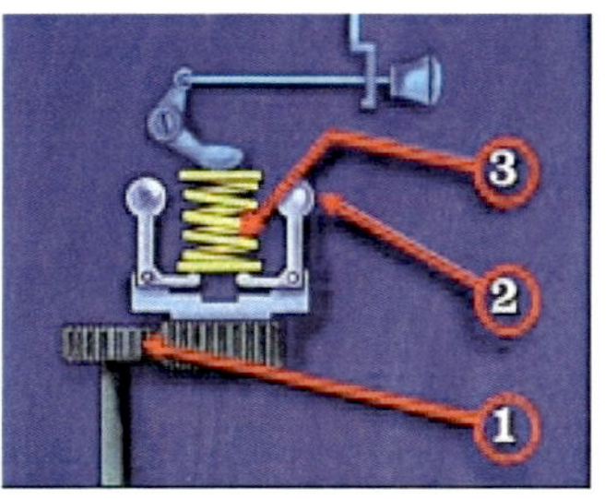

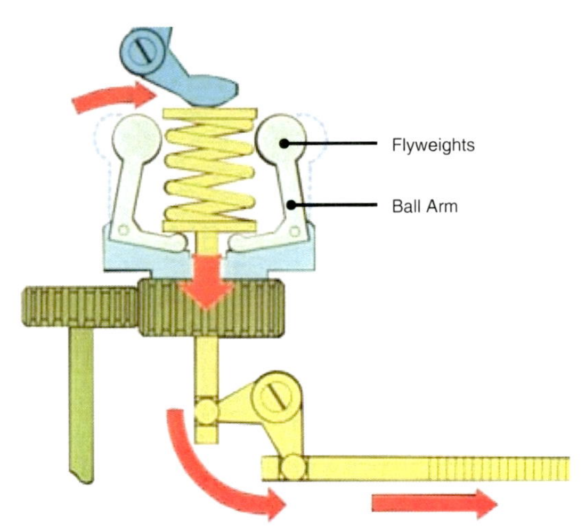

Centrifugal Force

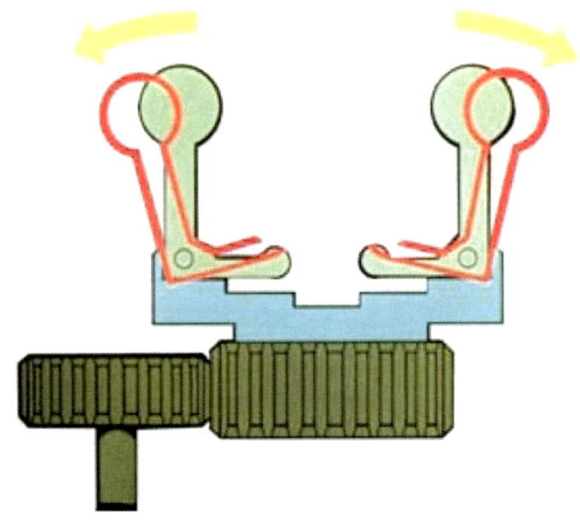

Governor Springs

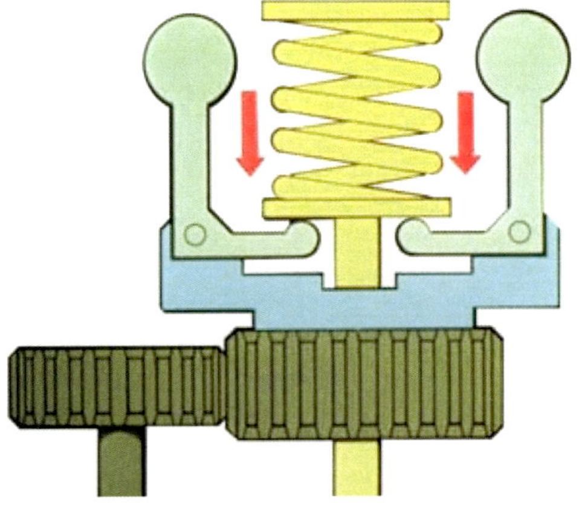

Governor Control

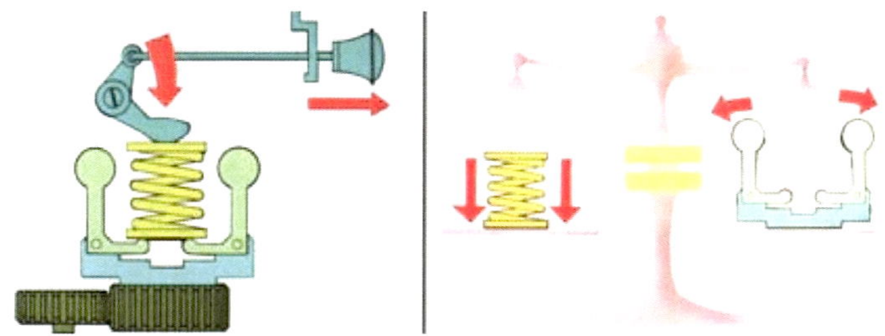

Fuel Changing Mechanism

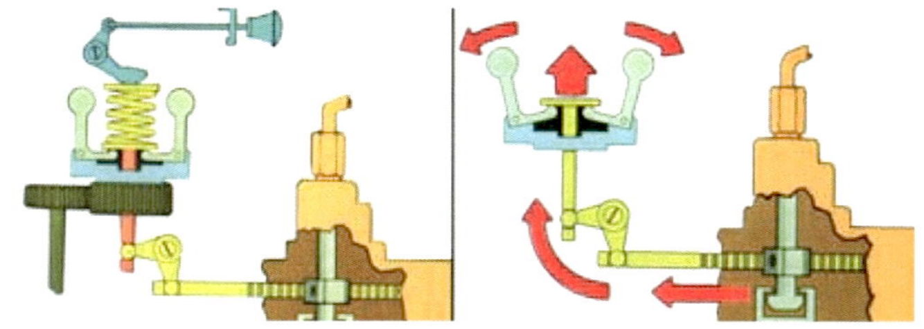

Adjusting Screws

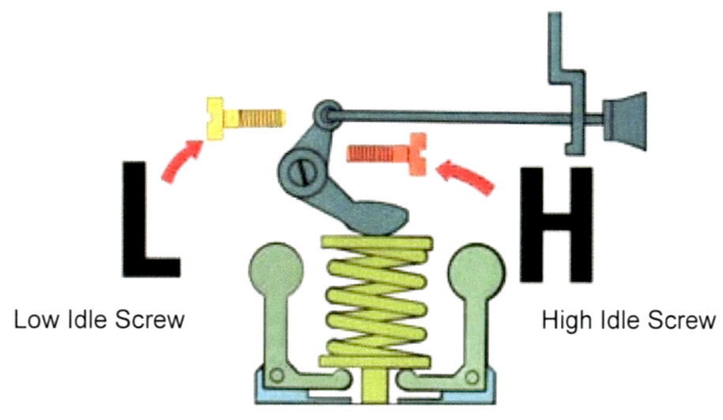

The control level adjustment

- Two adjusting screws limit the travel of the governor control lever between LOW IDLE position and the HIGH IDLE position

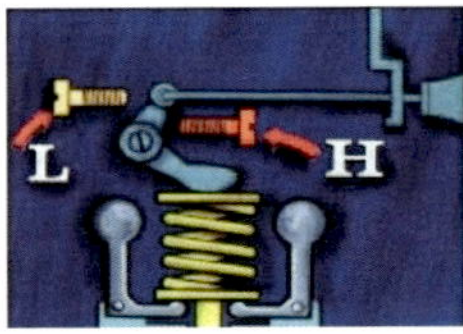

- The low idle stop and high idle stop are simply minimum and maximum engine RPM settings with no load on the engine

High and Low Idle adjustment

- The high and low idle adjusting screws are located under the cover on the governor.

- Notice that the holes in the cover are shaped to lock the screws and prevent them from turning after they are adjusted

High and Low Idle

- Looking, again, at the governor cross section see
- (1) the high idle adjusting screw and
- (2) the low idle adjusting screw. The lever is against the HIGH IDLE screw.
  The low idle and high idle screws, then limit minimum and maximum engine rpm with no load on the engine

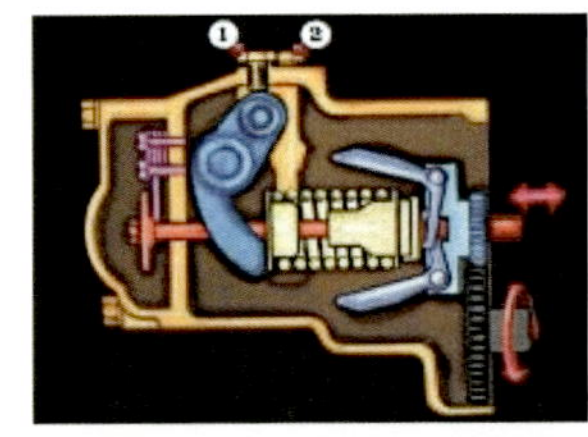

The Governor Lubrication

- The governor is lubricated by the engine lubricating system. Oil from the diesel engine oil manifold is directed to the governor drive bearing. All other governor parts are lubricated by **splash**
- The oil drains from the governor, through the fuel injection pump housing, back to the engine block.

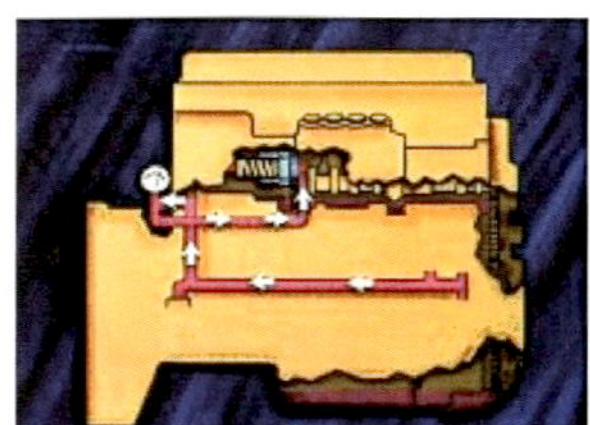

The speed measuring mechanism

- The flyweights and "L" shaped ballarms which pivot are mounted on the governor drive.

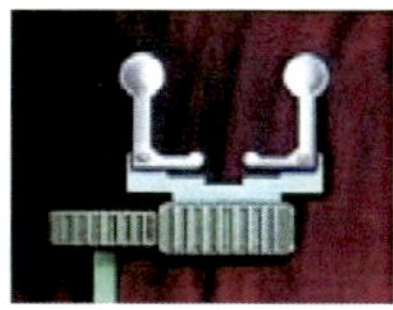

- The flyweights are rotated by the engine

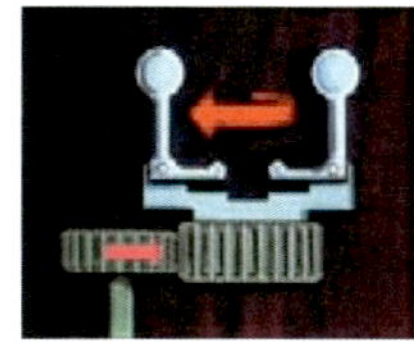

The speed measuring mechanism

- As the flyweights rotate, they exert a centrifugal force outward. The flyweights move outward pivoting the ballarms upward. The amount of outward force depends on the speed of rotation

- the greater the engine speed, the greater the centrifugal force and, therefore, the greater the movement of the flyweights and ball-arms

The speed measuring mechanism

- We need to control this centrifugal force, so we have the governor spring. The spring acts against the force of the rotating flyweights and tends to oppose them

- The force exerted by the spring depends on the governor control setting

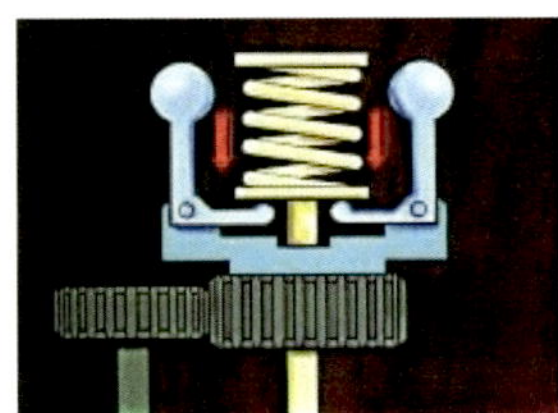

The speed measuring mechanism

- A lever connected to the governor control pushes on or compresses the spring. The spring force opposes the flyweights to regulate the desired engine speed setting
- The governor control, shown here as a simple push-pull knob, may be a hand operated control lever or a foot operated accelerator pedal

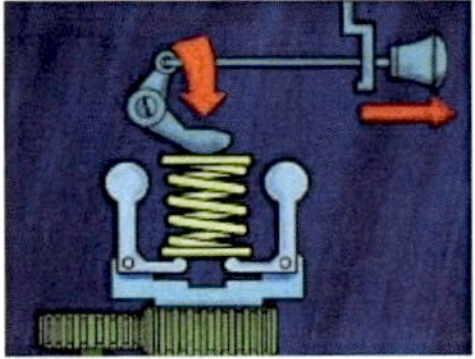

The speed measuring mechanism

The speed measuring mechanism, then, senses and measures engine speed changes. The fuel changing mechanism links the speed measuring mechanism with the fuel injection pumps to control engine

Fuel Changing Mechanism

- The fuel changing mechanism consists of the:
  - 1) connecting linkage,
  - 2) rack and
  - 3) the fuel injection pump.

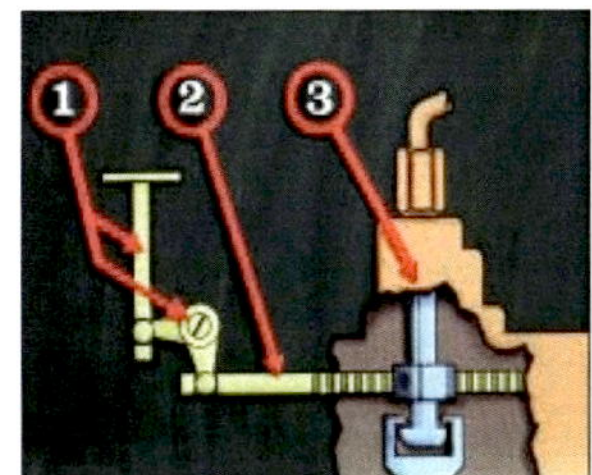

The Diesel Governor

- In Diesel engines air intake is not limited, and the cylinders always have more air than is needed to support combustion. **The amount of fuel injected into the cylinders controls engine speed**.

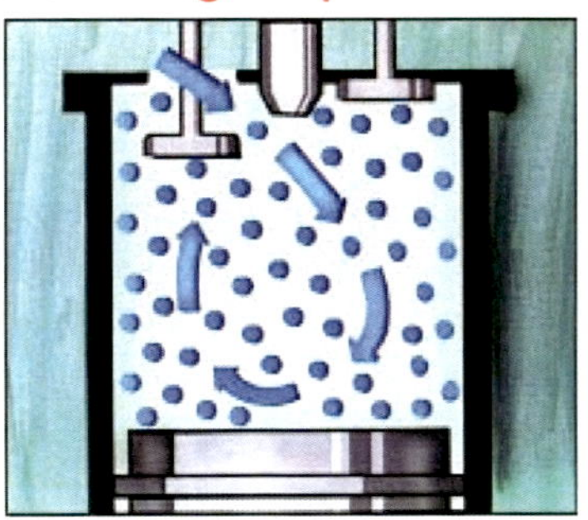

High Torque & Acceleration

- As the fuel is injected directly into the cylinders rather than into the air intake manifold, engine response is immediate. This , resulting greater power stroke, adds up to very rapid acceleration

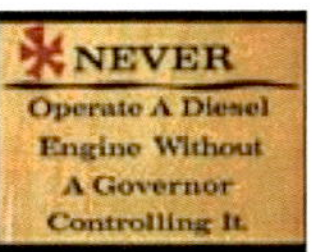

- As we said earlier, diesel engines can accelerate at a rate of more than 2000 revolutions per second. Because of this rapid acceleration, manual control is difficult, **if not impossible**.

Warning!!

- If you were to move the fuel rack of a diesel engine to the full "ON" position without a load and with the governor not connected, the engine speed might climb and exceed safe operating limits before
you could shut it down. One second...two seconds...before you knew what was happening, the engine may have been seriously damaged by overspeeding

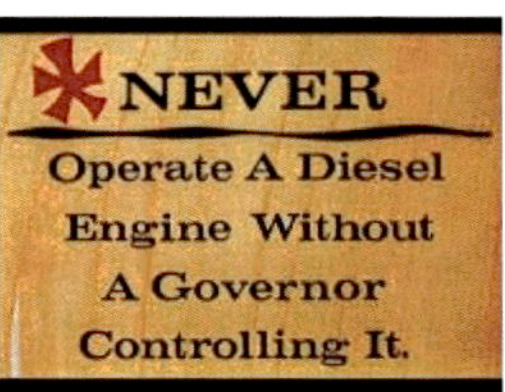

Rotary Fuel Injection Pump

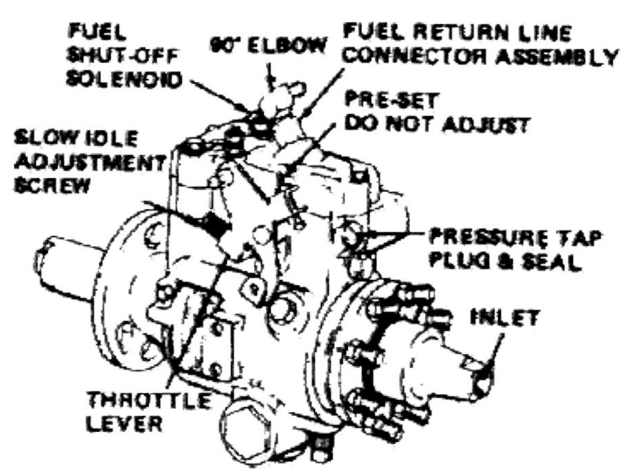

Rotary Vane-Pump

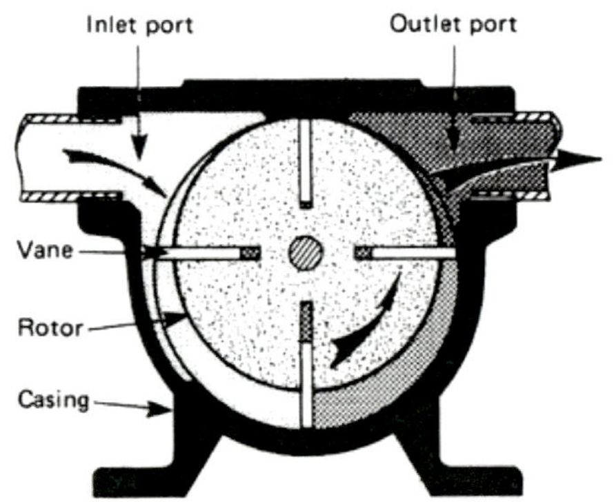

Rotary Central Piston Pump

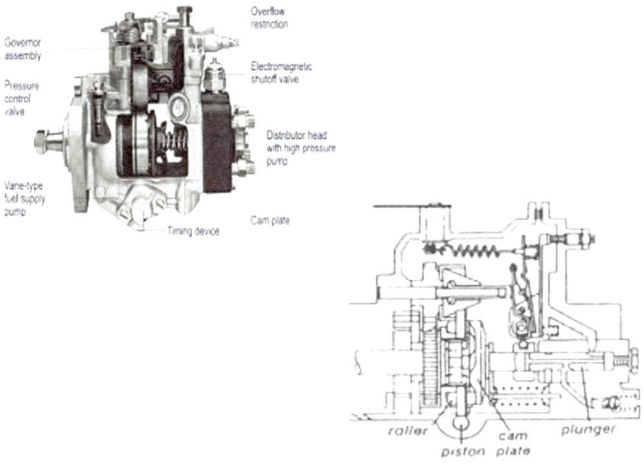

Hydromechanical Governor

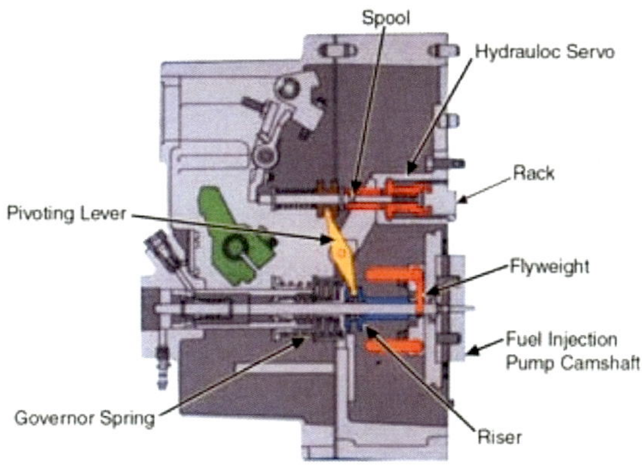

Electronic Fuel Injection Pump

The FSD at the left side of the Fuel Injection pump works together with the ECM of the engine to control the timing and pressure of fuel per each cylinder

Rotary Fuel Injection Pump

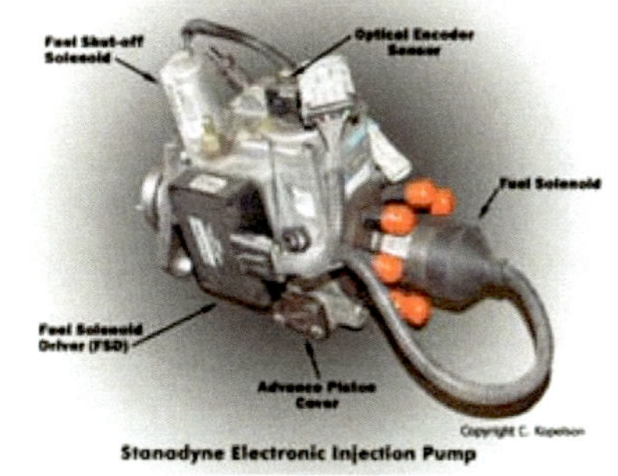

Electronic Rotary Fuel Injection Pump

This type of pump works similarly than the mechanical pump. In this case the timing and pressure is controlled by means of the PCM , FSD and Fuel solenoid

High Pressure Common Rail & Pump

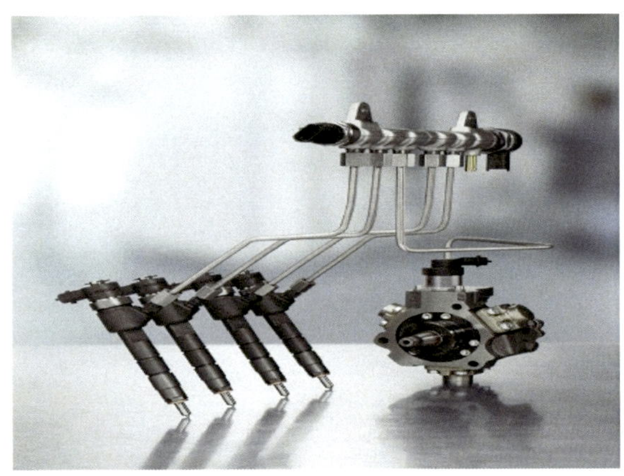

Common Rail Pump

- The supply pump is driven by the engine and produces high-pressure fuel

Common Rail Fuel Pump

Remove and replace the common rail pump is too simple because does not need take care with the timing marks . In other words the pump can be removed in any position

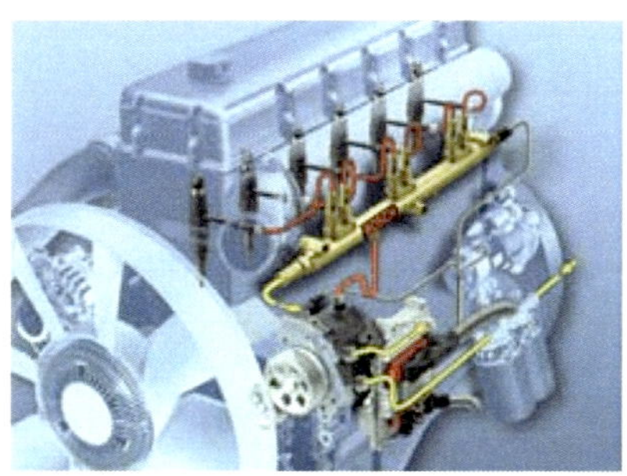

Common Rail Fuel Pump

This system have an input low pressure side and an output high pressure side to keep the rail pressurized

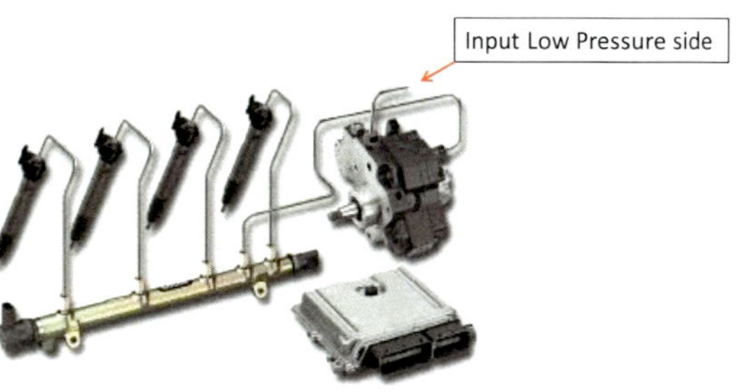

Common Rail System

The common rail system accumulates high-pressure fuel in the common rail and injects the fuel into the engine cylinder at timing controlled by the engine ECU

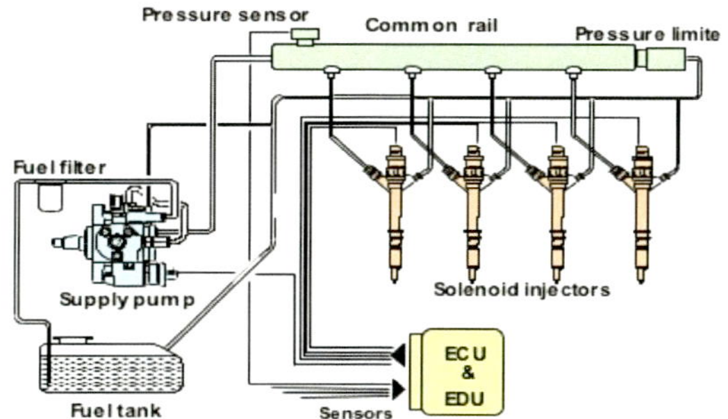

How the Common Rail System Works

The injectors that provide fuel to the cylinders are attached to a single tube (the common rail)

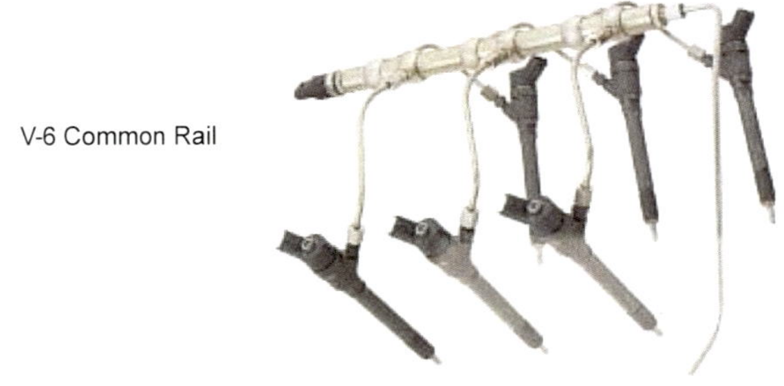

High Pressure Common Rail & Pump

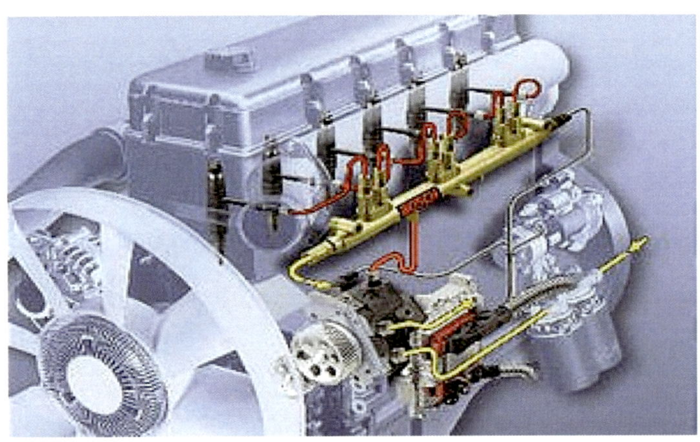

High Pressure Common Rail & Pump

- 1 = High Pressure Pump 2 = High Pressure Rail
- 3 = Electronic Control Unit
- 4 = Injector

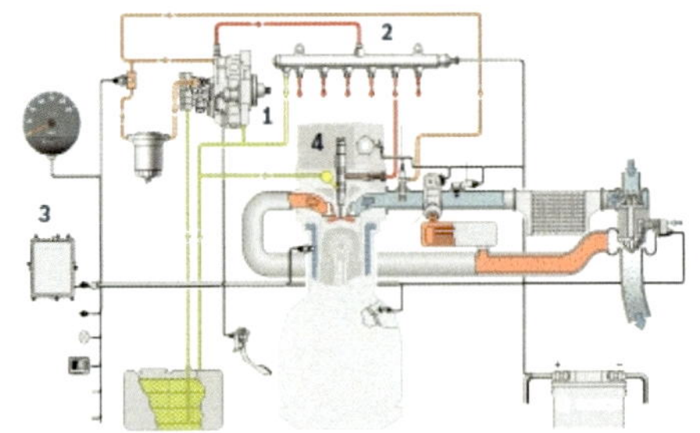

Common Rail

The way in which fuel is injected into the cylinders of diesel vehicles determines their torque, fuel consumption, emissions and noise level

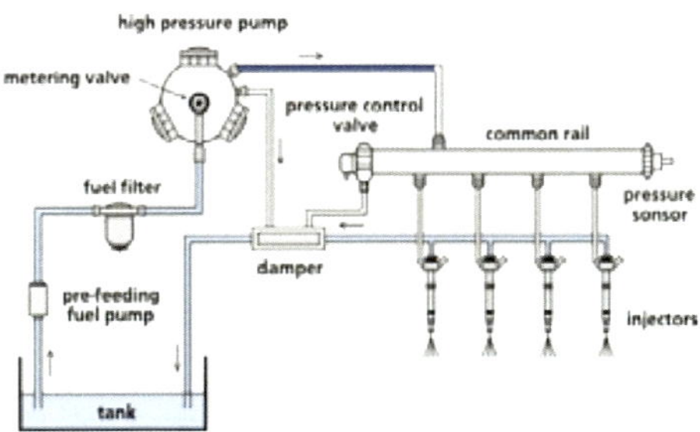

Common Rail

A common rail injection system separates these two functions (generating pressure and injecting) by first storing fuel under high pressure in a central accumulator rail and then delivering it to the individual electronically-controlled injection valves (injectors)

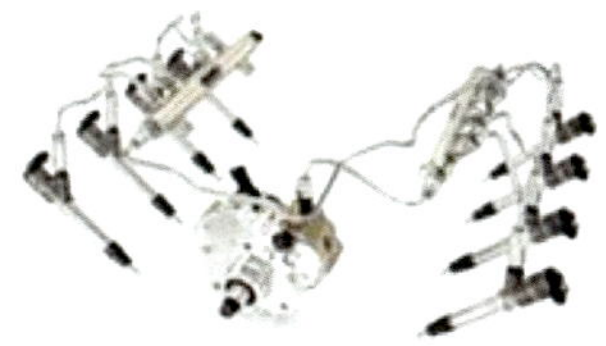

Common Rail Pressures

This ensures that incredibly high injection pressures (in some cases over 25,000 pounds per square inch) are available at all times

Pre Injection types

- The pilot injection, occurring well before ignition, provides time for fuel and air to mix

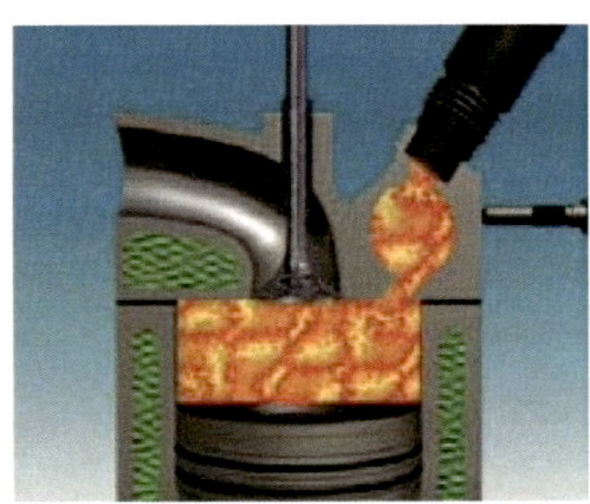

Rail System Injection types

- The "pre" injection shortens the ignition delay during the main injection and, as a result, reduces the generation of nitrogen oxide (NOx), noise and engine vibration

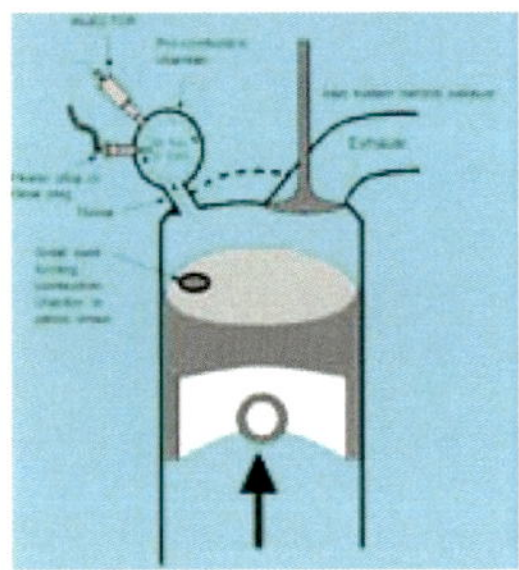

Feed Pumps

- The function of feed pump is to deliver the fuel from the fuel tank to fuel injection pump through the fuel filter
- The feed pump is attached to the injection pump and driven by its camshaft

302

Fuel Injection System

Diagnosis & Troubleshot

Engine Surges at Idle

Cause	Remedy
Tank empty **or** tank vent blocked	Fill tank / bleed system. check tank vent
Air in the fuel system	Bleed fuel system, eliminate air leaks
Pump rear support bracket loose	Replace as necessary
Overflow fitting interchanged with inlet fitting	Install fittings in their proper positions
Cold-start device malfunction	Check cold-start mechanism
Fuel injection pump defective or cannot be adjusted	Replace

Rough Idle When Engine is Warm

Cause	Remedy
Improper fuel (gasoline) in tank	Drain tank, flush system, fill with proper fuel
Air in the fuel system	Bleed fuel system, elimitate air leaks
Pump rear support bracket loose	Replace as necessary
Injection nozzle defective	Repair or replace
Injection sequence does not correspond to firing order	Install fuel injection lines in the correct order
Low or uneven engine compression	Repair as necessary
Fuel injection pump defective or cannot be adjusted	Replace

Low Power

Cause	Remedy
Improper fuel (gasoline) in tank	Drain tank, flush system, fill with proper fuel
Tank empty **or** tank vent blocked	Fill tank / bleed system. check tank vent
Fuel filter blocked	Replace fuel filter
Injection lines blocked / restricted	Drill to nominal I.D. or replace
Fuel-supply lines blocked / restricted	Test all fuel supply lines - flush or replace
Loose connections, injection lines leak or broken	Tighten the connection, eliminate the leak
Pump-to-engine timing incorrect	Readjust timing
Injection nozzle defective	Repair or replace
Engine air filter blocked	Replace air filter element
Injection sequence does not correspond to firing order	Install fuel injection lines in the correct order
Maximum speed misadjusted	Readjust maximum speed screw
Overflow fitting interchanged with inlet fitting	Install fittings in their proper positions
Low or uneven engine compression	Repair as necessary
Fuel injection pump defective or cannot be adjusted	Replace

Excessive Fuel Consumption

Cause	Remedy
Pump-to-engine timing incorrect	Readjust timing
Injection nozzle defective	Repair or replace
Engine air filter blocked	Replace air filter element
Injection sequence does not correspond to firing order	Install fuel injection lines in the correct order
Low idle misadjusted	Readjust idle stop screw
Cold-start device malfunction	Check cold-start mechanism
Fuel injection pump defective or cannot be adjusted	Replace

Fog-Like Exhaust in Full-Load Range (White or Blue)

Cause	Remedy
Improper fuel (gasoline) in tank	Drain tank, flush system, fill with proper fuel
Tank empty **or** tank vent blocked	Fill tank / bleed system. check tank vent
Air in the fuel system	Bleed fuel system, elimitate air leaks
Fuel filter blocked	Replace fuel filter
Injection lines blocked / restricted	Drill to nominal I.D. or replace
Fuel-supply lines blocked / restricted	Test all fuel supply lines - flush or replace
Paraffin deposit in fuel filter	Replace filter, use Winter Fuel
Pump-to-engine timing incorrect	Readjust timing
Overflow fitting interchanged with inlet fitting	Install fittings in their proper positions
Fuel injection pump defective or cannot be adjusted	Replace

Poor Performance, Black Smoke or Low Power

Cause	Remedy
Improper fuel (gasoline) in tank	Drain tank, flush system, fill with proper fuel
Air in the fuel system	Bleed fuel system, elimitate air leaks
Injection lines blocked / restricted	Drill to nominal I.D. or replace
Pump-to-engine timing incorrect	Readjust timing
Injection nozzle defective	Repair or replace
Engine air filter blocked	Replace air filter element
Injection sequence does not correspond to firing order	Install fuel injection lines in the correct order
Cold-start device malfunction	Check cold-start mechanism
Low or uneven engine compression	Repair as necessary
Fuel injection pump defective or cannot be adjusted	Replace

Engine Does Not Rev Up

Cause	Remedy
Improper fuel (gasoline) in tank	Drain tank, flush system, fill with proper fuel
Tank empty **or** tank vent blocked	Fill tank / bleed system. check tank vent
Air in the fuel system	Bleed fuel system, elimitate air leaks
Fuel filter blocked	Replace fuel filter
Fuel-supply lines blocked / restricted	Test all fuel supply lines - flush or replace
Loose connections, injection lines leak or broken	Tighten the connection, eliminate the leak
Pump-to-engine timing incorrect	Readjust timing
Overflow fitting interchanged with inlet fitting	Install fittings in their proper positions
Cold-start device malfunction	Check cold-start mechanism
Low or uneven engine compression	Repair as necessary
Fuel injection pump defective or cannot be adjusted	Replace

Incorrect Idle or Maximum Speed

Cause	Remedy
Low idle misadjusted	Readjust idle stop screw
Maximum speed misadjusted	Readjust maximum speed screw
Overflow fitting interchanged with inlet fitting	Install fittings in their proper positions
Cold-start device malfunction	Check cold-start mechanism
Fuel injection pump defective or cannot be adjusted	Replace

Air Intake System Components

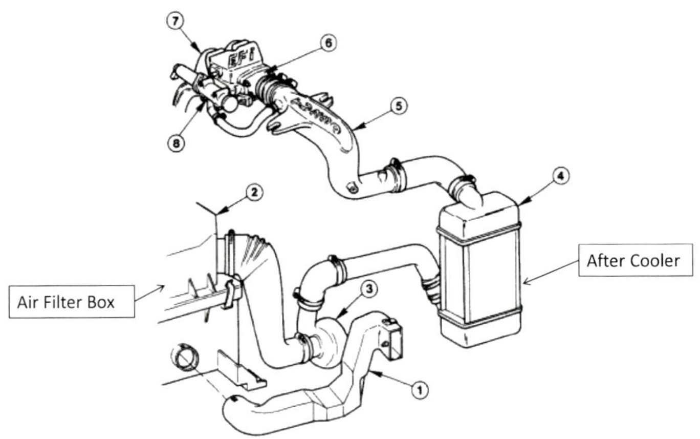

Air Intake

- Each cylinder should contain more air than required to produce combustion under ideal and controlled conditions
- Increasing the amount of fuel injected **not result** in more power if there is not enough air to combine with it

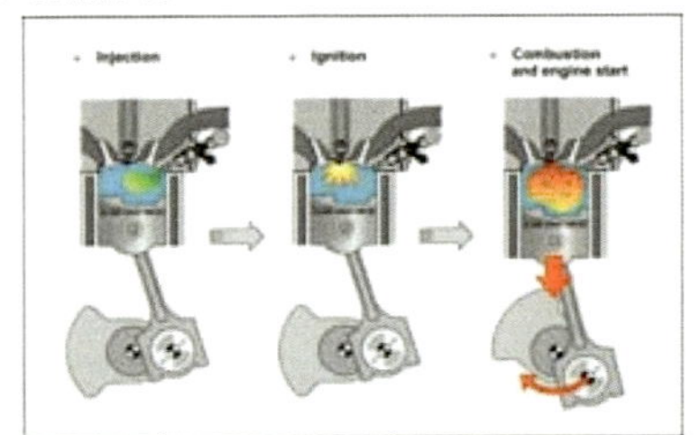

Volumetric Efficiency

Is a ratio (or percentage) of the quantity of air that is trapped by the cylinder during induction over the swept volume of the cylinder under static conditions

$$\eta_V = \frac{Volume\ of\ air\ taken\ into\ cylinder}{Maximum\ possible\ volume\ in\ the\ cylindre}$$

$$\eta_V = \frac{V_{air}}{V_c}$$

Volumetric Efficiency

Volumetric Efficiency can be improved by compressing the induction charge. In this case Volumetric Efficiency can exceed 100%

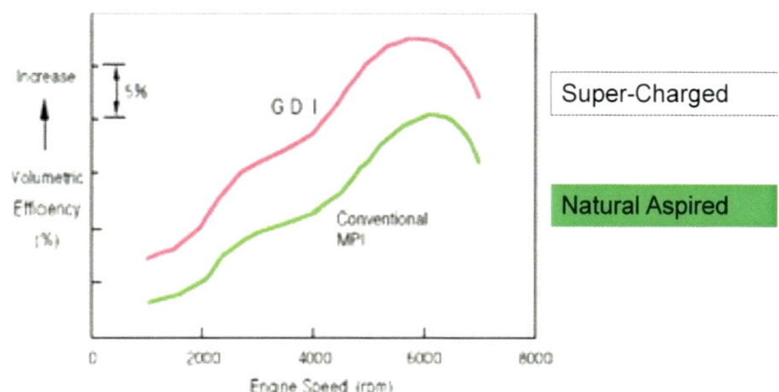

Types of Systems

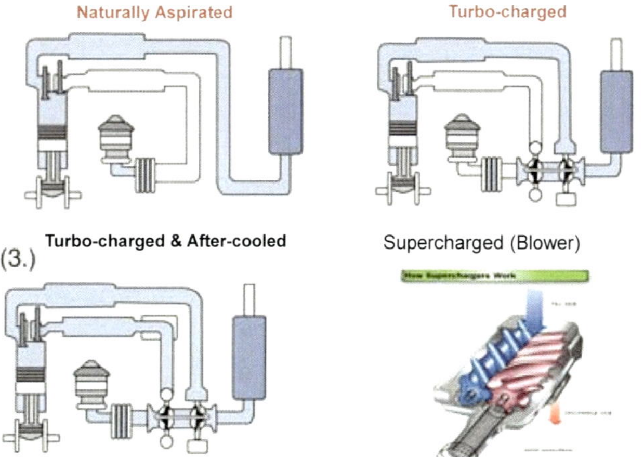

(3.)

Type No 1

Naturally Aspirated Intake and Exhaust Systems

922

Air Cleaner Filters

There are two types: Dry and Washable

Dry Air Cleaner Element

This efficiency of an air filter depends of the dust retained by the filter . For this reason a consumer should always use the air filter recommended by the engine manufacturer

924

308

Air Filter Defective

A defective air filter can damage the compressor wheel blades . The fractured pieces enter in the combustion chamber producing catastrophic consequences

Washable Air Cleaner Element

They have a VERY high pressure drop. The problem with many of them is that they have 3 layers of media. When you wash them out, it is practically impossible to dislodge the particles from the center layer

Air Intake Instruments

The air intake flow performance can be monitored with gauges at the dashboard or by the air filter sensor

Air Cleaner Service

When the air filter is clogged the mark at the sensor is located at the red zone. After the filter replacement the sensor could be reset

Air Cleaner Service

929

SCAVENGING

Scavenging is the process of removing exhaust gases from the cylinder after combustion and replenishing the cylinder with fresh air

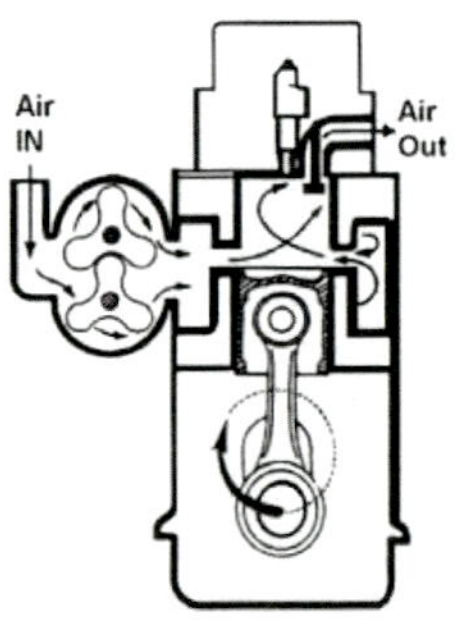

310

Scavenging in Marine Engines

Scavenging is generally provided by the engine's **super-charging** and turbo-charging system

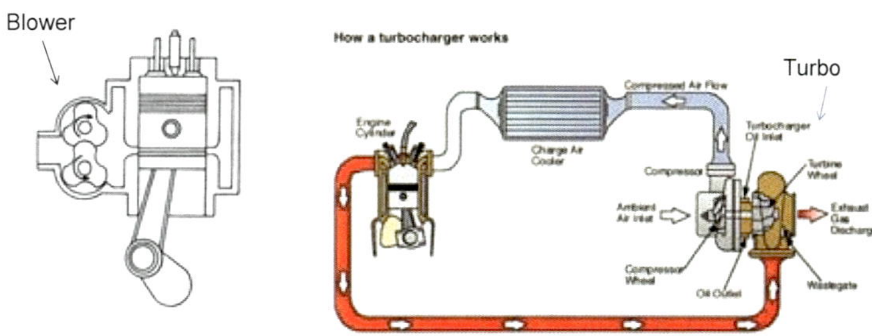

SCAVENGING

- Efficient scavenging is necessary for good combustion of fuel inside the engine cylinder.
- The passage of scavenge air will also assist cooling of the cylinder, piston and valves

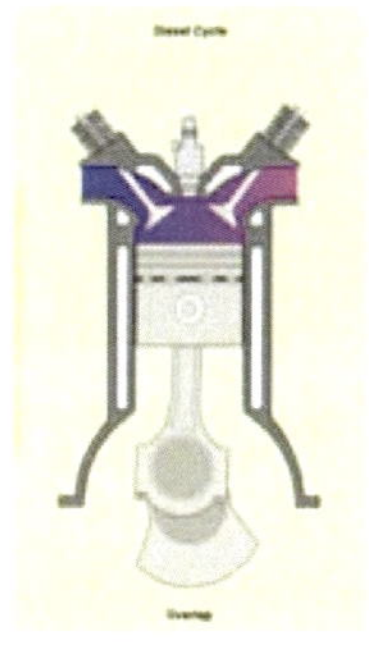

SCAVENGING

The time available for scavenging process in 2 stroke engine is less than 4 stroke engines

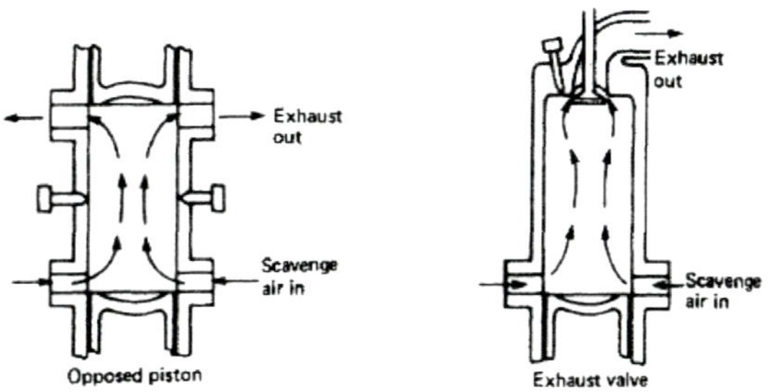

Scavenging in Marine Engines

- The more efficient the scavenging, the better is the fuel combustion and power output of the engine

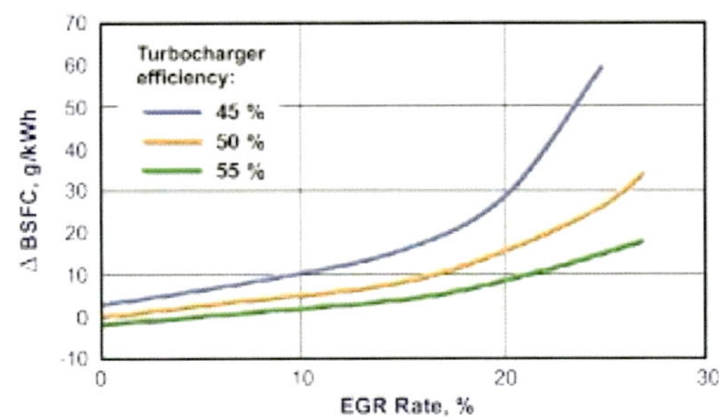

SCAVENGING METHODS

- The two principal methods by which this is accomplished are referred to as :

  - **PORT SCAVENGING**
    - Direct cross-flow
    - Uni-flow
  - **VALVE SCAVENGING**

Types of scavenging

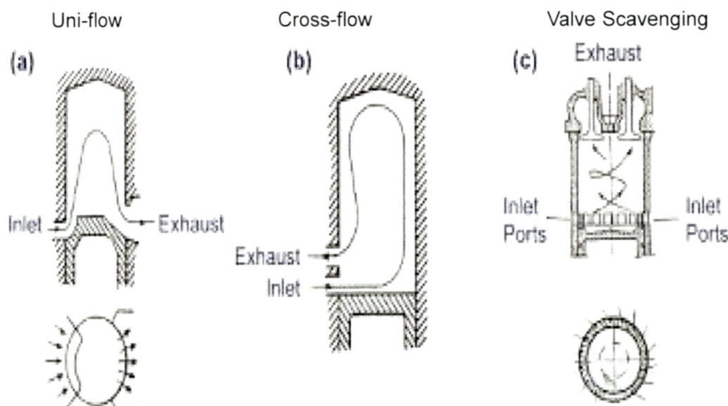

Valve Scavenging

- Valve Scavenging is of the **uni-flow** type

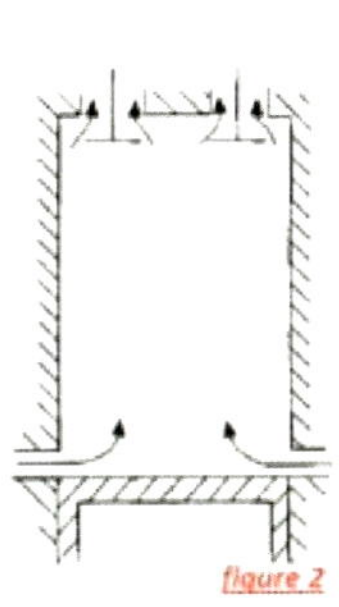

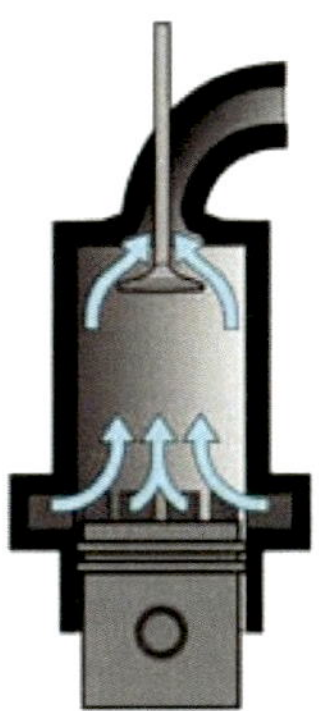

Two Stroke & Four Stroke

- Scavenging is not the same for both two stroke and four stroke engines.

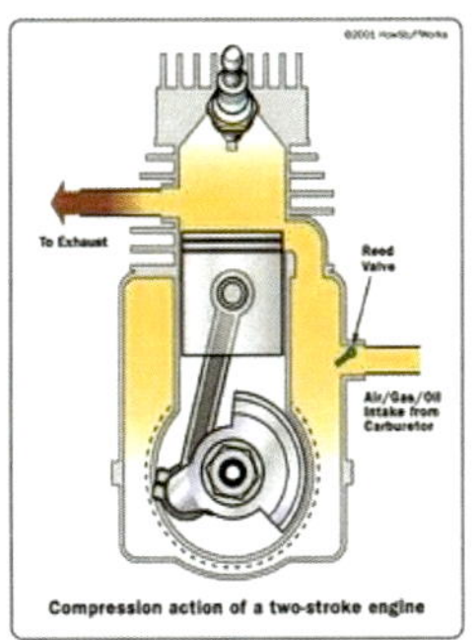

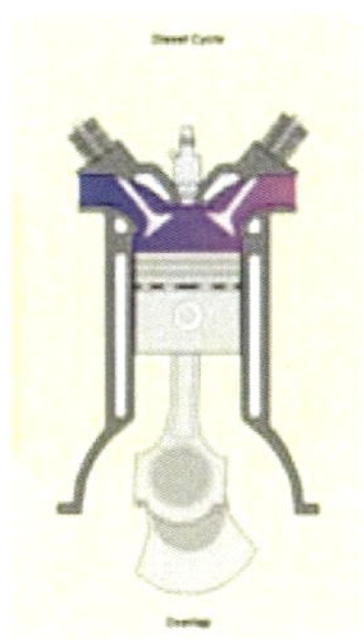

The Overlap Procedure

This is mainly due to the adequate overlap between the opening of the inlet valve and closing of the exhaust valve

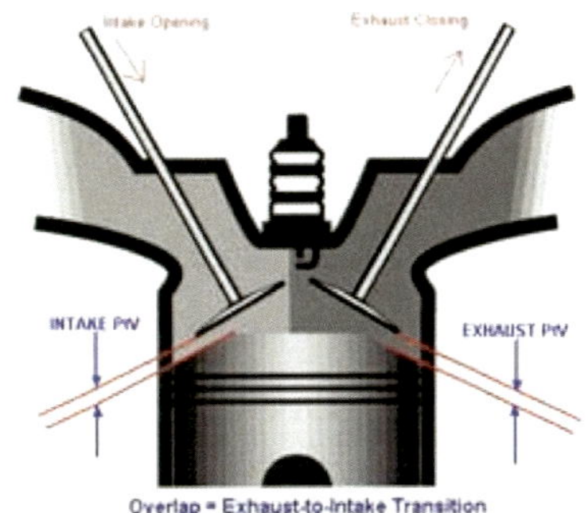

Overlap in Four Stroke Engine

The overlap of intake and exhaust permits the air from the blower to pass through the cylinder into the exhaust manifold, cleaning out the exhaust gases from the cylinder and, at the same time, cooling the hot engine parts

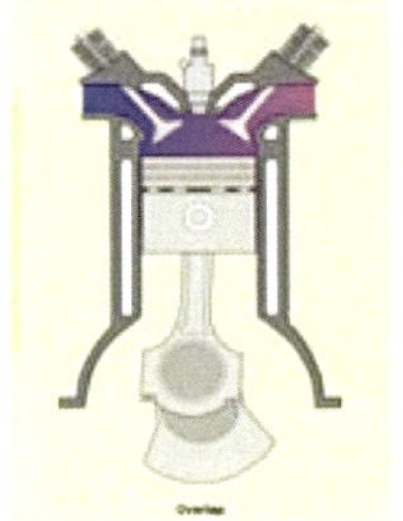

Overlap in Four Stroke Engine

When scavenging air enters the cylinder of an engine, it must be so directed that the waste gases are removed from the remote parts of the cylinder

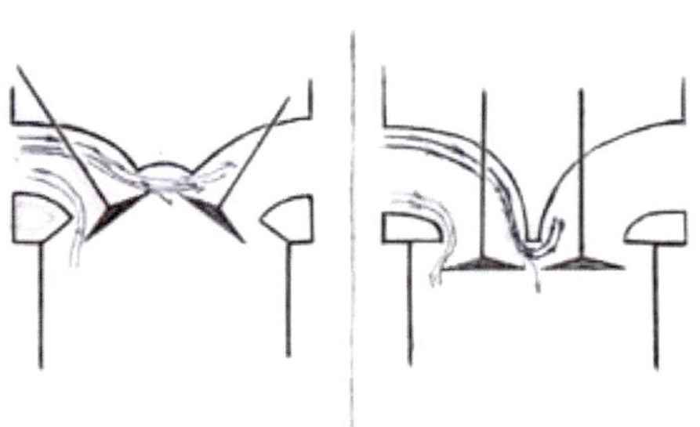

Two Stroke & Four Stroke

In a four stroke engine the air is induced during a downwards stroke of one of the two cycles, i.e per power stroke, and the exhaust gases are removed in the preceding stroke

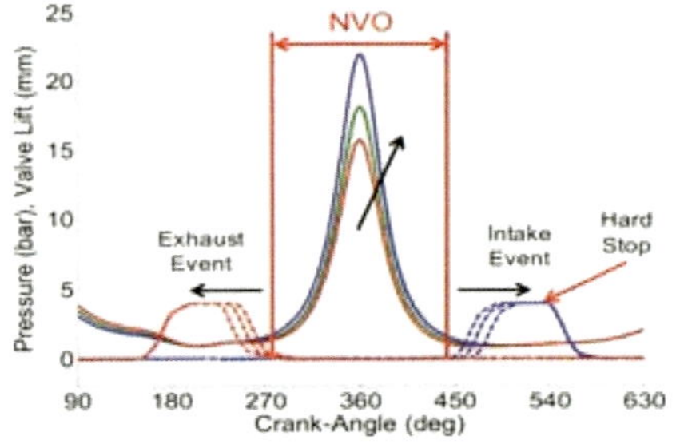

314

Scavenging in Two stroke

In a 2-stroke cycle engine, the process takes place during the latter part of the downstroke (expansion) and the early part of the upstroke (compression)

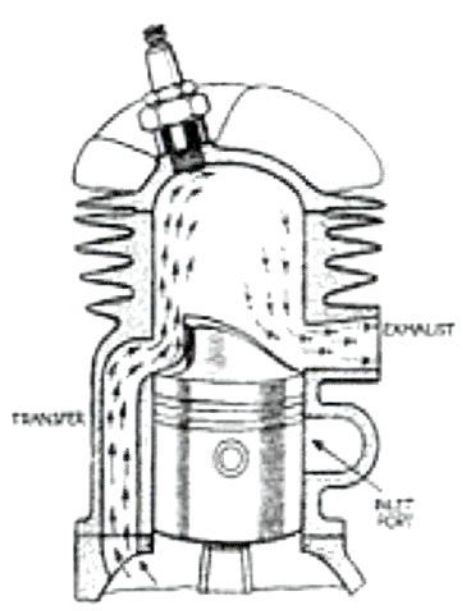

Scavenging in Four stroke

In a 4-stroke cycle engine, scavenging takes place when the piston is nearing and passing TDC during the latter part of an upstroke (exhaust) and the early part of a downstroke (intake). The intake and exhaust openings are both open during this interval of time

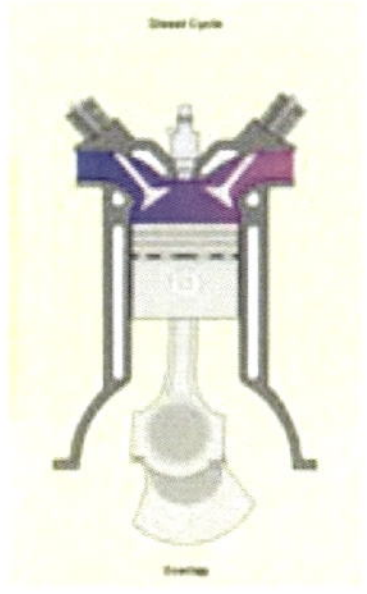

SCAVENGING AND SUPERCHARGING

Scavenging is done when air is admitted under low pressure into the cylinder while the exhaust valves or ports are open

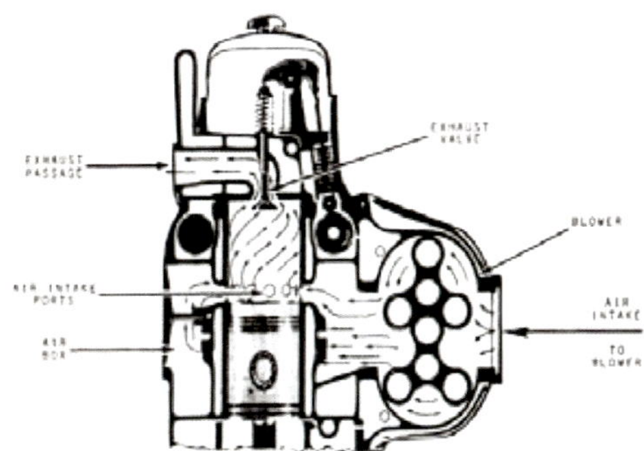

SCAVENGING AND SUPERCHARGING

Super-charging is done with the exhaust ports or **valves closed**, a condition that enables the blower to force air under pressure into the cylinder and thereby increase the amount of air available for combustion

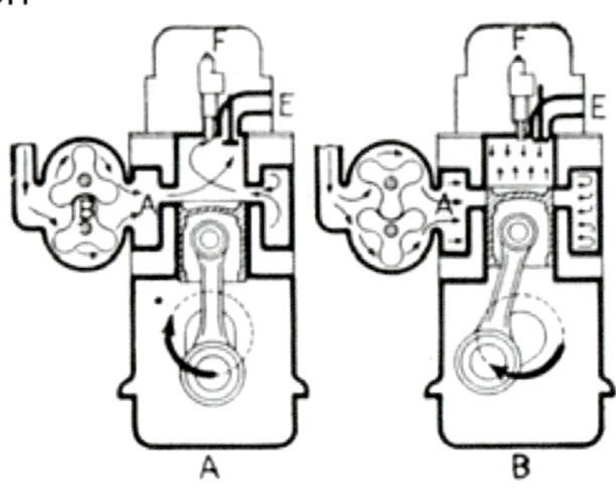

SCAVENGING AND SUPERCHARGING

- An engine is referred to as super-charged when the manifold pressure exceeds the <u>atmospheric pressure</u>
- The increase in pressure, resulting from the compression action of the blower, will depend on the type of installation
- In other words, an engine of a given size that is supercharged can develop more power than an engine of the same size that is not supercharged

Port Direct Scavenging

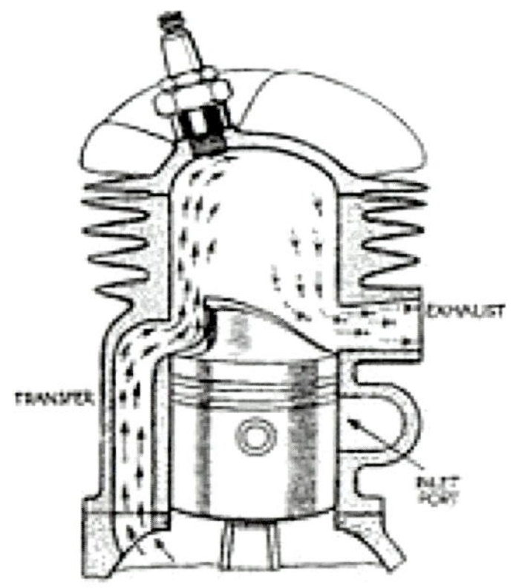

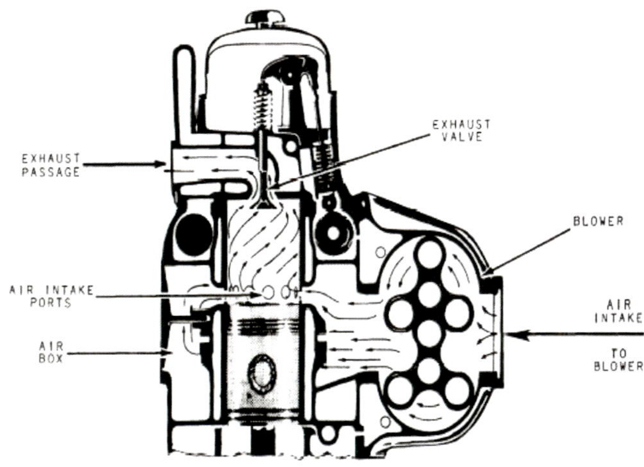

The Blower Function

- In the intake systems of all 2-stroke cycle diesel engines and some 4-stroke cycle diesel engines, a device, known as a blower, is installed to increase the flow of air into the cylinders

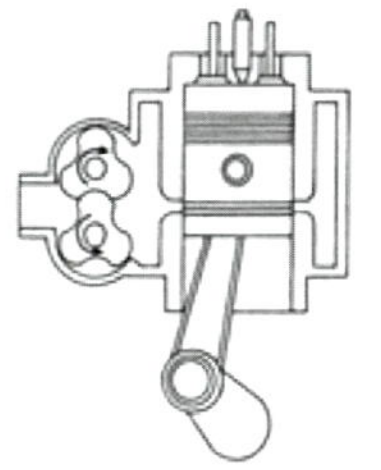 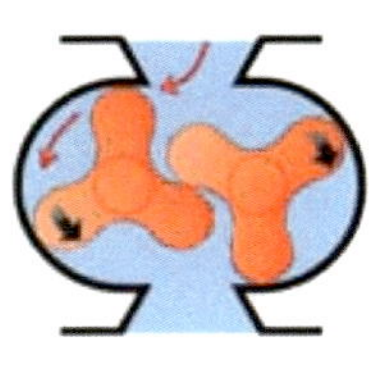

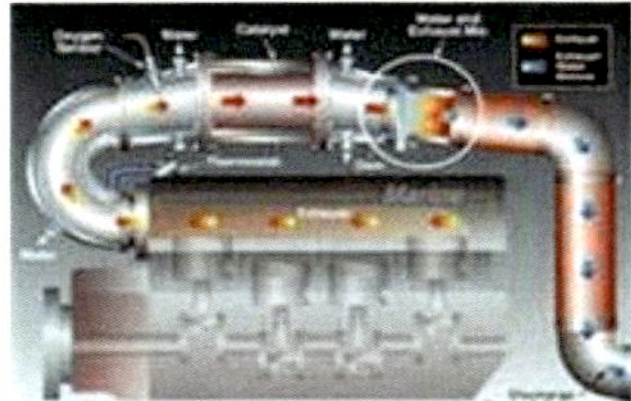

Marine Exhaust Systems

Wet & Dry

How they are Designed

Marine Exhaust Systems

- There are two basic types of marine exhaust systems, **wet** and **dry**
- The main purpose is, of course, to get rid of the hot exhaust gases. But the design of exhaust systems is also to cool the exhaust, muffle the noise and move toxic gases such as carbon monoxide, and noxious odors away from the boat and its occupants

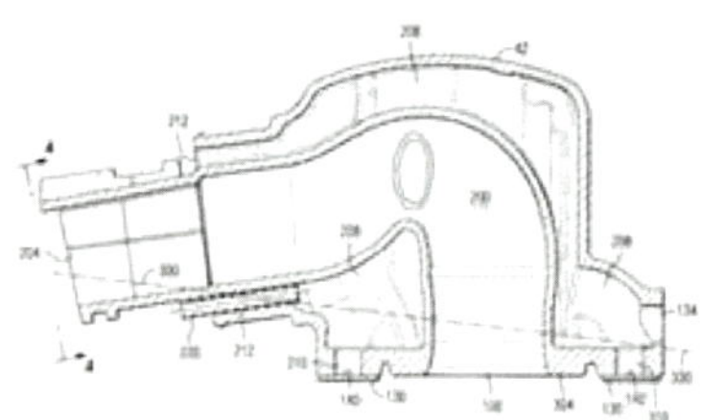

Marine Exhaust Systems

The design also needs to protect the interior of the boat from heat and to reduce back pressure that would keep the engine from running at maximum efficiency

Marine Exhaust Systems

The design has to be durable and safe, resistant to corrosion and corrosive gases, and not transmit noise and vibration into the boat structure

Wet Systems

Wet exhaust systems usually consist of a water cooled exhaust manifold, a riser off the manifold that lifts the exhaust up from the manifold and then down again

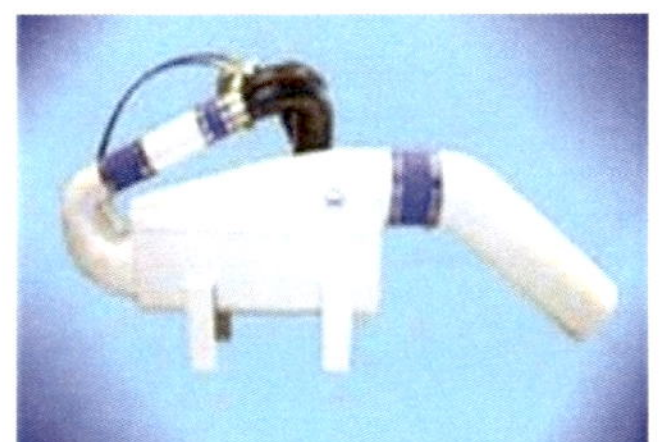 

The Risers

- The riser prevents water in the exhaust system from being ingested into the engine through the exhaust manifold
- At this point the cooling water is injected into the exhaust and blown out the exhaust opening along with the hot exhaust gases. This helps to muffle and cool the exha

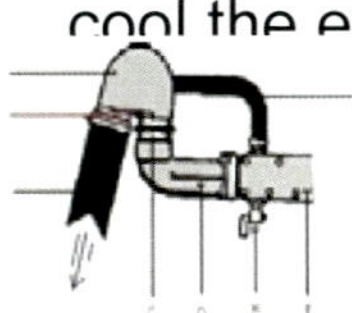  

The Muffler

- After the Raiser the system may be a muffler in the system to quiet the engine even more
- In the near future there may also be a catalytic converter in the system to reduce exhaust emissions

Dry Systems

Dry systems may or may not have water cooled manifolds. But after the exhaust exits the manifold, no water is injected into the system

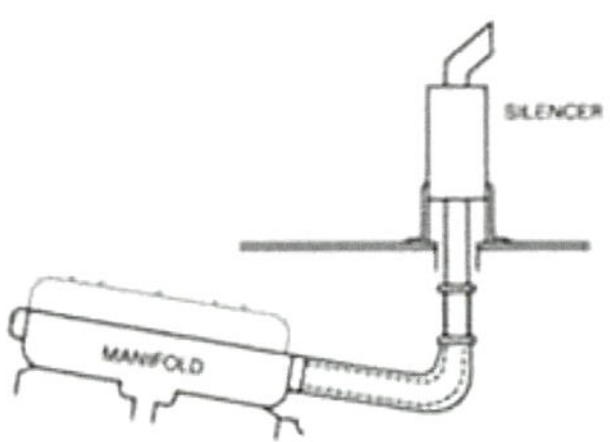

Dry Exhaust System

The exhaust pipe can then rise up vertically or go horizontally out of the boat. There is usually a muffler in the exhaust pipe

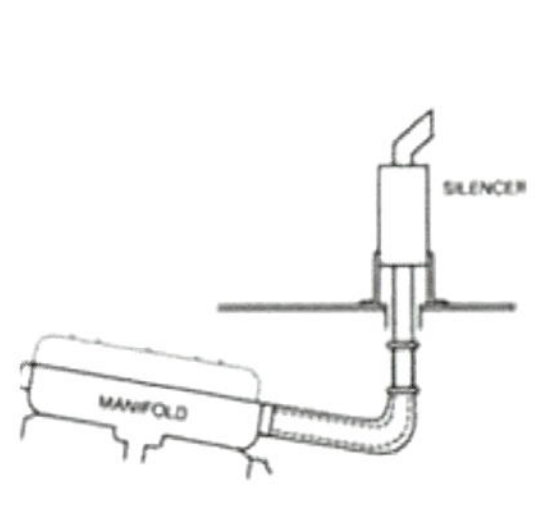

Dry Exhaust System

The pipes get very hot so usually they are wrapped with fiberglass lagging to insulate the interior of the boat from the heat and to prevent burns and fires

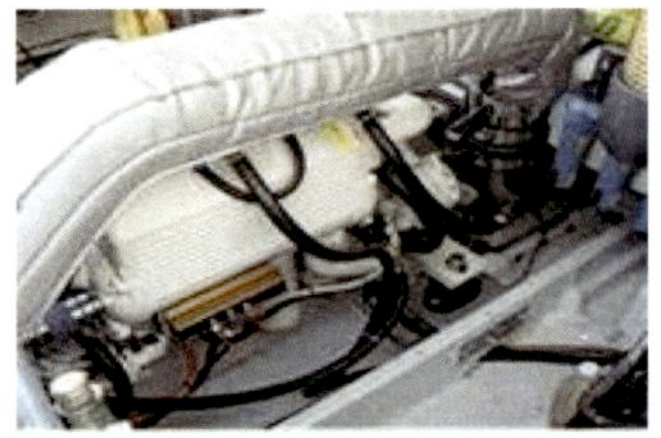

Dry exhaust Systems

Dry stacks, that is, tall vertical exhaust pipes, are commonly used on larger recreational and commercial boats to direct exhaust gases as far up and away from the boat as practical. This prevents carbon monoxide and other noxious gases from getting into the boat

Exhaust Back pressure

- One of the most misunderstood concepts in exhaust theory is back pressure.
- Your exhaust system is designed to evacuate gases from the combustion chamber quickly and efficiently.
- Exhaust gases are not produced in a smooth stream; exhaust gases originate in pulses. A 4 cylinder motor will have 4 distinct pulses per complete engine cycle

Exhaust Back pressure

- The more pulses that are produced, the more continuous the exhaust flow.
- Back pressure can be loosely defined as the resistance to positive flow - in this case, the resistance to positive flow of the exhaust stream

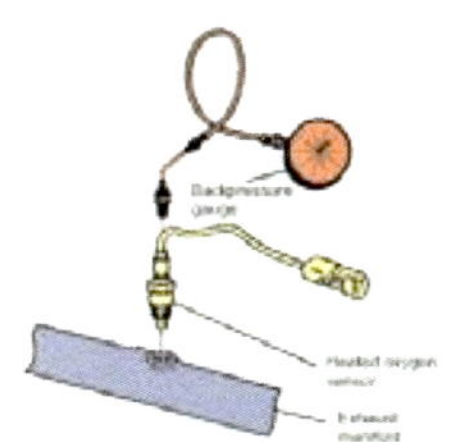

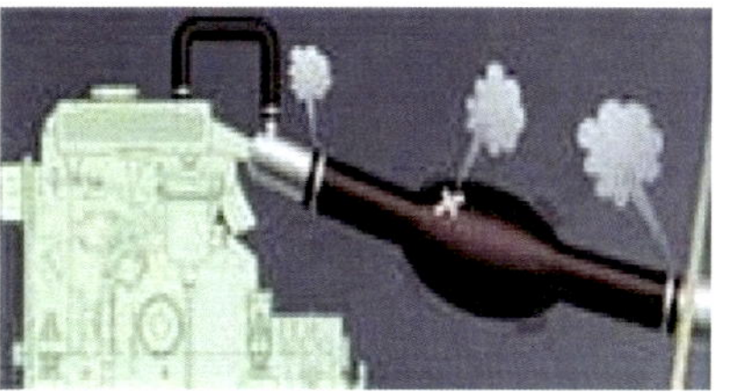

Back Pressure and Velocity

- Many people mistakenly believe that wider pipes are more effective at clearing the combustion chamber than narrower pipes.
- According with the VELOCITY concept. Here is an analogy. If you let the water just run unrestricted out of the hose it flows at a rather slow rate. However, if you take your finger and cover part of the opening, the water will spray out at a much faster rate

Exhaust Back Pressure

- You want the exhaust gases to exit the chamber and speed along at the highest velocity possible - you want a FAST exhaust stream
- If you have two exhaust pulses of equal volume, one in a 2" pipe and one in a 3" pipe, the pulse in the 2" pipe will be travelling considerably FASTER than the pulse in the 3" pipe

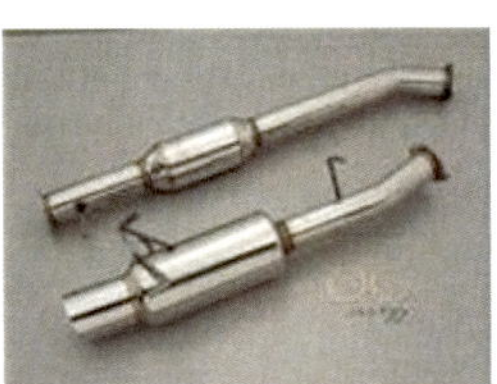 

Exhaust Back Pressure

- **Back pressure** refers to pressure opposed to the desired flow of a fluid in a confined place such as a pipe
- The trick is to have a pipe that that is as narrow as possible while having as close to zero back pressure as possible at the RPM range you want your power band to be located at

Exhaust Pipe Diameter

- Exhaust pipe diameters are best suited to a particular RPM range
- A smaller pipe diameter will produce higher exhaust velocities at a lower RPM but create unacceptably high amounts of back pressure at high rpm

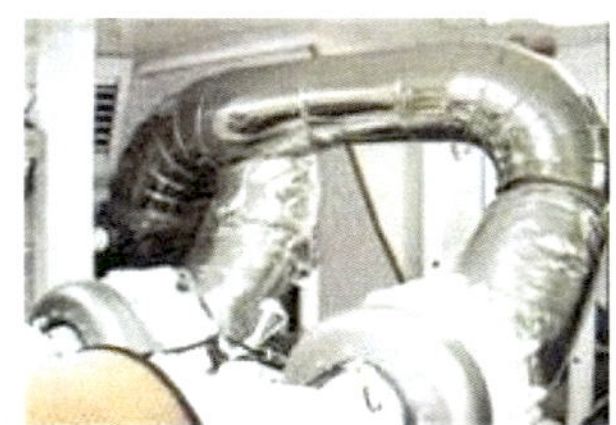

Back Pressure

- In four stroke engines the back pressure caused by the exhaust system (consisting of the exhaust manifold, muffler and connecting pipes) of a four stroke Diesel has a negative effect on engine efficiency resulting in a decrease of power output that must be compensated by increasing fuel consumption
- In a two stroke engine however, a certain amount of exhaust back pressure is needed to prevent unburned fuel/air mixture from passing right through the cylinders into the exhaust

Efficiency

- Dynamometer testing has proven that "hot" cylinders are a common cause of premature engine failure, especially in high performance marine applications
- The polishing of exhaust cavities improve scavenging of each cylinder to remove exhaust gasses that produce potentially damaging high cylinder head temperatures

  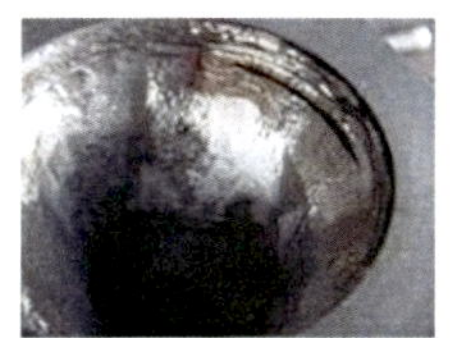

Flexible Exhaust Hose

- Hose used in wet exhaust systems shall comply with the performance requirements of UL1129

- Every exhaust hose connection shall be secured with at least two non-overlapping clamps at each end to produce a secure, liquid and vapor-tight joint

Hose and Clamps

- Clamps used for this purpose shall be entirely of stainless steel metal

- The bands shall be a minimum of one-half inch in width

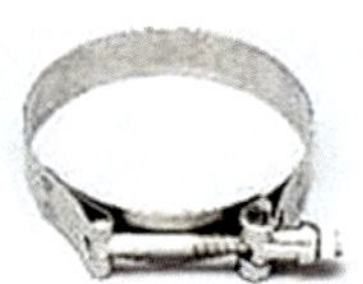  

Water Lift Silencers

'Water lift' or 'water-trap' silencers can be positioned well below the waterline, allowing the exhaust run from the engine to have a nice, safe, steep angle

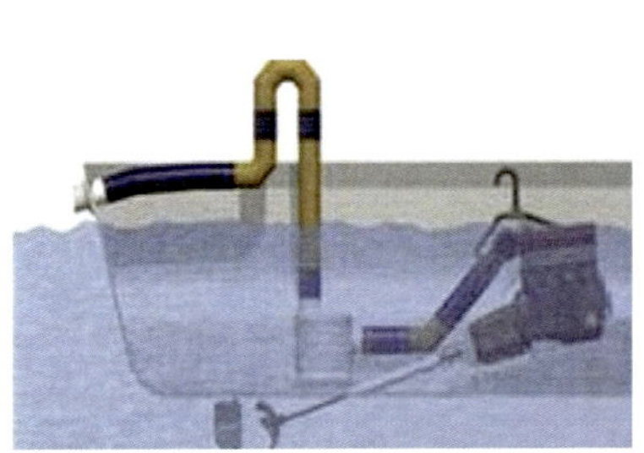

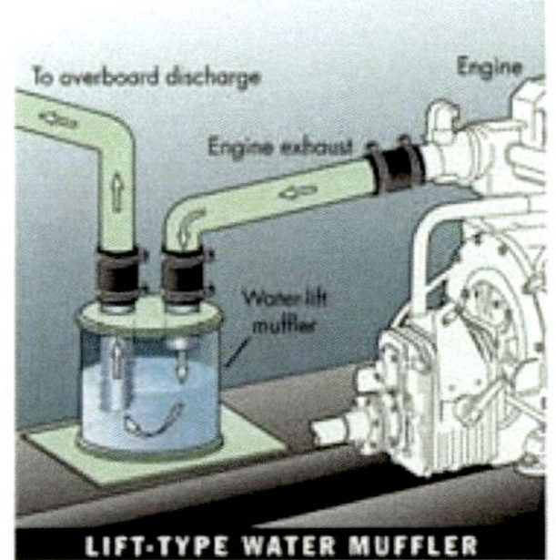

Water Lift Silencers

- You might think that such a device would rapidly fill with water and eventually block the exhaust completely, but this doesn't happen
- In a correctly designed unit, the water level stabilizes in the bottom of the chamber, just level with the bottom of the outlet pipe, with pure gas swirling around above it.

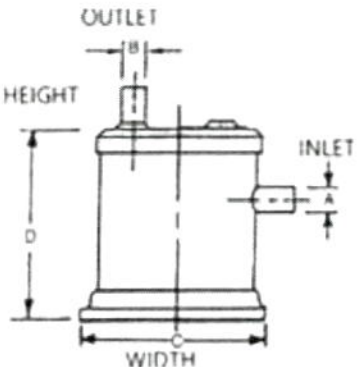

Waterlift Silencers

- Water is picked up by the gas on the way out and this further silences and cools the flow .
- Waterlift silencers should reduce exhaust noise by about 40%, compared with an unsilenced system .
- Most common is the vertical cylinder variety, which can have top inlets, horizontal or angled side inlets and a wide variety of outlets

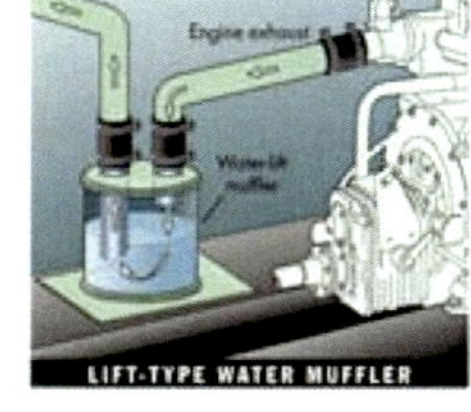

The Siphon Break

- The siphon break, or air break, is basically a small air valve which is fitted in the raw water hose line located high on the bulkhead above the engine
- This hose conveys the seawater from the engine's heat exchanger to the water injection elbow just aft of the engine exhaust manifold .

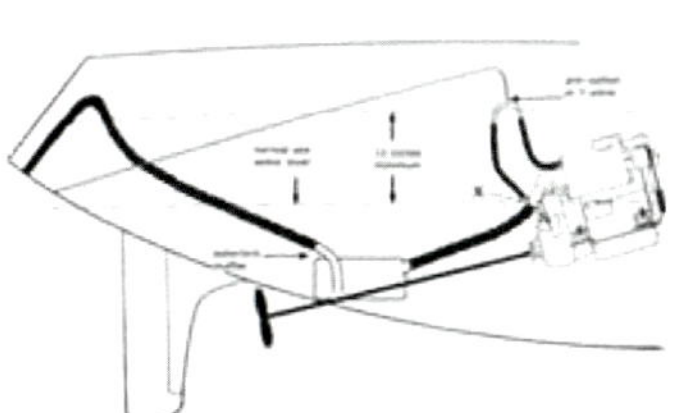

The Siphon Break

When the engine is running, sea water is drawn in via the thruhull valve [under the engine], passes through the intake strainer, the raw water pump, the heat exchanger, then up past the siphon break and down into the exhaust water-injection elbow, exiting via the exhaust muffler and hose

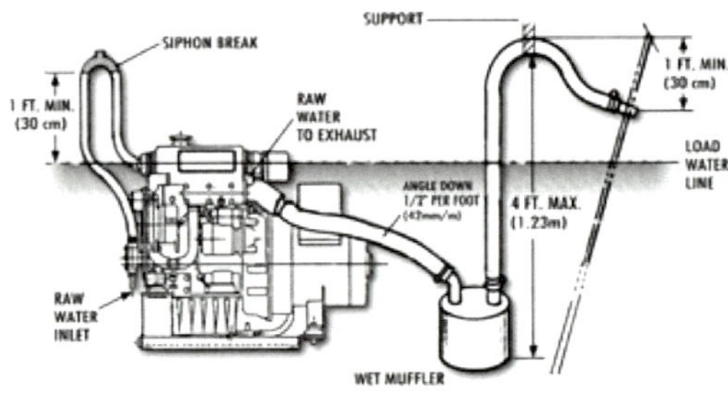

The Siphon Break

The air valve closes under water pressure with the engine running. When the engine shuts down, this valve opens and air is drawn into the hose, allowing the water within the hose to then fall away under gravity, down into the exhaust muffler

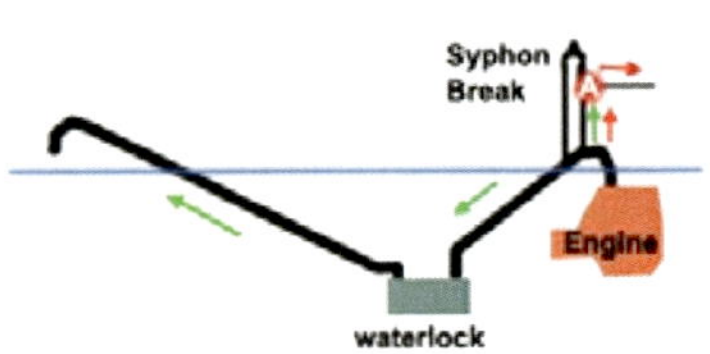

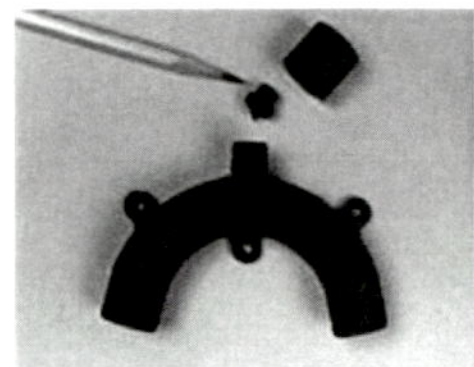

The Siphon Break

This is the critical function of the device: to prevent seawater from otherwise siphoning past the pump or being drawn in from the muffler, possibly filling the engine's exhaust manifold with sea water

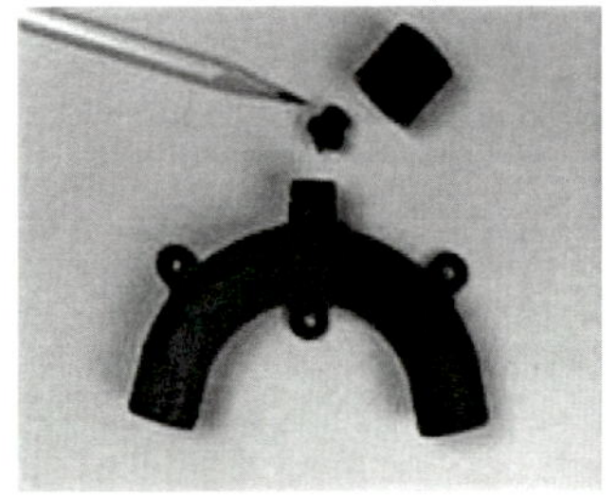

The Siphon Break

There are various types of air or siphon breaks available. The most common type consists of a lightweight ball, about the size of a pea, enclosed within a small plastic cage or basket on top [see Photo 1]. The ball presses up against a seat when water is flowing under pressure from the pump, and falls to allow air into the hose when the engine is shut down, stopping the pump

Water Separator

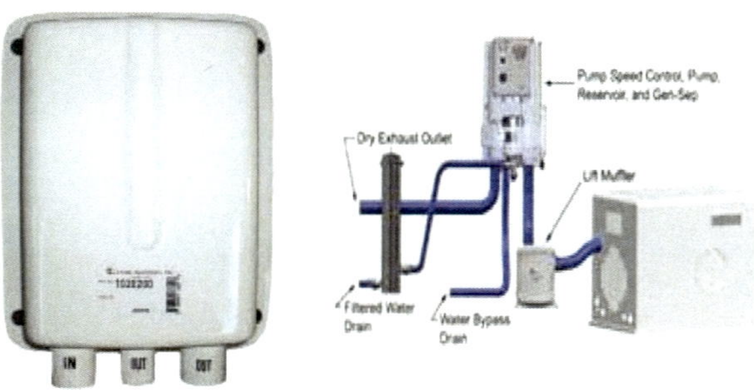

Installation with Water Separator

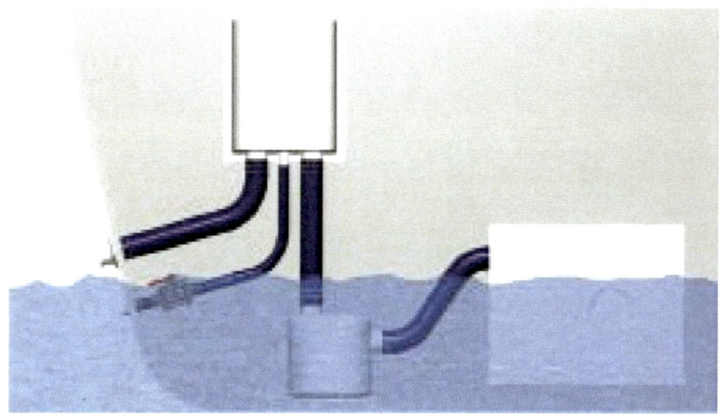

typical installation diagrams

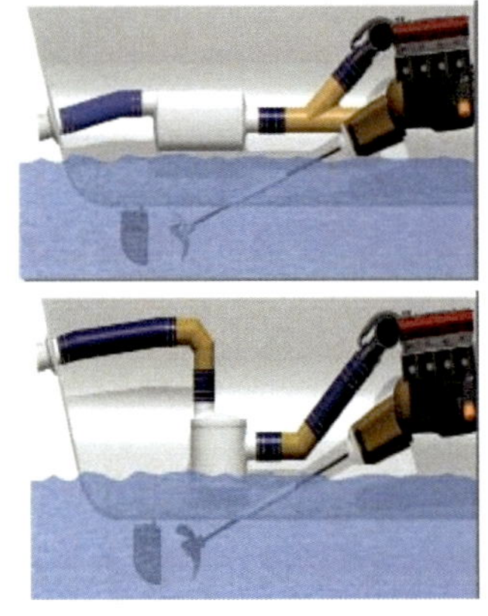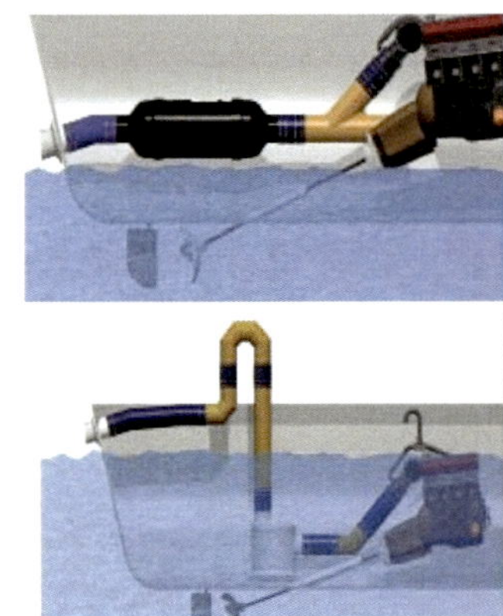

Exhaust Generator Installation

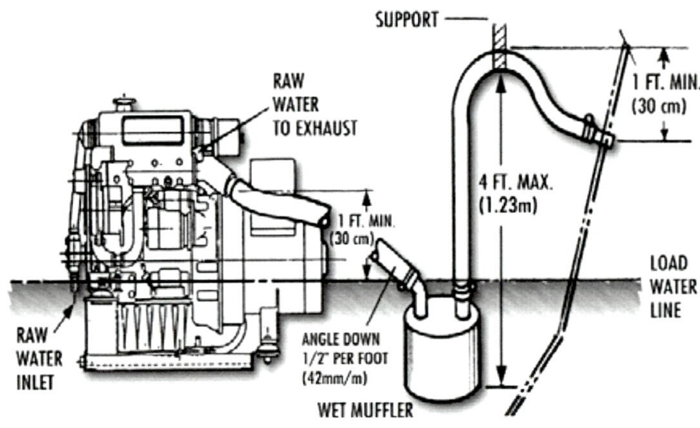

Exhaust Generator Installation

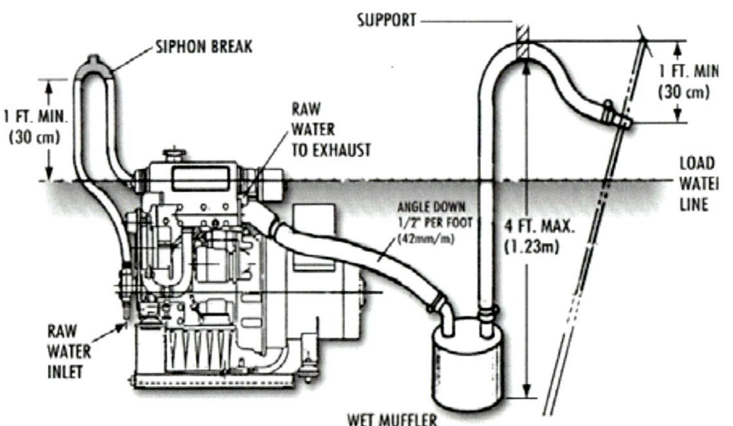

Turbocharger vs. Supercharger

- **Supercharges** use an air compressor to improve engine combustion.

- In other words a Supercharger is a fresh air compressor driven by a belt connected to crankshaft or simply by an electric motor activated by a sensor

- A **Turbocharged** engine recycles the engines exhaust.

- A turbocharger is just a fresh air compressor driven by the force of the exhaust gases

Turbocharger Terminology

- **Absolute Pressure:** Absolute pressure is the total pressure exerted by all elements in the system. This includes atmospheric pressure, vacuum created by the throttle blade, and pressurization due to forced induction.

- **Atmospheric Pressure:** The atmosphere has weight - approximately 14.7 psi (pounds per square inch) @ sea level. As you move to higher elevations, the pressure lessens .

- **Boost:** The term boost generally refers to any pressurization of the intake tract over that of the atmospheric pressure

Turbocharger Terminology

- **Blow Off Valve:** A blow off valve's primary purpose is to keep your turbo alive through sudden changes in pressure in the intake tract. A blow off valve generally is used in a MAP based system, as the valve dumps the excess pressure to the atmosphere, which would throw off the readings of a closed metered air flow system.
- In other words is a pressure drain valve to protect the life of the turbo and avoid over pressure in the combustion chamber

Turbocharger Terminology

- **Bypass Valve:** A bypass valve and a blow off valve have the same functions. The difference is the bypass valve is used in a metered air system, as the excess pressure is bled off from the intact tract, through the bypass valve, and dumped *after* the air flow meter, and *before* the turbo.
- The air has already been metered, and then flows through a circuit back into the turbo again. This keeps the turbo spinning more efficiently than just dumping the excess pressure to the atmosphere.

A diesel engine fitted with a turbocharger is called a **turbo-diesel**

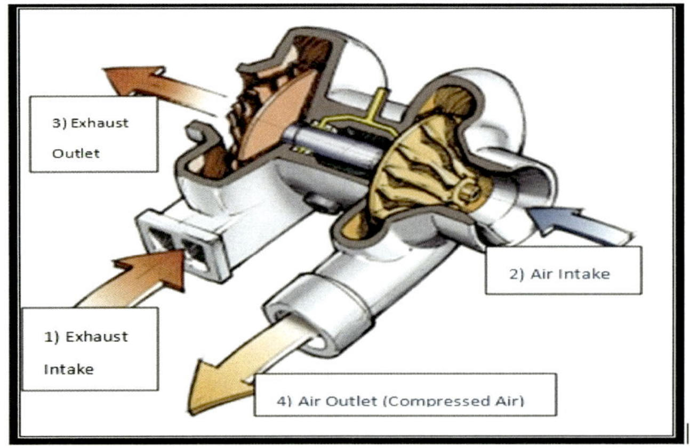

330

Working Principle

- A turbocharger is a small radial fan pump driven by the energy of the exhaust gases of an engine
- A turbocharger consists of a turbine and a compressor on a shared shaft

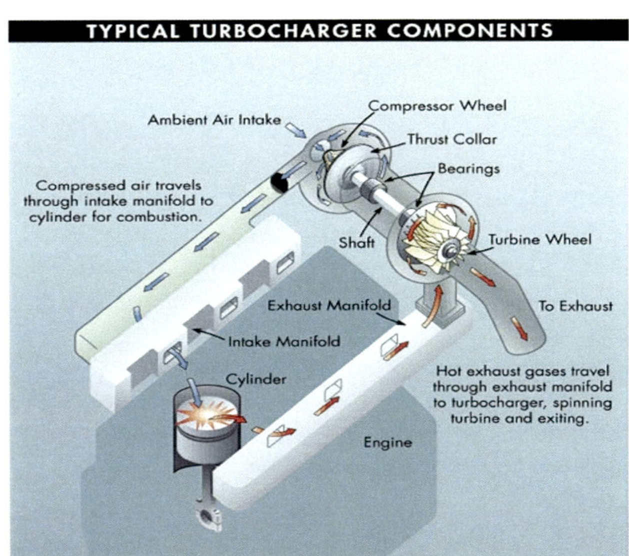

Turbocharger Terminology

- **Absolute Pressure:** Absolute pressure is the total pressure exerted by all elements in the system. This includes atmospheric pressure, vacuum created by the throttle blade, and pressurization due to forced induction.
- **Atmospheric Pressure:** The atmosphere has weight - approximately 14.7 psi (pounds per square inch) @ sea level. As you move to higher elevations, the pressure lessens .
- **Boost:** The term boost generally refers to any pressurization of the intake tract over that of the atmospheric pressure

Working Principle

The compressor draws in ambient air and pumps it in to the intake manifold at increased pressure, resulting in a greater mass of air entering the cylinders on each intake stroke

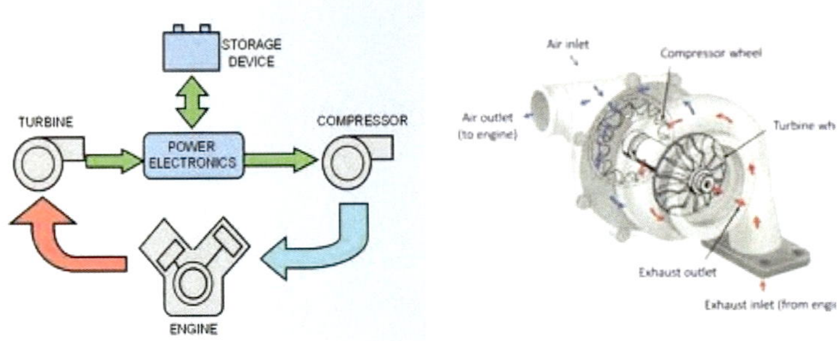

Working Principle

The turbine converts exhaust heat to rotational force, which is in turn used to drive the compressor

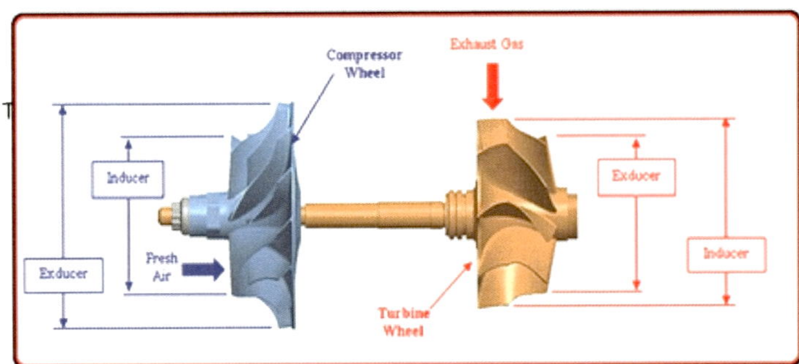

Working Principle

- The pressure in the atmosphere is no more than 1 atm (approximately 14.7 psi), there ultimately will be a limit to the pressure difference across the intake valves and thus the amount of airflow entering the combustion chamber.

- The turbocharger increases the pressure at the point where air is entering the cylinder, a greater mass of air (oxygen) will be forced in as the inlet manifold pressure increases.

Working Principle

- The additional air flow makes it possible to maintain the combustion chamber pressure and fuel/air load even at high engine revolution speeds, increasing the power and torque output of the engine
- Because the pressure in the cylinder must not go too high to avoid detonation and physical damage, the intake pressure must be controlled by venting excess gas. The control function is performed by a wastegate, which routes some of the exhaust flow away from the turbine. This regulates air pressure in the intake manifold

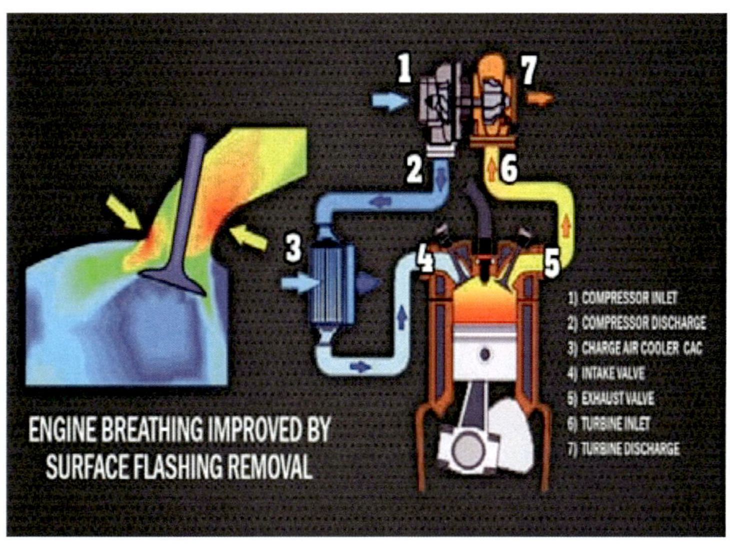

Turbochargers

A turbocharger is simply a device that pumps air and is designed to operate within a defined envelope of parameter ranges - namely boost pressure, air flow rate and turbine shaft rotational speed

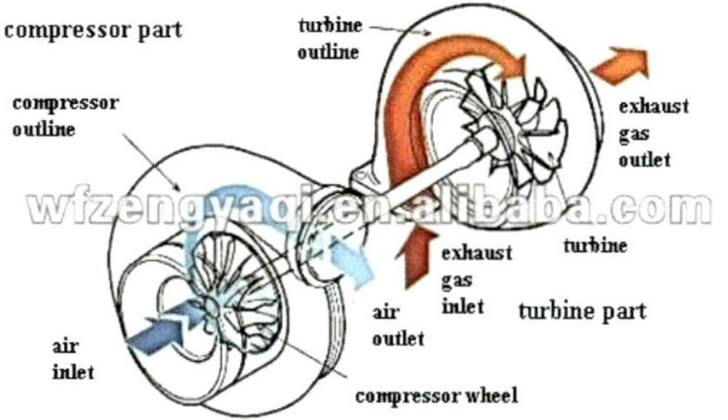

Turbochargers

- A combination of these parameters results in an **efficiency** that has a direct bearing on the temperature of air that the turbocharger compressor supplies
- The turbine in the turbocharger spins at speeds of up to **150,000 rotations per minute (rpm)**

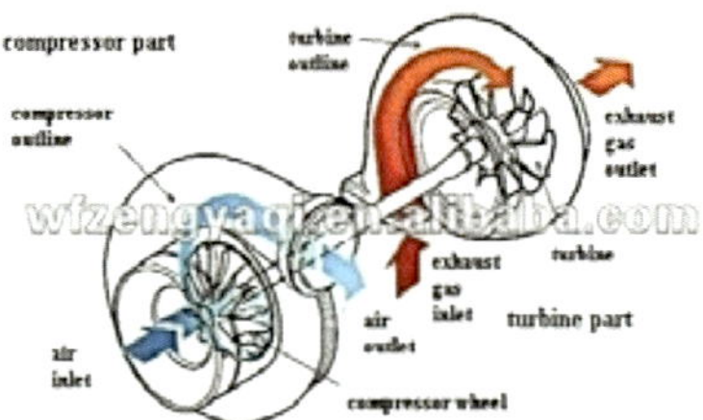

Turbocharger Components

Turbocharger Components

Common turbocharger components include the rotor assembly, bearing housing, and compressor housing

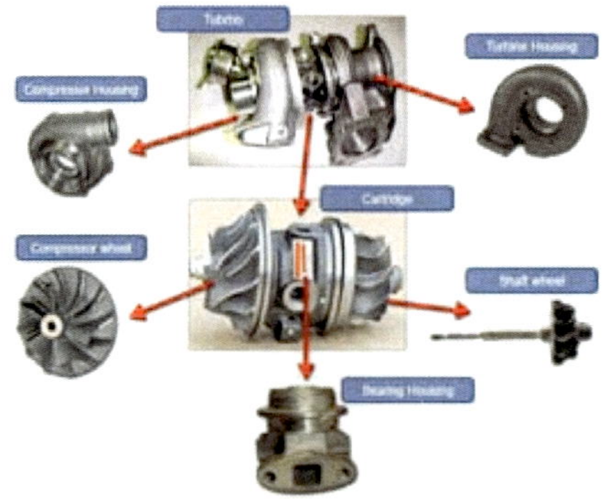

Turbocharger Components

The shaft bearings usually receive oil from the engine lubricating system. Engine coolant may circulate through the housing to aid in cooling

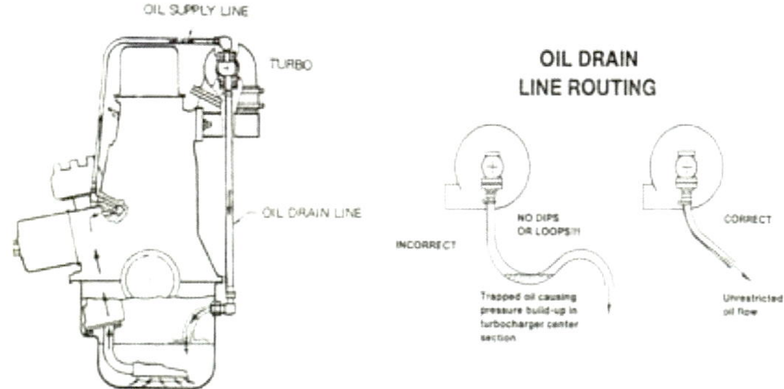

Turbocharger Basic parts

The center housing, or center hub rotating assembly (CHRA), is the part that houses the shaft and bearings that the two wheels spin on, and normally contains oil and coolant to lubricate it all

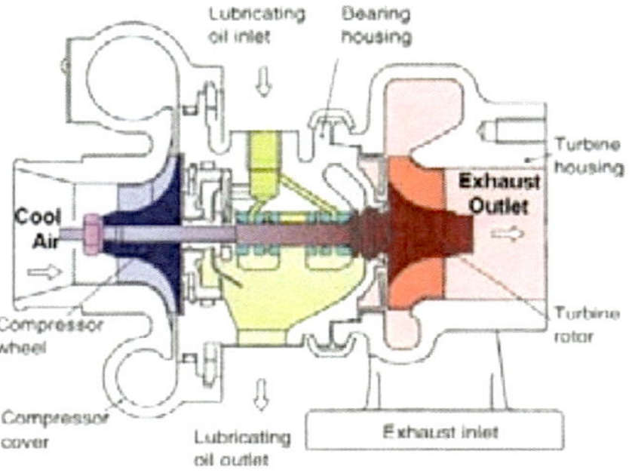

Turbocharger Basic parts

After engine shutdown, there is natural convection circulation in the turbo's oil and coolant lines from the turbo cooling off

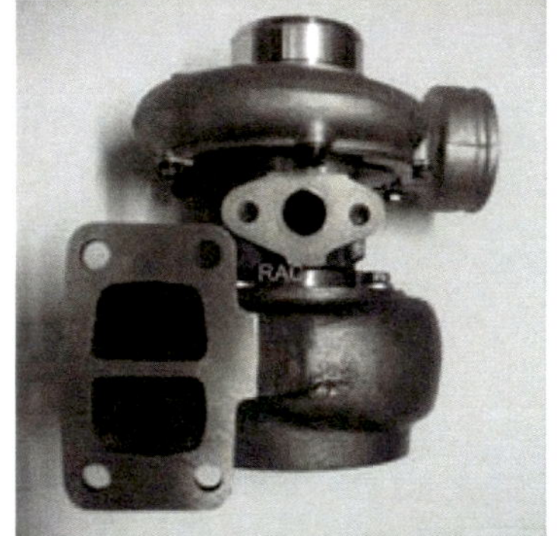

Turbine Housing (Hot Side)

Turbine housings are manufactured in various grades of spheroidal graphite iron to deal with thermal fatigue and wheel burst containment. As with the impeller, profile machining to suit turbine blade shape is carefully controlled for optimum performance

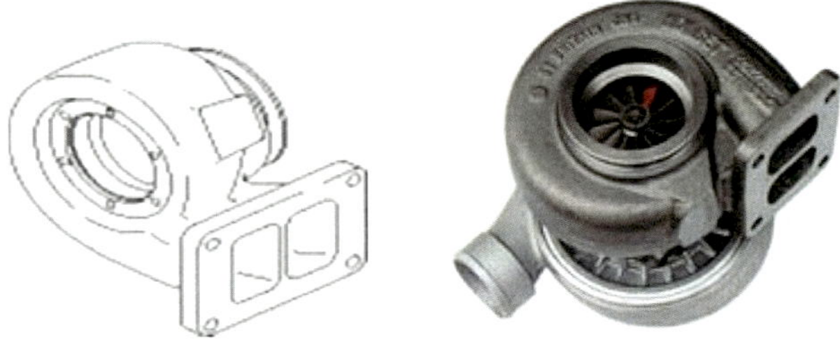

Turbine Wheel (Hot Side)

The Turbine Wheel is housed in the turbine casing and is connected to a shaft that in turn rotates the compressor wheel

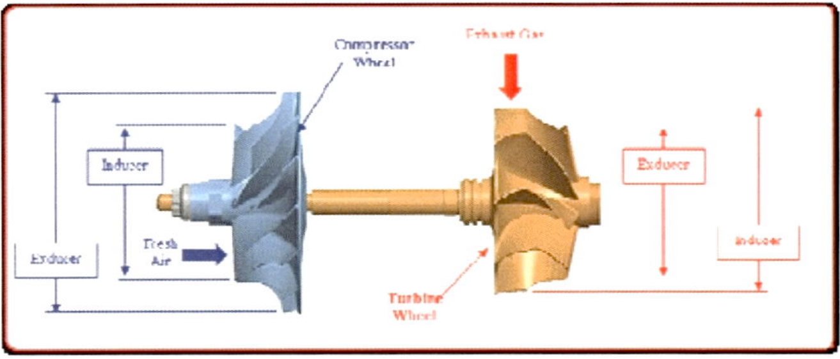

Compressor Cover (Cool Side)

Compressor housings are also made in cast aluminum. Various grades are used to suit the application. Both gravity die and sand casting techniques are used. Profile machining to match the developed compressor blade shape is important to achieve performance consistency

Compressor Wheel (Impeller)

- It is located on the same shaft with the impeller wheel and both are face to face
- An easy way to identify the compressor wheel is because it is always clean and close to the Air Filter

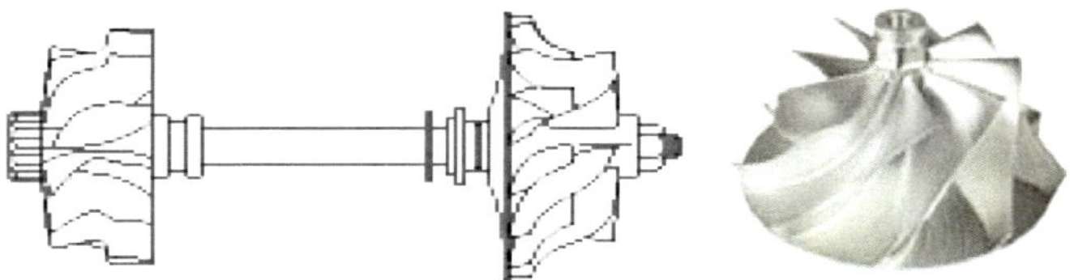

Bearing Housing

A grey cast iron bearing housing provides locations for a fully-floating bearing system for the shaft, turbine and compressor which can rotate at speeds up to 170,000 rev/min

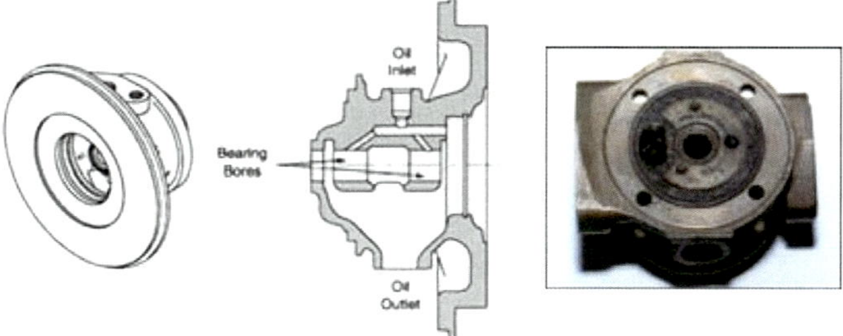

Bearing Systems

The bearing system has to withstand high temperatures, hot shut down, soot loading in the oil, contaminants, oil additives, dry starts

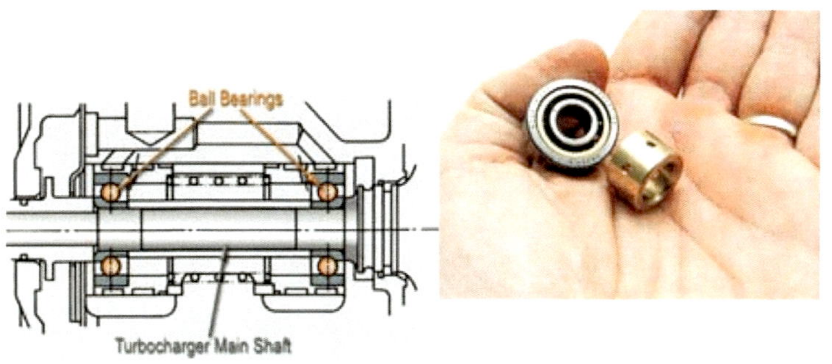

Bearing Systems

End thrust is absorbed in a bronze hydrodynamic thrust bearing located at the compressor end of the shaft assembly. Careful sizing provides adequate load bearing capacity without excessive losses

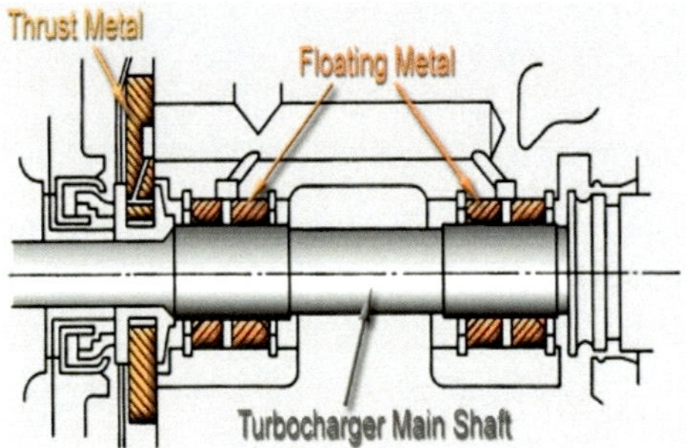

Turbo Wastegates

Internal

External

Electronically Controlled

Wastegate

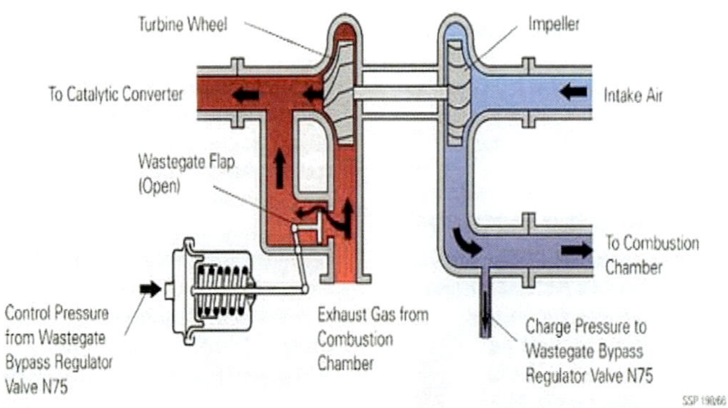

Waste-gate Turbo

- Without a wastegate, the amount of boost that a turbocharger creates varies with the pressure of the engine's exhaust. This happens because exhaust pressure varies with relation to the engine's speed (measured in RPM's)
- This implies that as an engine reaches higher RPM's, increasing amounts of boost will be created by the turbocharger
- The problem with this is that an engine can only accommodate a given amount of boost. Most stock engines are only meant to take about 10 PSI if not less.

Types of Wastegates

An internal wastegate is a component on the turbo unit itself. The gate is opened via an actuator which is a diaphragm type system

Waste-gate Turbo

- In order to regulate the amount of boost that comes into the engine, a wastegate acts as a door only allowing a given amount of exhaust to hit the turbocharger's exhaust turbine

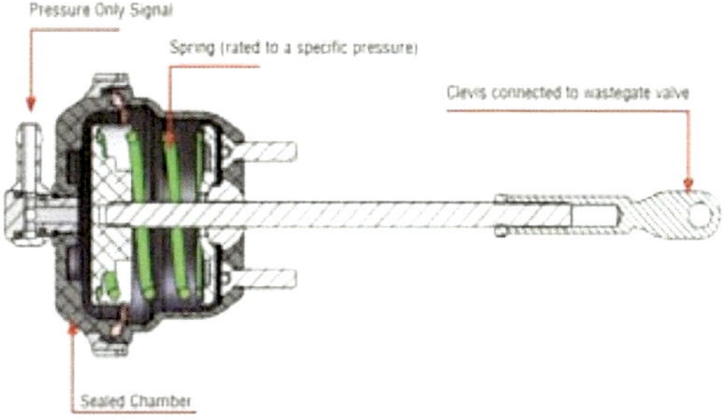

Waste-gate Turbo

Once the engine starts producing more exhaust pressure then the wastegate system will allow, a flap is opened to redirect excess exhaust away from the turbine blades. In turn, this is where a wastegate gets it's name

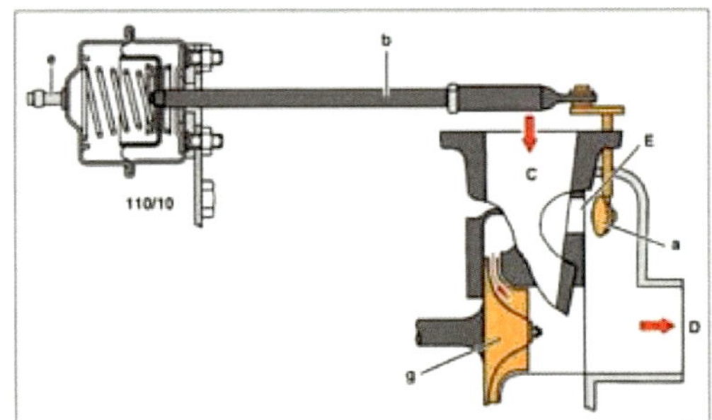

Waste-gate Turbo

It's a gate to carry away waste. In order to regulate when a wastegate opens, a boost conroller can be used

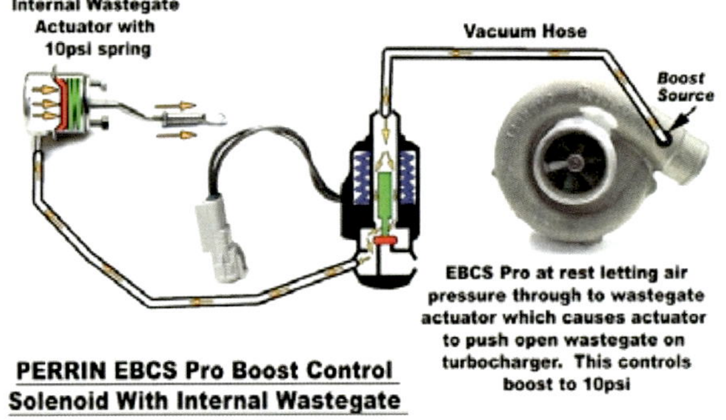

Wastegate

Built into the housing is an internal wastegate that lets excess exhaust gas and pressure out

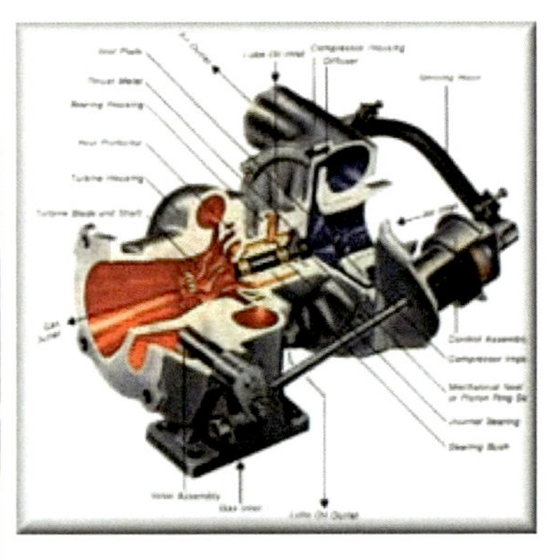

Internal Wastegate

Internal Wastegate

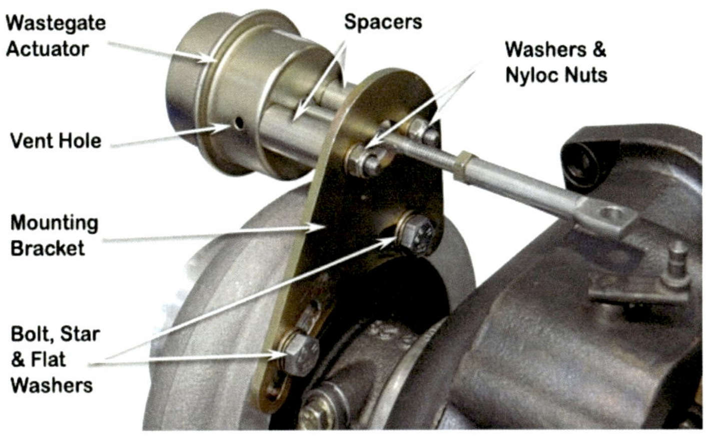

Types of Wastegate

The second one is called an external wastegate (left), unlike an internal wastegate, it is separate from the turbo unit and does not require an actuator. Excess exhaust can either be fed into the exhaust system or it can be vented straight out and into the atmosphere

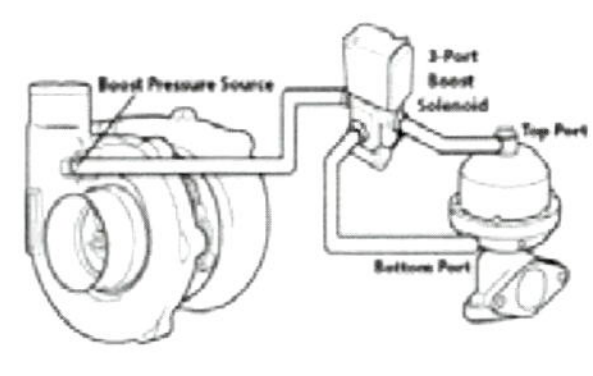

Wastegate Adjustment

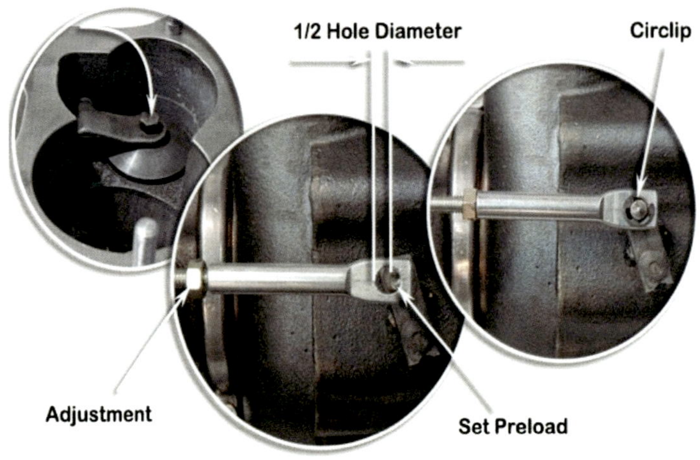

RPM Adjustment

Wastegate Inspection

- Levels of boost pressure indicate a problem:
  - Too high at full load conditions
  - Too low at all lug conditions
- Refer to specifications for wastegate

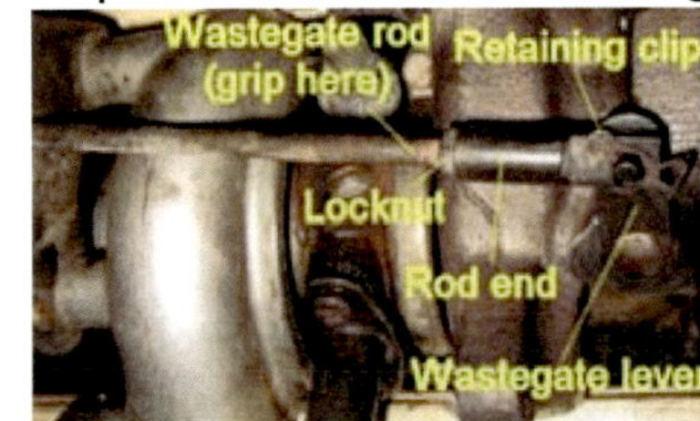

Wastegate Inspection

- Remove air line and apply corresponding amount of pressure to canister – DO NOT EXCEED 200 kPa (29 psi)
- The actuating lever should move
- NOTE: the housing assembly is preset at the factory and no adjustments can be made

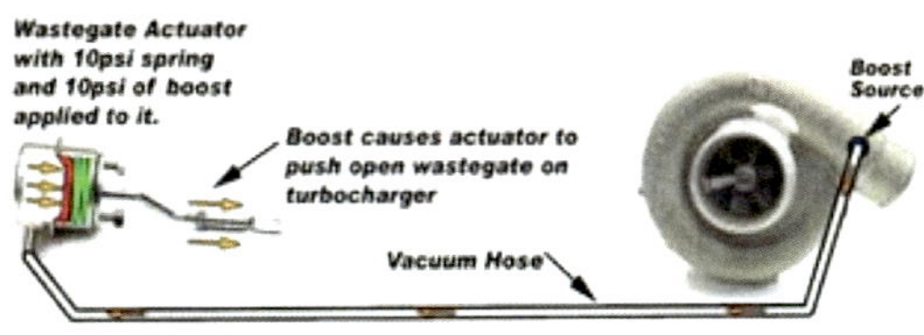

Pressure Control

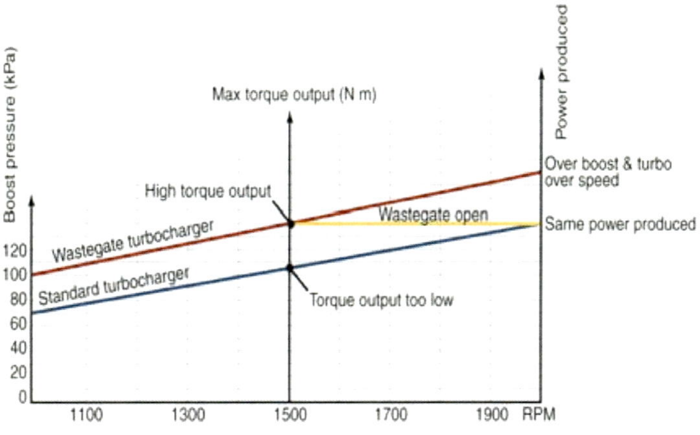

Electronically Controlled Wastegate

- Controlled by signal pressure to the ECM

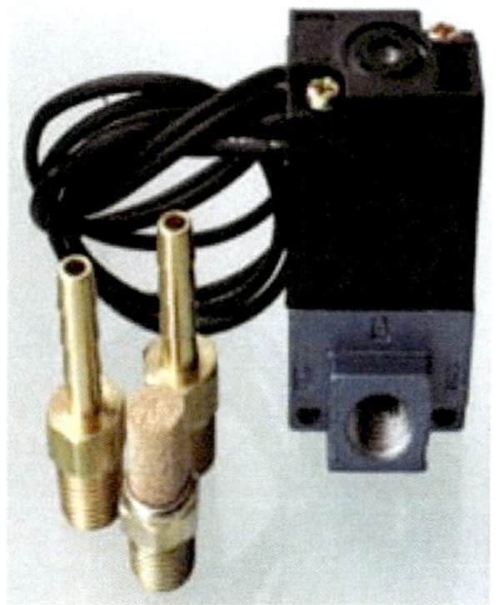

Superchargers

Scavenging & Supercharging

Blowers

Belt Driven Compressors

Electric Superchargers

Ducts & Lubrication

Types of Superchargers

There are two main types of superchargers defined according to the method of gas transfer: positive displacement and dynamic compressors

Positive displacement / Roots blower

Dynamic Compressor

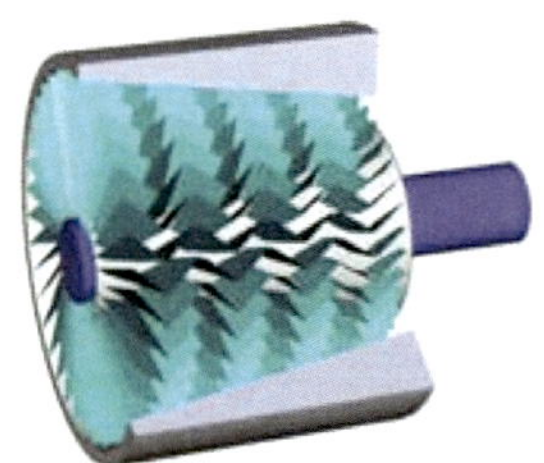

SUPERCHARGING

- An increase in airflow into the cylinders of an engine can serve to increase power output, in addition to being used for scavenging.

- Since the power of an engine comes from the burning of fuel, an increase in power requires more fuel; the increased fuel, in turn, requires more air since each pound of fuel requires a certain amount of air for combustion.

SUPERCHARGING

- The supplying of more air to the combustion spaces than can be supplied through the action of atmospheric pressure and piston action (in 4-stroke cycle engines) or scavenging air (in 2-stroke cycle engines) is called SUPERCHARGING

- In some 2-stroke cycle diesel engines, the cylinders are supercharged during the air intake simply by an increase in the pressure of scavenging air. The same blower is used for super-charging and scavenging

Valve Timing in a Supercharged Engine

- In the supercharged engine, the closing of the intake valve is slowed down so that the in-take valves or ports are open for a longer time after the exhaust valves close.

- The increased time that the intake valves are open (after the exhaust valves close) allows more air to be forced into the cylinder before the start of the compression event.

- The amount of additional air that is forced into the cylinder and the resulting increase in horse-power depends on the pressure in the air box or intake manifold

Valve Timing in a Supercharged Engine

- The increased overlap of the valve openings also permits the air pressure created by the blower to remove gases from the cylinder during the exhaust event.

- This overlap will increase power.

- The amount of the increase depends on the pressure the supercharger applies.

- The increased valve overlap allows the air pressure created by the blower to remove gases from the cylinder during the exhaust stroke

SCAVENGING AND SUPERCHARGING

- An engine is referred to as super-charged when the manifold pressure exceeds the <u>atmospheric pressure</u>

- The increase in pressure, resulting from the compression action of the blower, will depend on the type of installation

- In other words, an engine of a given size that is supercharged can develop more power than an engine of the same size that is not supercharged

Supercharger / Blower

A **supercharger** is an air compressor used for increasing the pressure, temperature , and density of air supplied to an internal combustion engine

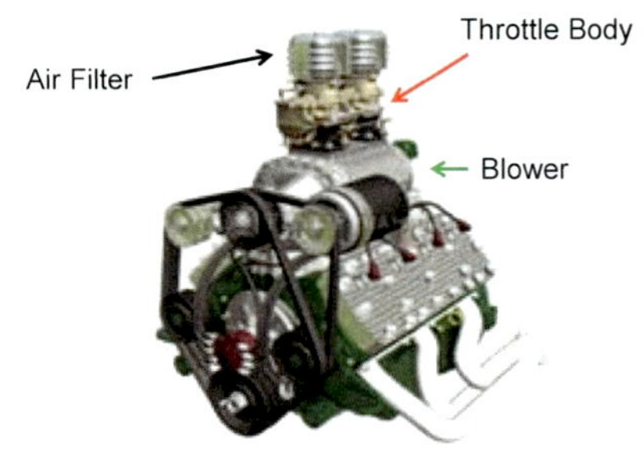

Supercharger / Blower

This compressed air supplies a greater mass of oxygen per cycle of the engine to support combustion than available to a naturally aspirated engine, enabling for more fuel to be burned and more work to be done per cycle, thus allowing to increase the power produced by the engine

The Blower Function

The blower compresses the air and forces it into an air box or manifold, which surrounds or is attached to the cylinders of an engine. Thus, more air under constant pressure is available as required during the cycle of operation

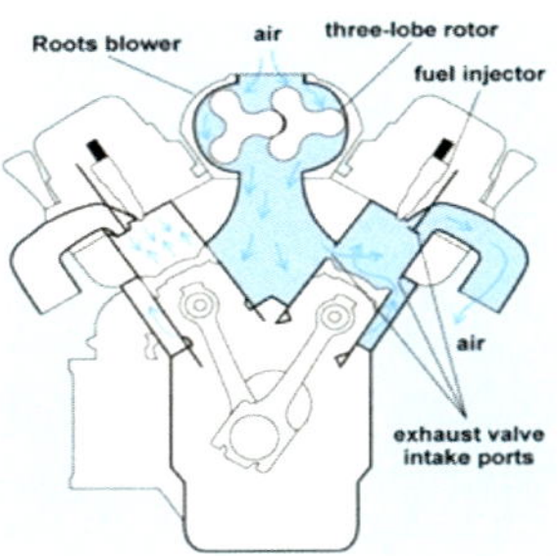

The Blower Function

- The increased amount of air, a result of blower action, fills the cylinder with a fresh charge of air
- During the process, the increased amount of air helps to clear the cylinder of the gases of combustion. The process is called SCAVENGING

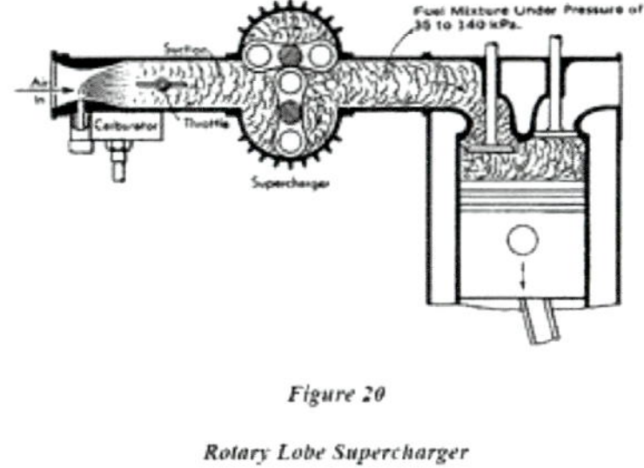

Figure 20

Rotary Lobe Supercharger

Positive Displacement Chargers

Positive displacement (Roots) blowers and compressors deliver an almost constant level of pressure increase at all engine speeds (RPM)

Dynamic Compressors

Dynamic compressors do not deliver any pressure at all at low speeds, past the threshold speed then increasing pressure with increasing speed

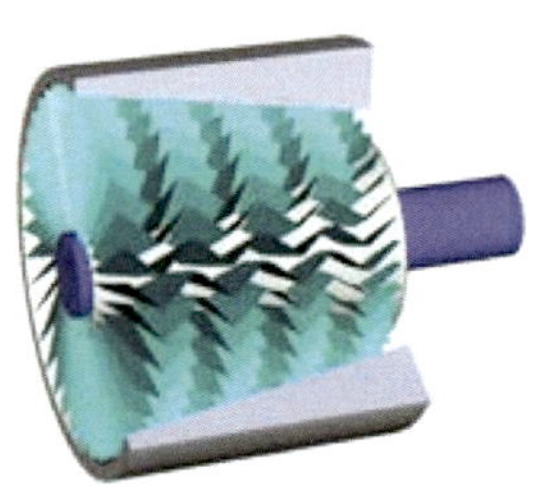

Roots Blower Parts

Roots Blower Components

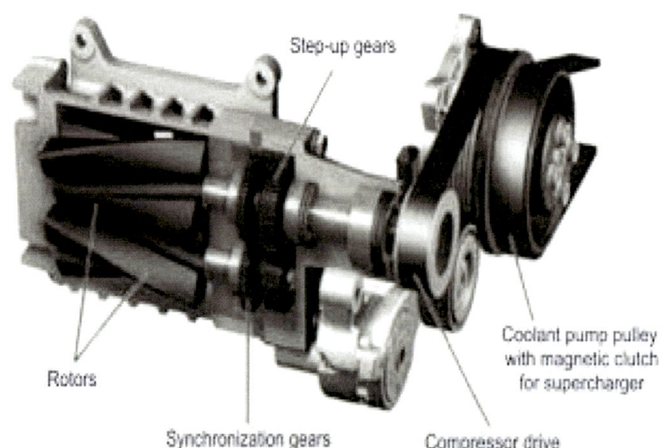

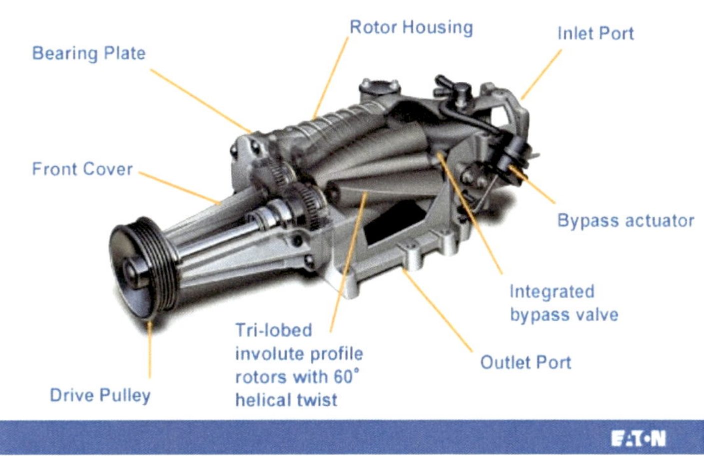

Belt Driven Supercharger

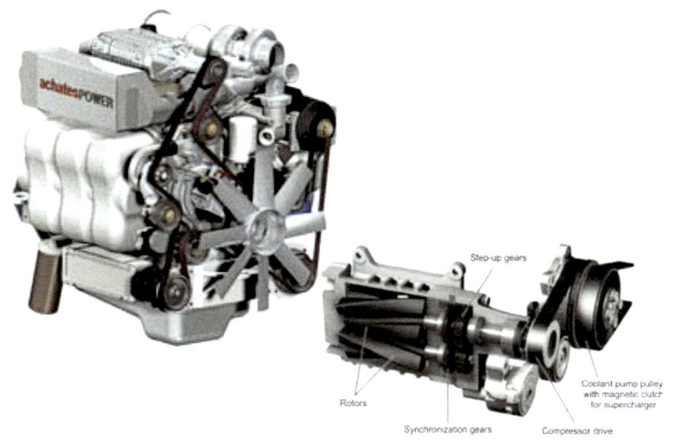

12/24V Turbo & Supercharger

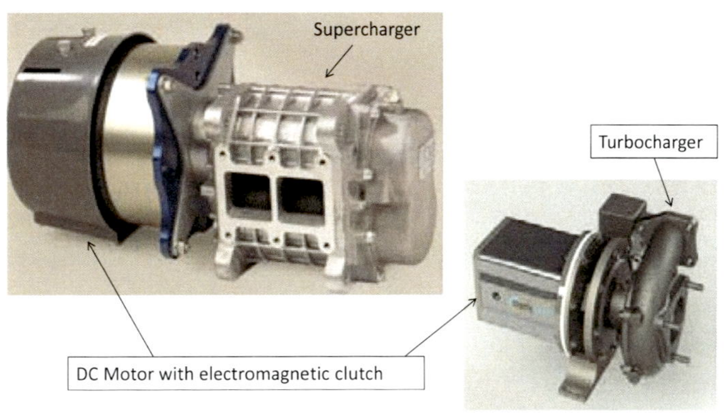

Blower Common Problems

- The main problems that will require repair include:
    - Losing power
    - Extreme noisiness
    - Blue & Black exhaust smoke
- The noise and power problems can normally be fixed easy. However, certain issues such as blue and black smoke will require specialized attention

Cleaning

- First clean the exterior. A dry cleaning solvent is the best
- Replace the air cleaning element. Carbon buildup can also occur in the housing unit, and this should be cleaned out

Duct Connections

- Because superchargers work by compressing air, any loose connections will cause problems with the air compression
- Most of the time the duct connections will have loosened and air will escape. This means that it is more difficult for the pressure to get high enough.

Air and Oil

- One of the most common problems is a dirty air filter. This can cause clogging and if left too long can cause oil leaks especially from the blower gear box

Turbocharger vs. Supercharger

- **Supercharges** use an air compressor to improve engine combustion.
- In other words a Supercharger is a fresh air compressor driven by a belt connected to crankshaft or simply by an electric motor activated by a sensor
- A **Turbocharged** engine recycles the engines exhaust.
- A turbocharger is just a fresh air compressor driven by the force of the exhaust gases

Turbocharger Selection

Altitude Vs Engine Performance

Boost Pressure & Pressure Ratio

Atmospheric Pressure

Turbo selection

Volumetric Efficiency

Sizing Turbocharger

Torque and Horsepower

Altitude Vs Efficiency

- High Altitude – quantity of air (oxygen) is reduced
- Without turbocharger
  - Engine performance falls significantly
- With turbocharger:
  - Power output reduced approx 1% per 300 M above sea level
  - Above 2000 M, fuel delivery must be decreased according to specifications

Turbocharger Efficiency

Turbocharger efficiency is essentially the ratio of useful work produced by the turbocharger from the work required to drive the turbocharger

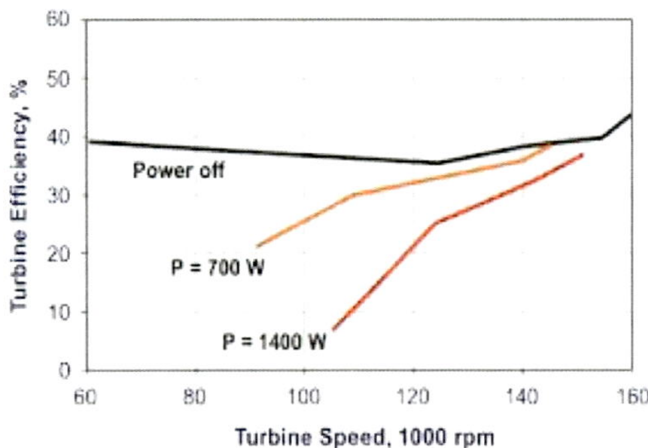

Efficiency Vs Air Density

- The efficiency is typically displayed as a fraction where an efficiency of 1.0 is ideal. It is typically accepted that an efficiency of 0.6 is the lowest operational limit.

- A form of supercharger, the purpose of a turbocharger is to increase the density of air entering the engine to create more power

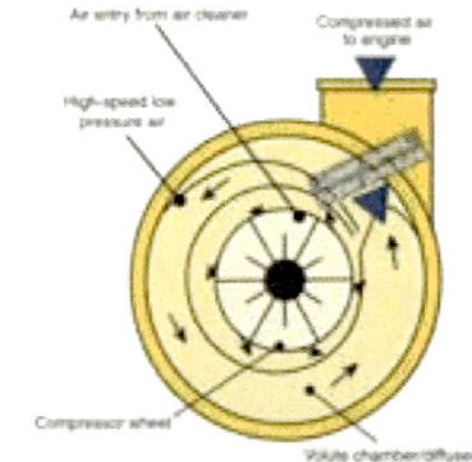

Boost Pressure Recommended

The typical boost pressure provided by either a turbocharger or a supercharger is 6 to 8 (psi) (gas Engines) and 10 to 17 PSI in diesel engines

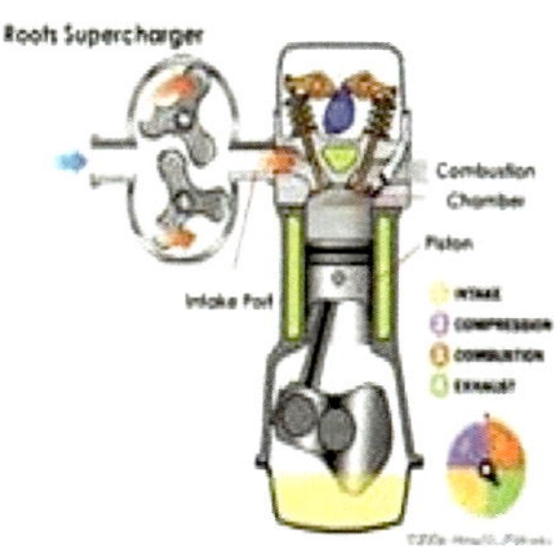

Atmospheric Pressure & Efficiency

- Since normal atmospheric pressure is 14.7 psi at sea-level, you can see that you are getting about 50-percent more air into the engine. Therefore, you would expect to get 50-percent more power
- It's not perfectly efficient, though, so you might get a 30-percent to 40-percent improvement instead

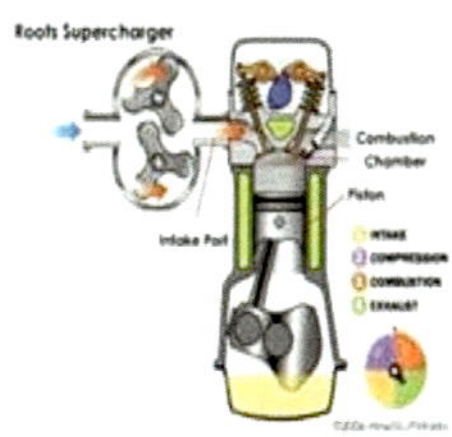

Turbochargers Efficiency

- Turbocharger efficiency is essentially the ratio of useful work produced by the turbocharger from the work required to drive the turbocharger
- An increase in temperature is of course a by product of compressing air and is unavoidable
- The efficiency is typically displayed as a fraction where an efficiency of 1.0 is ideal. It is typically accepted that an efficiency of 0.6 is the lowest operational limit
- Naturally the maximum rotational speed of the turbocharger rotating assembly (the turbocharger's red-line) is also a limiting factor

Turbo Vs. Supercharger

- The key difference between a turbocharger and a supercharger is its **power supply**
- Something has to supply the power to run the air compressor

Turbo selection & Volumetric efficiency

- If the exhaust housing is small but its turbine is large, the airflow will get choked.
- If the exhaust housing is large but the turbine is small, the airflow will not be efficiently directed at the turbine.
- The exhaust and intake side also should be in harmony.
- Like a water funnel, no matter how much water you put into the top, there is a range of how much water can come out the bottom

Turbo selection & Volumetric efficiency

- An inappropriately sized and/or matched component will prevent the components from working in the area of good efficiency, performance, and value.
- Each mod should have a set of supporting mods working towards an overall goal
- Before choosing components and modifying your engine, have an estimate of about how much power you want, then design the modifications around making that overall goal

Sizing Turbocharger

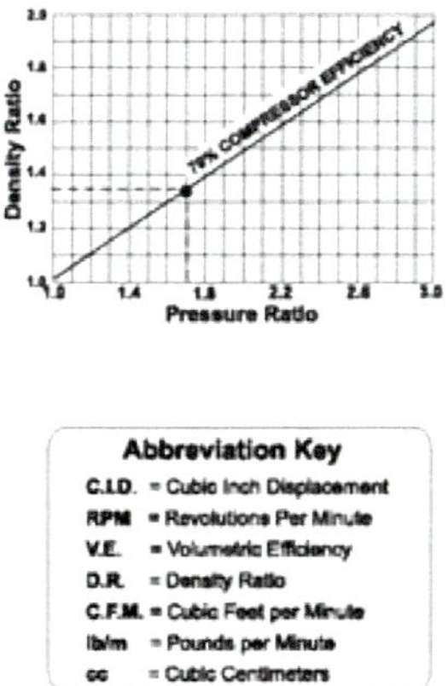

Abbreviation Key

C.I.D. = Cubic Inch Displacement
RPM = Revolutions Per Minute
V.E. = Volumetric Efficiency
D.R. = Density Ratio
C.F.M. = Cubic Feet per Minute
lb/m = Pounds per Minute
cc = Cubic Centimeters

Turbocharger Calculation

- In order to select the appropriate turbocharger for a custom application, you should calculate the following items
    - (1) The Boost Pressure Recommended
    - (2) Turbocharged AirFlow (CFM) (in lbs per minute)
    - (3) A copy of the gasket where the turbocharger should be bolted at the exhaust manifold

(CFM)   Air Flow Rate (Naturally Aspired)

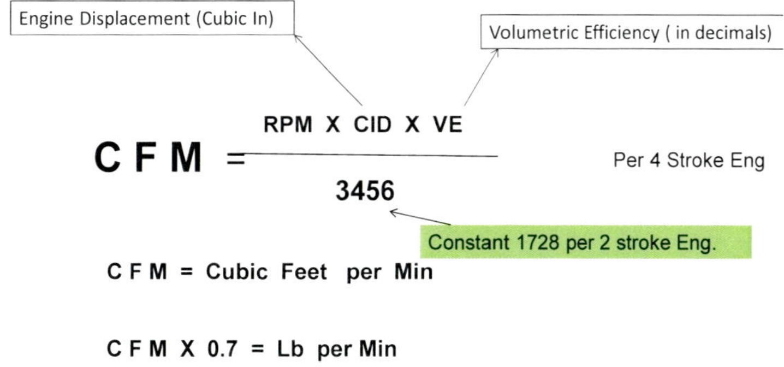

Volumetric Efficiency

Is a ratio (or percentage) of the quantity of air that is trapped by the cylinder during induction over the swept volume of the cylinder under static conditions

$$\eta_V = \frac{Volume\ of\ air\ taken\ into\ cylinder}{Maximum\ possible\ volume\ in\ the\ cylindre}$$

$$\eta_V = \frac{V_{air}}{V_c}$$

Volumetric Efficiency

Volumetric Efficiency can be improved by compressing the induction charge. In this case Volumetric Efficiency can exceed 100%

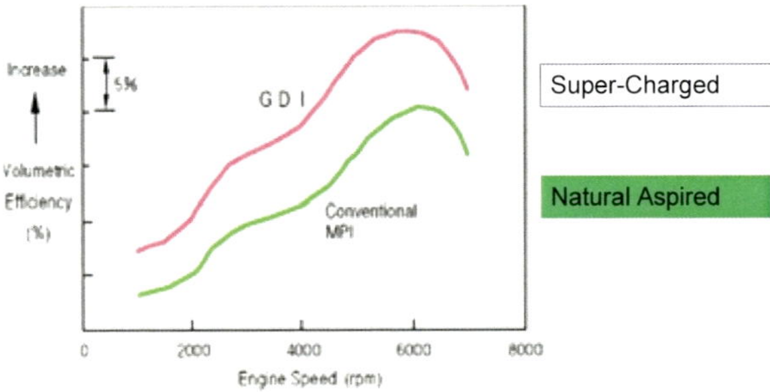

Volumetric efficiency

- The theoretical maximum *volume* of air that each cylinder can ingest during the intake cycle is equal to the swept volume of that cylinder [(3.1416/4) x bore x bore x stroke]
- The actual amount of air the engine ingests compared to the theoretical maximum is called volumetric efficiency (VE)
- An engine operating at 100% VE is ingesting its total displacement every two crankshaft revolutions

Volumetric efficiency

- If life was perfect, we could fill the cylinders completely with air
- If we had 17 psi boost in the intake manifold, we would open the intake valve and get 17 psi in the cylinder before the intake valve closed
- Unfortunately, this doesn't usually happen. With some exhaust remaining in the cylinder and the restriction offered by the intake ports and valves the actual amount of air that flows into the cylinder is somewhat less than ideal

Volumetric efficiency

- The amount that does flow divided by the ideal amount is called the volumetric efficiency.
- For your basic Cummins small block 6-cyl, this number is around 0.85 (or 85%)
- With tunnel rams some normally aspirated engines can get over 90% at certain rpms due to the ram effect
- To take this into account when we calculate flow into the engine, we multiply the ideal amount of air by the efficiency to get the actual amount of air

Turbo selection & Volumetric efficiency

- Also keep in mind that changing turbo components are only a part of increasing volumetric efficiency (VE).
- Adding camshafts, porting or tuning the intake manifold and cylinder heads, all change the volumetric efficiency and will further contribute to the efficiency of the engine.
- The most important characteristic of each turbo component is that they have to work well as part of a whole system

Volumetric Efficiency for Engine and Turbo

- The volumetric efficiency of a naturally aspirated engine is from 85% to 100% depending of the engine wear
- The volumetric efficiency of a turbocharger or Supercharger is between 50% and 60% depending on factors such as lubrication and restrictions on exhaust
- The maximum recommended Boost pressure for a gasoline engine is between **3 PSI and 7 PSI**
- The maximum recommended Boost pressure for a Diesel engine is between **10 PSI and 20 PSI**

Turbo (CFM) Produced

- Is the wanted (CFM's) at the output of the turbo

- CFM (Turbo) = CFM (naturally aspired) + 25% (no more than 30%)

Boost Pressure / Press Ratio

- **Boost Pressure** is the pressure in the induction system of an engine in excess of the standard sea-level atmospheric pressure
- **Pressure ratio (PR)** is the ratio between the wanted horsepower over the actual horsepower at the same RPM
- PR = HP (wanted) / HP (Actual Not Aspired)

Boost Pressure / PR

- For Example we wanted 200 HP after the turbo installation. But right now the power is only 132 HP at 2200 RPM
- PR = 200/132 = 1.47
- then   PR is = **1.47 : 1**
- This is <u>0.47 PSI</u> over the Atmospheric pressure
- **Boost Pressure = 14.7 X 0.47 = 7 PSI**
- You need 7 PSI of Boost to add 68 HP

Boost and Horsepower

- The general rule of thumb is that a turbo will increase horsepower by about 7 percent per pound of boost over a naturally aspirated configuration, and a supercharger will increase it by 5 or 6 percent per pound of boost
- The supercharger's return is a bit lower because it takes power from the crankshaft to turn the compressor, which a turbo does not
- For an engine with 150 HP then the gain with turbocharger will be 234 HP and with Supercharger will be 222 HP

Pressure Ratio / Over Sea Level

- Before calculating the compressor Pressure Ratio, you must decide what maximum boost
- Subtract .5 p.s.i. from 14.7 p.s.i. for each 1,000 ft. above sea-level to determine approximate atmospheric pressure at altitudes

$$\text{Pressure Ratio} = \frac{\text{Boost Press } + \text{Atmospheric Press}}{\text{Atmospheric Pressure}}$$

Atmospheric Pressure Vs Altitude

Subtract 0.5 p.s.i. from 14.7 p.s.i. for each 1,000 ft. above sea-level to determine approximate atmospheric pressure at altitudes

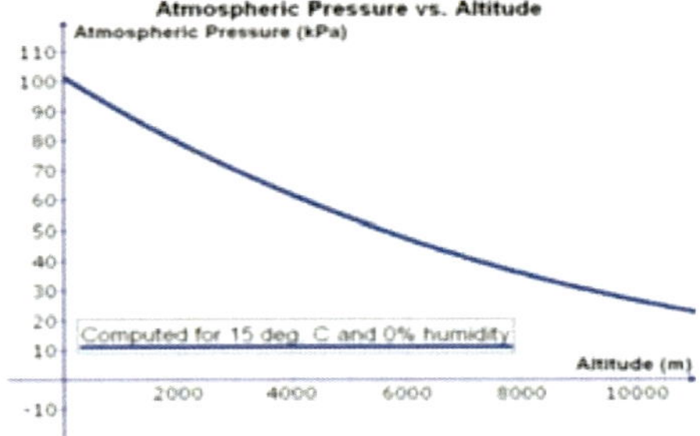

Marine Engine Torque Vs. Marine Engine Horsepower

- Most people make the common mistake of focusing on the marine engine horsepower rather than the marine engine torque.

- There is a common saying that "Horsepower sells a boat however Torque is what actually moves it". This could not be closer to the truth!

Marine Engine Torque Vs. Marine Engine Horsepower

- One should realize that horsepower is really a measure of the torque over a given period of time .

- This taken into account by the rpm variable in the specification. The following equation may help to shed some light as well.
    - Torque = Hp x 5252 / RPM   (5252 is a constant)
    - Torque (Max) = Hp x 5252 / RPM (Idle)

Marine Engine Torque Vs. Marine Engine Horsepower

- It is interesting to note that the formula also verifies the typical torque bell curve when the torque trails off as rpm increases at the top end.
- One may consider that engines making torque at a lower rpm tend to work better in marine applications due to the fact that "most" boats tend to plane in the range of 2,000 - 3,000 rpm.

Boost Pressure and Torque

- With this formula you can calculate the final torque produced for your engine according with the Horsepower calculated

- **Torque** $= \dfrac{5252 \times HP}{RPM}$

Now you can calculate the original the maximum torque at 132 HP (original) and the new torque with 200 HP (wanted)

Torque $_1$ = 770.29   Torque $_2$ = 1167.11 Ft-Lb

Note: Both torques shall be calculated at idle (900 RPM)

Summary

- In this particular example we can increase the Hp from 132 to 200 Hp , just with 7 PSI of boost pressure
- And the output torque produced by the engine will be increased in a 66%
- 770.29 Lb-in / 1167.11 Lb-in = 0.66
- The typical boost pressure provided by either a turbocharger or a supercharger is 6 to 8 (psi) (gas Engines) and 10 to 17 PSI in diesel engines.
- The highest values are recommended only for new engines

Tips to a Successful Turbo Rebuild

- if you were to pick up a turbo, without having any knowledge of how to rebuild a turbo, and simply begin taking it apart, I'm almost 100% positive you will break a part on your turbo.
- Please remember one thing, always be careful when you are attempting to rebuild a turbo. Basically, act like your turbo is a baby
- In order to help prevent any parts of your turbo from being damaged, it's best to place a rug, some rags, newspaper, cardboard, or some other type of material on the surface you are doing the rebuild on

Tips to a Successful Turbo Rebuild

- Once the part you are done working with is taken off the turbo, set it aside in a different area, to prevent it from getting accidentally hit and damaged.
- Finally, always keep your hands protected, especially when working around the compressor and exhaust wheels, which have the sharp blades to help fan exhaust out and fresh air in .
- The best thing to do is wear mechanics gloves and cover the wheels with a towel or rag, as you work around them

How to Troubleshoot Turbochargers

- Check for boost leaks by examining all hoses and clamps for any damage or cracks. Boost leaks can occur over time and cause the engine to run rich
- Remove the intake hose from the turbo compressor inlet and check the fan blades for any sign of damage. Additionally, try moving the main shaft in all directions to check for "shaft play." This can indicate worn bearings or other serious problems inside the core of the unit

How to Troubleshoot Turbochargers

- If you are unable to generate proper boost or have an excessive amount of oil in the upper intercooler pipe, this is a key indicator of internal bearing failure. The turbo will need to be removed and rebuilt or replaced.

- Examine the exhaust manifold and turbine housing for any signs of wear or cracking. This can cause a severe exhaust leak and prevent the turbo from properly spinning up and creating boost. Symptoms of an exhaust leak include hesitation, poor acceleration and excessive noise